Philosophy of Engineering and Technology

Volume 29

The Philosophy of Engineering and Technology book series provides the multifaceted and rapidly growing discipline of philosophy of technology with a central overarching and integrative platform. Specifically it publishes edited volumes and monographs in: the phenomenology, anthropology and socio-politics of technology and engineering the emergent fields of the ontology and epistemology of artifacts, design, knowledge bases, and instrumentation engineering ethics and the ethics of specific technologies ranging from nuclear technologies to the converging nano-, bio-, information and cognitive technologies written from philosophical and practitioners' perspectives and authored by philosophers and practitioners. The series also welcomes proposals that bring these fields together or advance philosophy of engineering and technology in other integrative ways. Proposals should include: A short synopsis of the work or the introduction chapter. The proposed Table of Contents The CV of the lead author(s). If available: one sample chapter. We aim to make a first decision within 1 month of submission. In case of a positive first decision the work will be provisionally contracted: the final decision about publication will depend upon the result of the anonymous peer review of the complete manuscript. We aim to have the complete work peer-reviewed within 3 months of submission. The series discourages the submission of manuscripts that contain reprints of previous published material and/or manuscripts that are below 150 pages/75,000 words. For inquiries and submission of proposals authors can contact the editor-in-chief Pieter Vermaas via: p.e.vermaas@tudelft.nl, or contact one of the associate editors.

More information about this series at http://www.springer.com/series/8657

Sacha Loeve • Xavier Guchet
Bernadette Bensaude Vincent
Editors

French Philosophy of Technology

Classical Readings and Contemporary Approaches

Springer

Editors
Sacha Loeve
Faculté de Philosophie/Institut de Recherches Philosophiques de Lyon (IRPhiL - EA 4187)
Université Lyon 3 Jean Moulin
Lyon, France

Xavier Guchet
Connaissance, Organisation et Systèmes TECHniques (COSTECH)
Université de technologie de Compiègne (UTC)
Compiègne, France

Bernadette Bensaude Vincent
UFR de Philosophie/Centre d'Etude des Techniques, des COnnaissances et des PRAtiques (CETCOPRA - EA2483)
Université Paris 1 Panthéon - Sorbonne
Paris, France

ISSN 1879-7202 ISSN 1879-7210 (electronic)
Philosophy of Engineering and Technology
ISBN 978-3-030-07791-4 ISBN 978-3-319-89518-5 (eBook)
https://doi.org/10.1007/978-3-319-89518-5

Printed on acid-free paper

This Springer imprint is published by the registered company Springer International Publishing AG part of Springer Nature.
The registered company address is: Gewerbestrasse 11, 6330 Cham, Switzerland

Contents

1 **Is There a French Philosophy of Technology? General Introduction** 1
Sacha Loeve, Xavier Guchet, and Bernadette Bensaude Vincent

Part I Negotiating a Cultural Heritage

2 **Philosophy and Technology in the French Tradition. The Legacy of François Dagognet** 23
Gérard Chazal

3 **On the Current Uses of Simondon's Philosophy of Technology** 37
Vincent Bontems

4 **Christianity and the Philosophy of Technology in France** 51
Daniel Cérézuelle

5 **Rise of the Machines: Challenging Comte's Legacy with Mechanology, Cybernetics, and the Heuristic Values of Technology** 65
Ronan Le Roux

6 **We Have Never Been Wild: Towards an Ecology of the Technical Milieu** 81
Victor Petit and Bertrand Guillaume

7 **The Role of the Philosophy of Technology in French-Language Studies of Video Games** 101
Mathieu Triclot

Part II Coining and Reconfiguring Technoscience

8 **Technoscience: From the Origin of the Word to Its Current Uses** 121
Gilbert Hottois

9 **Cybernetics Is an Antihumanism. Technoscience and the Rebellion Against the Human Condition** ... 139
Jean-Pierre Dupuy

10 **Lyotard on the (In)Humanity of Technoscience** ... 157
François-David Sebbah

11 **Toward a Philosophy of Technosciences** ... 169
Bernadette Bensaude Vincent and Sacha Loeve

Part III Revisiting Anthropological Categories

12 **Technology and Nature** ... 189
Catherine Larrère and Raphaël Larrère

13 **Leroi-Gourhan: Technical Trends and Human Cognition** ... 209
Charles Lenay

14 **The Anthropocene: Where Are We Going?** ... 227
John Stewart

15 **Toward an Object-Oriented Philosophy of Technology** ... 237
Xavier Guchet

16 **Cyborgs, Between Organology and Phenomenology: Two Perspectives on Artifacts and Life** ... 257
Thierry Hoquet

17 **Of Times and Things. Technology and Durability** ... 279
Bernadette Bensaude Vincent

18 **Transcendental Imagination in a Thousand Points** ... 299
Bernard Stiegler

Part IV Innovating in Ethics, Design and Aesthetics

19 **Ethics of Technology in France** ... 317
Jean-Yves Goffi

20 **Wisdom in the Technosphere** ... 331
Michel Puech

21 **Design Made in France: Perspectives on "Industrial Aesthetics" (1951–1984)** ... 345
Vincent Beaubois and Victor Petit

22 **Design and Aesthetics in Nanotechnology** ... 361
Sacha Loeve

23 **Ontophany Theory: Historical Phenomenology of Technology and the Digital Age** ... 385
Stéphane Vial

Contributors

Vincent Beaubois Histoire des Arts et des Représentations (HAR), Université Paris Nanterre, Nanterre Cedex, France

Bernadette Bensaude Vincent Centre d'Étude des Techniques, des Connaissances et des Pratiques (CETCOPRA), Université Paris 1 Panthéon-Sorbonne, Paris Cedex 05, France

Vincent Bontems Laboratoire de Recherche sur les Sciences de la Matière (LARSIM), Commissariat à l'Énergie Atomique et aux Énergies Alternatives (CEA), Gif-sur-Yvette Cedex, France

Daniel Cérézuelle Programme Autoproduction et Développement Social (PADES), Paris, France

Gérard Chazal Centre Georges Chevrier, Université de Bourgogne, Dijon, France

Jean-Pierre Dupuy Ecole Polytechnique, Stanford University, Paris, France

Jean-Yves Goffi Université Grenoble Alpes, PPL, Grenoble, France

Xavier Guchet Connaissance, Organisation et Systèmes TECHniques (COSTECH), Université de technologie de Compiègne (UTC), Compiègne, France

Bertrand Guillaume Centre de recherches et d'études interdisciplinaires sur le développement durable (CREIDD), Université de technologie de Troyes (UTT), Troyes, France

Thierry Hoquet Institut de Recherches Philosophiques (IREPH) (EA 373), Université Paris Nanterre, Nanterre Cedex, France

Gilbert Hottois Université Libre de Bruxelles, Bruxelles, Belgium

Catherine Larrère University of Paris 1 Panthéon-Sorbonne, Paris Cedex 05, France

Raphaël Larrère Institut National de la Recherche Agronomique (INRA), Paris Cedex 07, France

Ronan Le Roux EA 4395 "Lettres, Idées, Savoirs" (LIS), Université Paris-Est Créteil Val de Marne, Créteil Cedex, France

Charles Lenay Connaissance, Organisation et Systèmes TECHniques (COSTECH), Université de technologie de Compiègne (UTC), Compiègne, France

Sacha Loeve Faculté de Philosophie/Institut de Recherches Philosophiques de Lyon (IRPhiL - EA 4187), Université Lyon 3 Jean Moulin, Lyon, France

Victor Petit Connaissance, Organisation et Systèmes TECHniques (COSTECH), Université de technologie de Compiègne (UTC), Compiègne, France

Michel Puech Département de Philosophie, Sorbonne Université, Paris, France

François-David Sebbah Département de philosophie, University of Paris Nanterre, Nanterre Cedex, France

John Stewart Connaissance, Organisation et Systèmes TECHniques (COSTECH), Université de technologie de Compiègne (UTC), Compiègne, France

Bernard Stiegler Institute for Research and Innovation (IRI, Paris Center of Art and Culture Georges Pompidou), Paris, France

Ars Industralis, Paris, France

Connaissance, Organisation et Systèmes TECHniques (COSTECH), Université de technologie de Compiègne (UTC), Compiègne, France

Nanjing University, Nanjing, China

Mathieu Triclot FEMTO-ST Institute, Besançon, France

Recherches et Études sur le Changement Industriel, Technologique et Sociétal (RÉCITS), Université de Bourgogne Franche Comte, Besançon, France

Université de Technologie de Belfort Montbéliard (UTBM), Belfort, France

Stéphane Vial PROJEKT Design Lab (EA 7447), University of Nîmes, Nîmes, France

ACTE Institute, Sorbonne Paris 1 University, Paris, France

Abbreviations

AFM	Atomic Force Microscope
C-K	Concept-Knowledge theory
CRCT	Centre de Recherche sur la Culture Technique
ELSI	Ethical, Legal land Societal Implications/Impacts
ICSID	International Council of Societies of Industrial Design
JCVI	John Craig Venter Institute
IHST	Institut d'Histoire des Sciences et des Techniques
GMO	Genetically Modified Organism
MRI	Magnetic Resonance Imaging
NBIC	Nanotechnology, Biotechnology, Information technology and Cognitive science
NHTSA	National Highway Travel Safety Administration
OECD	Organisation for Economic Co-operation and Development
OECE	Organisation Européenne de Coopération Economique (Organisation for European Economic Co-operation)
R&D	Research and Development
STM	Scanning Tunneling Microscope
STS	Science and Technology Studies
TAC	Technology is Anthropologically Constitutive/Constituent
TRIZ	Teorija Reshenija Izobretateliskih Zadatch (Theory of the resolution of invention-related tasks)
TSRD	Techno-Scientific Research and Development
UCAV	Unmanned Combat Air Vehicle
UTC	Université de Technologie de Compiègne
WWII	Second World War

Chapter 1
Is There a French Philosophy of Technology? General Introduction

Sacha Loeve, Xavier Guchet, and Bernadette Bensaude Vincent

Abstract The existence of a French philosophy of technology is a matter of debate. Technology has long remained invisible in French philosophy, due to cultural circumstances and linguistic specificities. Even though a number of French philosophers have developed views and concepts about technology during the twentieth century, "philosophy of technology" has never been established as a legitimate branch of philosophy in the French academic landscape so far. This book, however, demonstrates that a community of philosophers dealing with various issues related to technology and built up on the legacy of the previous generations has emerged. In gathering scholars with quite diverse theoretical backgrounds and matters of concern, this volume outlines a coherent, albeit heterogeneous, philosophical trend. Five chief characteristics are identified in this introduction: (i) a close connection between history and philosophy, with a focus on the *temporalities* of technology, (ii) the prevalence of the *anthropological* approach to technology whether it be social anthropology or paleoanthropology, (iii) a focus on technological *objects* that we characterize as a "thing turn" *à la française*, (iv) the dignification of *technoscience* as a philosophical category, and (v) a pervading concern with ethical issues based on the anthropological interpretation of technology and quite distinct from current trends in applied ethics.

S. Loeve (✉)
Faculté de Philosophie/Institut de Recherches Philosophiques de Lyon (IRPhiL - EA 4187), Université Lyon 3 Jean Moulin, Lyon, France
e-mail: sacha.loeve@univ-lyon3.fr

X. Guchet
Connaissance, Organisation et Systèmes TECHniques (COSTECH), Université de technologie de Compiègne (UTC), Compiègne, France
e-mail: xavier.guchet@utc.fr

B. Bensaude Vincent
Centre d'Étude des Techniques, des Connaissances et des Pratiques (CETCOPRA), Université Paris 1 Panthéon-Sorbonne, Paris Cedex 05, France
e-mail: bvincent@univ-paris1.fr

S. Loeve et al. (eds.), *French Philosophy of Technology*, Philosophy of Engineering and Technology 29,
https://doi.org/10.1007/978-3-319-89518-5_1

Keywords Artifacts · Ethics of technology · History of technology · Paleoanthropology · Philosophy of technology · Social anthropology · Technology · Technological objects · Technoscience · Thing turn

"Is there, actually, a French philosophy of technology?" The question was phrased in these terms a few years ago by French philosopher Daniel Parrochia, in a paper devoted to examine the setting up and the development of a philosophy of technology in the French-speaking area (including Belgium in particular), from Descartes to nowadays. (Parrochia 2009) In this ambitious paper published in 2009, Parrochia featured and mentioned none of the contributors to the present volume apart from Jean-Yves Goffi and Gérard Chazal. Would it mean that in less than a decade, a new generation of French philosophers of technology has come into existence? And if so, what would be the distinctive features of this emerging community with regard to the previous generations mentioned in Parrochia's survey?

1.1 We Have Never Been Philosophers of Technology

At first glance, and despite Parrochia's focus on seminal French contributions to the philosophy of technology that still prove relevant for current research worldwide, the existence of a specifically French tradition is still questionable. As a matter of fact, the philosophy of technology has no institutional existence in the French academic area. Unlike other countries (Germany, the Netherlands, the United States in particular) where the philosophy of technology is established in university chairs, institutes, departments, academic journals, and learned societies, in France there is no academic infrastructure. Until recently, there were no university chairs labelled 'philosophy of technology', and no journals devoted to this topic. Admittedly, a French Society for the Philosophy of Technology (Société pour la Philosophie de la Technique) was created at the initiative of Gilbert Hottois and Daniel Cérézuelle in the early 1990s, on the model of the (North-American and North-European) Society for Philosophy and Technology, but for the last 10 years the Society has been more or less dormant.

Yet the situation seems to be changing: two chairs have been recently created, in Compiègne and Lyons universities, a book series "*L'évolution des machines*" of philosophical studies of technology has been launched by Hermann publishing house, and the French Society for the philosophy of technology is restarting its activities.[1] Interestingly, one might also mention the topic set for the French *Agrégation de philosophie* in 2018: *Travail – Techniques – Production* (work,

[1] Société Francophone de Philosophie de la Technique (SFPT).

technologies, production). Despite these encouraging signs however, and despite the long history of philosophical concerns with technology since the seventeenth century in France, there is strictly speaking no such thing as a French tradition of the philosophy of technology.

It does not mean that there were no philosophical considerations about technology in France. A number of individual philosophers dedicated part of their scholarship to technology but they did not attract disciples. Gilbert Simondon and Jacques Ellul, for instance, remain relatively isolated in their lifetimes. For decades, Simondon's works were more familiar to architects, ergonomics engineers, designers and promoters of technological education than to philosophers. And ironically Ellul was more famous in the United States than in France.

While no research school emerged throughout the twentieth century, a number of collectives of thought (something like Fleck's *Denk Kollectives*) are clearly identifiable:

- Within the Personalist movement founded by Emmanuel Mounier around the journal *Esprit*, Ellul and Bernard Charbonneau developed significant views about technology in an attempt to define a third way between Marxism and Liberalism along Christian lines (see Cérézuelle in this volume).
- Invited in the early 1950es at the Sorbonne Institute for the history of science by Pierre Ducassé (maybe the first french philosopher of technology), the "Cercle d'études cybernétiques" was a point of convergence for theorists interested in creating a science of machines (Lafitte, Couffignal, Russo), with a direct influence on Simondon and Canguilhem (see Le Roux in this volume).
- The Institute for Industrial Aesthetics (l'*Institut d'Esthétique Industrielle*) pioneered a vision of design theories and practices different from those that later prevailed in the USA, under the leadership of Jacques Viénot and with contributions by Etienne Souriau, Georges Friedmann and Simondon (see Beaubois and Petit in this volume).
- The University of Lyon, where François Dagognet initiated a sort of "materiological" tradition (Dagognet 1985), on the basis of Bachelard's and Canguilhem's epistemologies (see Chazal in this volume).
- The laboratory COSTECH (Connaissance, Organisation et Systèmes Techniques) founded by Bernard Stiegler in 1993 at the University of technology of Compiègne, which prompted a distinctively 'technologically constitutive' approach to the philosophy of cognition (Steiner 2010).

Occasionally French scholars talk about the "Lyon school" or the "Compiègne School." Such labels are, however, highly questionable. The Compiègne School is known as such mainly in Compiègne University of Technology and by extension in the small area of Humanities researchers working in the two other French Universities of Technology. As for the Lyon School, Jean-Claude Beaune, Parrochia and Chazal significantly pursued Dagognet's work in Lyon.[2] All of them share

---

[2] In the 1980s, Beaune created and directed an important collection of philosophy of technology, still active today, "*Milieux*," in the publishing house Champ Vallon.

common references to the French historical epistemology, but their concepts and matters of interest remain quite different. Each of them has developed his own specific approach, so that this community is not strictly speaking a 'research school'. It is worth noting that recently, the university of Lyons welcomed new scholars with great interest in the philosophy of technology (namely Thierry Hoquet and Sacha Loeve, both contributors to this volume), but their theoretical orientations could hardly be viewed as the continuation of Dagognet's works.

To be sure, many contributors to the present volume explicitly shape their views with reference to the authors mentioned above (Cérézuelle Dupuy and Le Roux are deeply interested in the French cybernetic movement, Beaubois and Petit focus on the industrial aesthetics community, Bontems, Guchet and Loeve often quote Simondon in their respective works on nanotechnology, synthetic biology, and big scientific instrumentation such as the Large Hadron Collider or high-throughput genome sequencing). However, none of them would dare introduce himself as a heir or a disciple of these prestigious predecessors. In this respect, there is a striking contrast with the development of philosophy of technology in the USA or the Netherlands. While (Don Ihde 1995, 2008, 2009; Selinger 2006) and Peter-Paul Verbeek (2006, 2008a) federate a strong community under the label "postphenomenology," and scholars at Delft University promote the label "The Dual Nature of Technical Artefacts," (Kroes and Meijers 2006) no identifiable philosophical label could serve as an umbrella term to characterize the French community.

Last but not least, the leading figures in the French tradition never claimed to write about the "philosophy of technology." Henri Bergson, for instance, was continuously concerned with technology throughout his works. Yet since he considered technology as coextensive to life, and closely connected to science, ethics and politics, he did not identify technology as a specific domain of philosophy. Similarly Canguilhem developed views about technology and life in a Bergsonian mode in his recently published earlier works (Canguilhem 2011; Braunstein 2000). Nevertheless he is viewed as a philosopher of biology rather than as a philosopher of technology. Simondon – whose complementary thesis on "technical objects" (1958) was supervised by Canguilhem – conducted his academic career in departments of psychology. After graduating as an engineer from École Polytechnique, Jean-Pierre Dupuy turned to the epistemology of complex modelling in cognitive, economic and social sciences, not to "philosophy of technology." Nowadays Bernard Stiegler is first and foremost viewed as a political philosopher. For all of them, technology is obviously a major topic of philosophical investigations, and its study leaves no field of philosophy unscathed – a view shared by most of the contributors to this book. But this did not lead to delineate a specific field of philosophy.

In brief: the philosophy of technology has never been established as an academic sub-discipline in France due to the overarching presence of technology in philosophical reflections rather than to a lack of concern.

1.2 Technique and Technology

This paradoxical situation is rooted in historical and linguistic specificities of French culture. The phrase "philosophy of technology" is a German invention, not a French one. Not only German scholars have made extensive use of "*Philosophie der Technik*" and "*Technikphilosophie*" ever since Ernst Kapp's *Grundlinien einer Philosophie der Technik* (1877),[3] but the current generic term of "technology" denoting a specific category of human activities is itself a German invention. As historians of the phrase "technology" have argued, its current meaning derives from a translation and adaptation of the German term "*die Technik*" rather than from the eighteenth-and-nineteenth-centuries notion of "technology." (Schatzberg 2006, Marx 2010)

While *die Technik* referred to industrial arts and their material means of production, *die Technologie* (*la technologie* in French)[4] was concerned with the *study* of these activities. It literally meant the *logos* of *technè*, the "discourse on arts." This tradition of "techno-logy"[5] initiated in the European Encyclopedist movement (Ames 1629; Alsted 1630; Wolff 1735) and institutionalized in the German universities during the cameralist period (Beckmann 1777, 1806) is instantiated in the innumerable volumes of dictionaries and encyclopedias published in Germany and in France between 1700 and 1900. (Krünitz 1773–1858; Lenormand 1819, 1822–1835; Leca-Tsiomis 1999; Stalnaker 2010) By the end of the nineteenth century, this use of the term *Technologie* declined. Technology was despised as useless literature and sacrificed on the altar of progress. (Mertens 2010, 2011) It was supplanted by a variety of fashionable notions such as "applied sciences," the political economy of machinery, championed by Andrew Ure's and Charles Babbage's in

[3] See for instance Franz Reuleaux, *Kultur und Technik* (1884) Eberhard Zschimmer, *Philosophie der Technik: vom Sinn der Technik und Kritik des Unsinns über die Technik* (1914), *Technik und Idealismus* (1920), *Deutsche Philosophen der Technik* (1937); Friedrich Dessauer, *Philosophie der Technik. Das Problem der Realisierung* (1927); Otwald Spengler, *Der Mensch und die Technik* (1931/1932); Manfred Schröter, *Philosophie der Technik* (1934), *Deutscher Geist in der Technik* (1935), and Eugen Diesel, *Das Phänomen der Technik* (1939).

[4] A similar use of "technology" is also attested in English until the late nineteenth century. Thomas Blount's *Glossographia* defines technology as "a description of crafts, arts and workmanship" (Blount 1670), The New World of Words as "a description of arts, especially the mechanical." (Phillips 1706) See also Jacob Bigelow's Elements of Technology (1829), where "Technology" refers to his book and "the useful arts" to its subject matter. The term had also a similar meaning when the Massachusetts Institute of Technology was founded in 1861.

[5] For the sake of readability, we here use the term "technology" in the loose meaning it has acquired in today English language. However, the distinction between "*technique*" and "*technologie*" continues to make sense in France for a number of scholars as exemplified in a number of chapters in this volume. In some cases, "technology" denotes a higher degree of systematicity than technique, and will translate for instance Ellul's use of "*la technique*" or "*le système technique*," while "technics" or "techniques" stand for the plural "*les techniques*." When it means "the *study* of technologies," as in Leroi-Gourhan's or Simondon's uses, some contributors have chosen to render the French "*technologie*" by "techno-logy" and some by "Technology," with the capital "T" denoting the discipline rather than its subject matter.

England, by the scientific organization of work prompted by Frederick Taylor in the US, and last but not least, by *die Technik* in Germany.

Die Technik was the object of engineering. It became a keyword of German engineering culture in a context of promotion of engineers as a distinctive social class, and a matter of lively debates stirring the *Verein Deutscher Ingenieure* (VDI, the Association of German Engineers, founded in 1856). These discussions at the VDI gathered together engineers with philosophers such as Eberhard Zschimmer or Friedrich Dessauer and revolved around the spiritual meaning of *die Technik* and its relation with *Kultur* (Hård 1998). The relation between *die Technik* and modern capitalism was also a recurrent topic in the early twentieth century German school of economic sociology (Schmoller, Simmel, Sombart, Weber).

In this period, there was no equivalent term available in French to discuss the spiritual meaning of the industrial arts as a whole. Significantly, while Bergson's *Creative evolution* (1907) is full of terms such like "artificial," "constructions," "industry," "instruments," "inventions," "fabrication," "functional," "machine," "manipulation," "manufacture," "mechanisms," or "tools," it contains not a single occurrence of "*la technique*," whereas one finds two occurrences of "*la technique*" in *The Two Sources of Morality and Religion* (1932).

The substantive denomination "*la technique*" has not been used in French language until the interwar period. "Technique" was used either as adjective qualifier, as in "*les termes techniques*" ("technical terms"), or as a term calling for a complement of the noun, as in "*la technique de la peinture*" ("the technique of painting") or "*la technique de la maçonnerie*" ("the technique of masonry"). In the latter cases "technique" is oblivious of its Greek roots *techné* that prevailed in "*technologie*" and closer to the Late Latin *techna* (trick, deceit), with a pejorative connotation of pedantry and sophistication. Thus, whilst Germans philosophers and engineers considered *die Technik*, in France the term of "arts" still prevailed. In the absence of an appropriate term it was unlikely that technology could emerge as a specific field of philosophical investigation.

It was Alfred Espinas, who pointed out this linguistic lacuna, which he characterized as a major limitation, in *Les origines de la technologie* (1897) – "*technologie*" meaning "philosophy of technology" or "technical thought" for Espinas:

> We could give to the useful arts the name of "techniques" so as to distinguish them from the arts meant to produce aesthetic emotion. This word of "technique" has unfortunately a rather restricted meaning among us. We say *the technique* of education, the technique of such or such fabrication, and we so designate the operating processes or, in general, the special parts of the industrial arts (…) [i.e. the methods of fabrication] rather than these arts themselves; we shall have the greatest difficulties to say *les techniques* instead of the "useful arts," especially if (…) the groups of superior rules (…) such as politics and morality, must be counted among the arts and thus become techniques. There would be nevertheless some advantage in being able to designate so, as the Greeks did, the conscious and thoughtful practices that are to a certain point in opposition to the simple practices or customs that establish themselves spontaneously before any analysis. (Espinas 1897: 7–8)[6]

[6] Espinas could read German, and few pages below, he referred to Kapp's theory of "organic projection" (On Espinas see Goffi in this volume; on Kapp see Hoquet).

Beyond language limitations, another reason why French philosophers were not interested in the promotion of a philosophy of technology is their strong attachment to the project of a techno-logy instantiated in the *Encyclopédie* edited by Diderot and D'Alembert. The encyclopedic ideal flourished during the interwar period in the multi-volume project of the *Encyclopédie française*, edited by French historian Lucien Febvre. (Poirot-Delpech 1988) And to a certain extent, this project together with the various conferences organized by the Centre International de Synthèse (Gemelli 1987) can be viewed as an alternative to the philosophy of technology promoted by German supporters of the Nazi regime such as Eugen Diesel and Manfred Schröter (Bontems 2009). In France, the multidisciplinary science of techniques has been rejuvenated in the 1930s through a convergence of initiatives: Social scientist Marcel Mauss developed views and methodological insights about "*les techniques et la technologie*" (1927–1928, 1948) and energetically campaigned for increasing the place of Technology in the "miscellaneous" section of the journal *L'année sociologique* (Schlanger 2012). Marc Bloch and Febvre, who founded the journal *Annales d'histoire*, dedicated a special issue to "Technology, History and Life" (Bloch 1936), where André-Georges Haudricourt, a linguist, botanist and ethnologist, published the first of a series of technological papers. Their endeavors have been reinforced by André Leroi-Gourhan in ethnology and prehistory (Leroi-Gourhan 1943, 1945, 1949, see Lenay in this volume), and later on by the proponents of "*la technologie comme science humaine*" ("techno-logy as human science"),[7] namely François Sigaut, an agronomist and historian, together with Haudricourt (Sigaut 1985; Haudricourt 1987). Even though these disciplines were claiming a systematic and general techno-logy, their leading figures knew little about the history of technological thought[8] until its rediscovery by a group of young historians and philosophers working at the Sorbonne's Institut d'Histoire des Sciences et des Techniques from 1963 to 1965 under the supervision of Canguilhem. (Guillerme and Sebestik 1966; Morère 1966; Sebestik 2007).

One major distinctive feature of these French approaches to techno-logy compared to other twentieth-century traditions of the philosophy of technology is the rejection of a functionalist view of technology and, instead, an effort to understand and to evaluate technology *per se*. Tools, objects, machines, operations, and gestures, are scrutinized for their own sake rather than as means for external ends or for the purpose of the moral evaluation of these ends. However the concept of technological objects is promoted as a necessary mediation *to understand the Human*, in keeping with the tradition of Encyclopedism, which ambitioned to constitute a human *paideia* through the circular survey of all the arts and fields of knowledge.

[7] It is also partly along these lines that the Universities of Technology (Universités de Technologie) were created in France in the 1970s. The project modeled after the Massachusetts Institute of Technology consisted in combining engineering sciences with the Humanities in order to overcome the French divide between engineering schools and universities. These universities were meant for training a new generation of "engineers-philosophers," also called "technologues" (Deforge 1985; Lamard and Lequin 2005).

[8] To our knowledge, Mauss never referred to Johann Beckmann, for instance.

Simondon in particular was a big fan of Diderot's *Encyclopédie* and claimed to promote a new techno-logical Encylopedism. (Simondon 1958: 105) And Dagognet adopted the profile of an encyclopedist as he wrote dozens of volumes about a broad variety of topics. (Parrochia 2011) In a nutshell, there was no room for a philosophy of technology in the French culture and language. Significantly, André Lalande's *Vocabulaire technique et critique de la philosophie,* a dictionary that remained the reference work for all philosophy students in the French academic system throughout the twentieth century, explicitly states that "*La technologie est la théorie ou la philosophie des techniques*" ("techno-logy is the theory or the philosophy of techniques"). (Lalande 1902–1947: 1107) In this case, a philosophy of technology would just be a redundancy.

1.3 An Emerging Community

This book, however, aims to demonstrate that, in the absence of institutional status and international visibility, a new community of philosophers focused on issues related to technology has emerged. To be more precise, this volume is less a mirror image of this emerging community, than a sort of catalyst. It will make the works achieved within this community (more) accessible to an international readership, and at the same time help the constitution of this community. In gathering various scholars with quite diverse theoretical backgrounds and matters of concern this collective volume displays a coherent, albeit, heterogeneous, philosophical trend. Due to the limited format of this volume, and to the demand of the series editors, we did not solicit contributions from authors who were already well known among English-speaking scholars such as Bruno Latour or Michel Serres for instance. Priority has been given in particular to young scholars who explore original pathways in questioning technology. We only included translations of two seminal papers by Bernard Stiegler and Gilbert Hottois (in an augmented version) that were not hitherto accessible in English, and a republication of a paper by Jean-Pierre Dupuy because of his structuring role in the French studies of contemporary technoscience.

What are the distinctive features of this emerging community? Let us go back to Parrochia's 2009 characterization in order to identify potential changes. He mentioned two major trends: A descriptive and historical approach and a social-anthropological approach to techniques.

- *Concerning the former,* contemporary French philosophers are more or less following the historical pathways opened up by Bertrand Gille (1978, 1979) and Maurice Daumas (1962, 1965, 1971), or the historical epistemology approach developed by Bachelard, Canguilhem and Serres among others. In stark contrast to the analytical approach carried out in a rather ahistorical manner within the framework of the Dual Nature of Technical Artefacts program or to the post-phenomenological approach, many chapters in this volume still keep Canguilhem's historical epistemology in the backstage. However, the traditional focus on the history of technological concepts and objects is shifting toward a concern for the genesis and lifetime of objects which results in the central notion of the multiple temporalities of objects (see in particular Bensaude-Vincent), with a strong reference to evolutionary models. Here, the pioneers are Bergson, André Leroi-Gourhan and above all Simondon, but also Serres and Stiegler. Contrary to Heidegger claiming that modern technology prevents us from authentically experiencing our temporal condition, several contributors to this volume rather claim that technological objects display temporalities of their own – new "timescapes" according to Bensaude-Vincent. They challenge evolutionary approaches and open the way to an investigation of the regimes of temporality in technology (see also debates around the concept of Anthropocene, Stewart in this volume).

– *Concerning the sociological and anthropological approach to the philosophy of technology*. French scholars have undoubtedly been sensitive to sociological works, in particular to the Actor-Network Theory (ANT) carried out by Madeleine Akrich, Michel Callon and Bruno Latour at the Centre de Sociologie de l'Innovation of the École des Mines de Paris. In this respect, the emerging French community is following the international mainstream, given that the ANT and its US counterpart – the Social Construction of Technology (SCOT) (Pinch and Bijker 1984, Law et al. 1987) – inspired numerous philosophical works over the past decades, in particular Verbeek, Andrew Feenberg (2000, 2003), and the "empirical turn in the philosophy of technology." (Brey 2010)

A more distinctive French characteristic is the importance of anthropology, and especially of paleoanthropology. Most French scholars share the conviction that technology has been constitutive of the human condition. This anthropological orientation goes on with persisting references to Leroi-Gourhan – an ethnologist and prehistorian who still suffers from little international fame. He was nevertheless a major reference in the works of French scholars ranging from Simondon, Gilles Deleuze, Stiegler (1994); Serres, Franck Tinland (1997a, b); Dominique Bourg (1996), and Guchet (2005), or, in this volume, Lenay). Even Latour, who used to mock the "third-rate biology" of technological evolutionists (1992: 2), came to take on a sort of "paleoanthropological turn" with laudatory reference to Leroi-Gourhan, in particular in *Cogitamus* (Latour 2010). Quite surprisingly Leroi-Gourhan is as influential as Simondon among French scholars, but while the latter enjoys international fame (despite the fact that English translations of his major works are not available yet), the former is still scarcely quoted in English papers – even though a translation of Leroi-Gourhan's major opus, *Gesture and Speech* (1964, 1965), has been available since 1993! Hopefully this volume, and in particular Lenay's chapter, will remediate to this regrettable lack of international visibility.

This "anthropological turn" should not be mistaken for a comeback of "human nature." Dupuy'chapter for instance speaks of human *condition*, not of human *nature*. It is definitely not an essentialist approach that would assume an ontological distinction between human beings and the rest of the world. On the contrary, in interweaving technology and anthropology French scholars try to avoid both essentialism in their account of what is human and "anthropologism" in their account of what is technology. As they assume that technology is not specifically human and begins with animals, they do not draw a clear-cut boundary between animals and humans. It is on this non-essentialist basis that they address the thorny issue of what distinguishes Humans *among* other living beings – the so-called "anthropological difference." (Tinland 1997a) Baboons for instance do have technologies as well as a complex social life. But baboons' technologies and baboons' societies remain rather disconnected. (Latour 2001) In other terms it is not technology *per se* but the *way* it supports collective life, which seems to be constitutive of the Human.[9] Leroi-Gourhan, Simondon, Serres, Stiegler, Lenay among others share the view that technology is "anthropologically constitutive," to quote Lenay's "*these de la Technologie Anthropologiquement Constitutive/Constituante*." (Lenay et al. 2002) This "TAC thesis" underlies several approaches presented in this volume together with the widely shared conviction that technology is not neutral. While the non-neutrality of artifacts has been asserted on the basis of political considerations (Winner 1986) or moral arguments (Verbeek 2006), French philosophers tend to give a paleoanthropological significance to this non-neutrality. Technological evolution shapes the conditions of possibility of human experience in ways that could have been otherwise (see Stiegler in this volume). In this respect, the TAC thesis should not be mistaken for technological determinism. (Steiner 2010)

Once again, there is a striking and pervading influence of Bergson on the French community. Bergson claimed that technology articulates two dimensions of the human condition, the biological and the socio-political ones: tools continue the motion of life through non-organic means, but industrial machines open novel ways in the evolution of man, thus prompting great moral and political challenges. Canguilhem argued that Bergson was a major conveyor of a "biological philosophy of technology" in the French area (Canguilhem 1952), which is instantiated in Canguilhem's approach, in Jacques Lafitte's (1932), Leroi-Gourhan's, Simondon's and even, to some extent, in Serres and Lenay.[10] In this respect, French scholars alongside German philosophers of technology distinguished themselves by the importance they confer to biology in their accounts of technology whereas biology is *not* the core of reflections on technology worldwide, with the exception of Tim Ingold, who is more an anthropologist than a philosopher, by the way.

In addition to its historical and anthropological approach to technology, the French emerging community can be distinguished by two specific themes of investigation: *technological objects* on the one hand and *technoscience* on the other hand.

- A "thing turn" *à la française*. The phrase "thing turn" refers to the impressive number of recent publications in various fields of scholarship that have been dealing with things, artifacts, instruments, collections … Mundane things such

[9] Or else for Dupuy it is not technology *per se* but *its metaphysics* which is constitutive of the Human.

[10] The younger generation of scholars (for instance Thierry Hoquet in this volume) – intends to overcome this biologically-rooted approach to technology, but in so-doing they acknowledge the overarching importance of biology in the French tradition of the philosophy of technology.

as zippers (Friedel 1994) or speed bumps (Latour 1994) are dignified as research objects: they deserve detailed empirical studies, which sometimes raise big philosophical issues. In this world of worldwide phenomenon, things have put on considerable weight. Since Langdon Winner's influential article about the bridges on Long Island in New York (Winner 1986) artifacts have a political load and an entry ticket into the field of moral philosophy (Latour 2002; Verbeek 2008b), even though Winner's example proved factually inadequate (Woolgar and Cooper 1999). Anthropologists such as Arjun Appadurai (1986) and Igor Kopytoff (1986) have loaded commodities with social potential and cultural meanings in response to their treatment by Marxist philosophers, and a number of scholars reconsidered the activity of design in light of the dispositions and contrivances of materials. (Ingold 2013) In addition to their cultural and social loads artifacts carry emotions and affects, embedded in their sensory properties and appearance as well as in the stuff they are made of. (Miller 1998; Daston 2004) The prescriptive power carried by artifacts has raised interesting issues about their intentionality since they co-shape human behaviors and regulate their relation to the world. (Idhe 1993; Verbeek 2008a) Scientific objects are credited with more special powers: Instruments not only carry the theoretical knowledge embedded in their design and construction but they also generate knowledge and meaning. (Baird 2004) According to Daston (2000) historical entities scientific objects do not only make up the world, they also have the potential to subvert classical metaphysics because they undermine the classical dilemma between realism and constructivism. Their historicity determines a specific ontology, which is not amenable to the either/or, real/unreal categories of classical metaphysics. And the object-oriented ontology developed by Graham Harman (2011) suggests that this is not a privilege of scientific objects only.

Edmund Husserl's famous injunction "back to the things themselves" (Husserl 1900–1901: 168) seems to come to fruition when things compete with the transcendental subject in the academic world! How and to what extent does the French emerging community participates in this silent Copernican revolution?

First the "thing turn" has been initiated by Simondon with his concept of "technical object" (1958) and further carried on by Dagognet (1985, 1989) who claimed that the objects better revealed the nature of the human mind than the subjects themselves. Dagognet invited philosophers to study the most mundane objects of daily life (1985, 1993). In a different perspective, social anthropologists developed extensive studies of mundane objects in eco-museums (Bonnot 2002, 2014) even providing a fine-grained description of their "situatedness." (Julien and Rosselin 2009; Montjaret 2014) The chapters in this volume display a wide range of such objects: Video games (Triclot), plastics (Bensaude-Vincent), huge scientific instruments (Bontems), objects of industrial design (Beaubois and Petit, Vial), objects of technoscience (Bensaude-Vincent et al.), nano-objects (Guchet, Loeve), bio-objects (Guchet), screens (Vial). Most of them shape their philosophical insights on the basis of a close empirical examination of objects: how they come into being, and are maintained, what they do, how they shape social relationships, how they interact with the world around them.[11] It is worth noting that most of these French contributors to the "thing turn" *à la française* attach importance to the conceptual distinction between thing and object – thus the right expression to label their empirical orientation would rather be "object turn," provided that the "object" is not reduced to a mere *vis-à-vis* of the "subject." Equating this divide and Heidegger's opposition between "*Ding*" and "*Gegenstand*" would be misleading. Mostly epistemological considerations have motivated the "object turn" in the French philosophy of technology. Indeed, due in part to the perennial pregnancy of Bachelardian epistemology, which has no sympathy for the concept of thing, French philosophers of technology mainly state that philosophical studies on technology should begin with the shaping of suitable concepts of technological object. More specifically, the "thing turn" *à la française* is characterized by the attempt to shape a robust concept of technological *object* distinct from the concept of *artifact*. What is at stake in the distinction between objects and artifacts?

Artifacts, i.e. man-made objects, are usually defined in contrast to natural entities. In the post-phenomenological framework as in the analytical framework of the dual nature (both physical and intentional) of technical artifacts, artifacts are viewed as "mediations" between humans and technology, and the main focus is on how users and designers relate to technology. To be sure, these contemporary developments challenge the metaphysical divide between subjects and objects: in Verbeek's approach, subjects are distributed, and they are shaped by non-human entities. However, the post-phenomenology framework remains overly anthropometric in the sense of a man-measure of all things. The French emerging community is by contrast kin to investigate how technology relates to non-human entities in the world. A nanomachine for instance is much more than an artifact that mediates humans-world relationships: it comes into being in the material world and interacts with other material entities at a length scale out of reach of direct human perception and action. Plastics interfere with temporalities that extend far beyond their lifetime as human commodities. (see Bensaude-Vincent in this volume) So technological objects should not be considered only as *arti-facts*, i.e. artificially made. They have to be described in a worldly perspective as inhabitants of the planet. It is plausible that the relevance of Simondon in current research is due to his attempt at reshaping the concept of technological object as a being of its own that has its own way of relating to the world. (see Guchet in this volume).

- *Technoscience as a philosophical category.* Early in the 1980s, Gilbert Hottois has drawn the philosophers' attention toward the practices of Big Science. He coined the compound term "technoscience" precisely to refer to scientific

[11] For some, such as Charles Lenay, Vincent Bontems or Mathieu Triclot, doing philosophy of technology even goes through *making technological objects*. Triclot practices machine learning and game design, and Lenay is working for years on the conception of tactile communication systems, both in French Universities of Technology while Bontems has developed with Vincent Minier some digital devices of scientific mediation in a technological research organisation (CEA). Their developments are not some practical applications of their theoretical views, but operative ways of addressing philosophical issues.

research where technology is embedded both as a milieu and as a driving force. But this concept escaped to its author as Hottois reminds us in this volume. (Hottois) The concept of technoscience became extremely fashionable, when used in various contexts and with various meanings by scholars who have been associated with postmodernism in the 1980s, – such as Latour, Jean-François Lyotard, and Donna Haraway. As the concept gained in extension its comprehension shrunk. Among philosophers the phrase "technoscience" lose most of its load of meanings and came to be more or less reduced to Bachelard's notion of "phénoménotechnique." (Rheinberger 2005) However in the 2000s, a number of European philosophers of science including some of the contributors to this volume have reinvested a notion whose meaning was diluted through its multiple re-appropriations. Their philosophy of *technoscience* invites a revision of both our epistemological and metaphysical categories. In particular, by unfolding the multiple modes of existence of technoscientific objects it suggests a shift from epistemic pluralism to ontological pluralism. (Bensaude-Vincent et al. in this volume);

– Finally, there is a strong tradition of *ethics of technology* in the French speaking area instantiated in Belgium by Hottois (1982, 2004), Geneviève Pinsart (2003), and Céline Kermisch (2011), in Switzerland by Bernard Baertschi (2005), in Quebec by Marie-Hélène Parizeau (2010). In this volume, ethics is a core topic in Dupuy's Goffi's, Catherine and Raphaël Larrère's, Michel Puech's contributions, and a peripheral issue in other chapters (Bensaude-Vincent, Guchet, Hottois, Loeve). However, the ethical questions raised by technology do not delineate a specific subfield that could be labeled "applied ethics" (i.e. an ethics applied to technology). In this respect, French scholars converge with the current international trend to challenge ELSI approaches (Ethical, Legal and Societal Impacts/Implications) that became mainstream in ethical debates (especially around the Human Genome Project and nanotechnology initiatives worldwide). Just as a number of their colleagues in other countries, the contributors to this volume do not express their ethical concerns in the anticipation of potential consequences of current technological research and development. They rather consider what a flourishing human existence could be in a technological environment (Puech in this volume), or try to disentangle the ethical implications of different concepts of technological action (Dupuy, Larrère and Larrère). Their ethical views are closely related to anthropological issues: cybernetics (yesterday) and converging technologies (today), Dupuy argues, offer nothing else than deconstructing the subject of humanism, his place in the world and his responsibility. Biotechnology and nanotechnology, Larrère and Larrère argue, blurs the divide between two views of technology (fabricating and piloting) that used to carry major anthropological significance in human history. Bensaude-Vincent raises ethical questions about how the multiple temporalities of technological objects interfere with human temporalities – a topic of anthropological relevance, undoubtedly. Guchet addresses the question of norms in technology through a critical analysis of current approaches that overlook its anthropological dimension, namely approaches that fall under the label "philosophy of artifacts." Also note that ethical concerns are never divorced from political issues (see for instance Loeve). So, the contributions gathered in this volume address ethical issues from anthropological or historical standpoints, together with a close examination of the ontological status of technological objects. Here, ontology, anthropology and ethics are closely intertwined: The entanglement of these three components characterizes the French contributions to the philosophy of technology. This is at least the main message carried out by this volume.

Let us end with a few remarks about the limited scope of this volume. It does not make full justice to a number of philosophical trends, known for having provided valuable insights on technology over the past decades.

In particular, our coverage of the phenomenological tradition is relatively modest given the continuous importance of Heideggerian thought in French approaches to technology, whose concept of *Gestell* has raised a lot of comments and reinterpretations (Janicaud 1985; Henri 1987; Milet 2000). French phenomenologists such as Maurice Merleau-Ponty or even Jean-Paul Sartre did not directly contribute to the philosophy of technology.[12] Sartre's reflections on the mundane use of things in l'*Etre et le néant* (1943) were largely inspired by the "tool analysis" of *Sein und Zeit*. Heidegger remains an inescapable reference, but he did not generate original views about technology in France as he did in the North-American and Dutch post-phenomenological trends (Ihde 1979; Borgmann and Mitcham 1987).[13] However, this volume features some innovative *uses* of phenomenology by Hoquet (in this volume), Stiegler (conversing with the German tradition as well as with Derrida), or Vial (who hybridizes phenomenology with Bachelardian "phenomenotechnics" and historical epistemology). In his *Ethics of Ordinary Technology* Michel Puech is also interested in the way "the grandiose structures of intentionality that were deployed in abstract compositions (Husserl) or gloomy destiny tales (Heidegger) have been redeployed by post-phenomenology within the fabric of daily life" (Puech 2016: 99, see also this volume) and some of the contributors to this book also practice a post-phenomenology in their own way.[14] We also care to emphasize that French phenomenologists who are concerned with technology take inspiration from the deconstruction movement, in particular Jacques Derrida and Jean-François Lyotard (see Sebbah in this volume).

Similarly, there is no chapter on the French Marxist tradition illustrated in the works of Georges Sorel (1908), Georges Friedmann (1946), Henri Lefebvre (1964),

[12] Conversely, some of the French philosophers who contributed the most originally to reflections on technology during the Second War and post-war periods such as Etienne Souriau (1943, 1956), Pierre Ducassé (1958), Sebbah and Wilson (2014) or even Simondon (who was close to Merleau-Ponty and Mikel Dufrenne), where quite critical of phenomenology in their times.

[13] For a more recent stance on Heidegger (and Marx) see Vioulac (2009, 2016).

[14] One may mention the works conducted at COSTECH around the TAC thesis with a phenomenological (and enactive) approach to cognitive science (Steiner and Stewart 2009; Lenay 2013). Their researches on the technologically constituted conditions of perception are very close to the current Dutch school of postphenomenology. Unfortunately, they have too few interactions, although both take place in Universities of Technology (Compiègne, Twente, Eindhoven).

Serge Moscovici (1968), Cornelius Castoriadis (1978), or André Gorz (1988).[15] However this tradition is not totally discarded since Petit and Guillaume argue in their chapter that the French Marxists interested in technology (Moscovici, Gorz in particular) ended up developing a political ecology. So Marxism appears in this concept as a legacy of the concept of *technical milieu*.

Finally we apologize for the "Franco-French" tone of this volume. Given the limited space allowed we could not include contributions by colleagues from Belgium, Switzerland, Quebec and other French-speaking countries. Their inspiring contributions, especially in the domain of the ethics of technology, would need another volume.

Acknowledgements We would like to express our gratitude to the Groupement d'Intérêt Scientifique pour l'Unité de la Technologie et des Sciences de l'Homme (GiS UTSH), a network gathering research teams in Social Sciences and Humanities from the three French Universities of Technology around the promotion of "technological research in the Humanities." UTSH sponsored the book and funded the translations of several papers. Warms thanks are due to John Stewart, for his translations of numerous chapters, as well as to other translators Virginia Coulon, George Collins, and Daniel J. Palumbo†, who left us in 2017.

References

Alsted, J. H. (1630). *Encyclopaedia*. Herborn: Corvini.

Ames, W. (1629). *Technometria, Omnium & singularum Artium fines adæquatè circumscribens*. English edition: Ames, W. (1979). *Technometry* (L. W. Gibbs, Trans.). Philadelphia: University of Pennsylvania Press.

Appadurai, A. (Ed.). (1986). *The social life of things: Commodities in cultural perspective*. New York: Cambridge University Press.

Baertschi, B. (2005). *Enquête philosophique sur la dignité: Anthropologie et éthique des biotechnologies* (No. 43). Geneva: Labor et Fides.

Baird, D. (2004). *Thing knowledge: A philosophy of scientific instruments*. Berkeley: University of California Press.

Beckmann, J. (1777). *Anleitung zur Technologie, oder zur Kentniß der Handwerke, Fabriken und Manufacturen, vornehmlich derer, welche mit der Landwirtschaft, Polizey und Cameralwissenschaft in nächster Verbindung stehn*. Göttingen: Wittwe Vandenkoek.

Beckmann, J. (1806). *Entwurf der algemeinen Technologie*. Göttingen: Johann Friedrich Röwer.

Bergson, H. (1907). *L'évolution créatrice*. Paris: Alcan. English edition: Bergson, H. (1911). *Creative evolution* (A. Mitchell, Trans.). Lanham: University Press of America.

Bergson, H. (1932). *Les deux sources de la morale et de la religion*. Paris: Presses Universitaires de France. English edition: Bergson, H. (1935). *The two sources of morality and religion* (R. A. Andrà, C. Brereton, & W. H. Carter, Trans.). New York: Henry Holt and Co., Inc.

[15] Neither do we deal with the controversy about the moral issues of "machinism" ("*la querelle du machinisme*") sparkled by writers such as Georges Duhamel and André Maurois in the 1930s, and which involved philosophers such as Pierre-Maxime Schuhl (1938) and Alexandre Koyré (1948). Historians Bertrand Gille and Jean-Pierre Vernant took also part in the debate, both attracting attention over the importance of machines and technological thought in the Middle Age or in Antique Greece.

Bigelow, J. (1829). *Elements of technology, taken chiefly from a course of lectures delivered at Cambrigde on the Aplication of the sciences to the useful arts*. Boston: Hilliard, Gray, Little, and Wilkins.

Bloch, M. (1936). Les techniques, l'histoire et la vie: Note sur un grand problème d'influences. *Annales d'histoire économique et sociale, 8*(42), 513–515.

Blount, T. (1670). Technology. In *Glossographia: Or, a dictionary interpreting all such hard words as are now used in our refined English Tongue* (3rd ed.). London: Tho. Newcomb.

Bonnot, T. (2002). *La vie des objets*. Paris: Éditions de la Maison des sciences de l'homme.

Bonnot, T. (2014). *L'attachement aux choses*. Paris: CNRS éditions.

Bontems, V. (2009). Simondon, les nanotechnologies et les hadrons. Actualité d'une philosophie des machines. *Revue de Synthèse, 1*, 37–66.

Borgmann, A., & Mitcham, C. (1987). Works about Heidegger. *Philosophy Today, 31*(2), 167–191.

Bourg, D. (1996). *L'homme-artifice. Le sens de la technique*. Paris: Gaillimard.

Braunstein, J.-F. (2000). Canguilhem avant Canguilhem. *Revue d'histoire des sciences, 53*(1), 9–26.

Brey, P. (2010). Philosophy of technology after the empirical turn. *Techné: Research in Philosophy and Technology, 14*(1), 36–48.

Canguilhem, G. (1952). Machine et organisme. In *La connaissance de la vie* (pp. 101–127). Paris: Hachette. English edition: Canguilhem G. (2008). Machine and organism. In *Knowledge of life* (pp. 75–97, S. Geroulanos & D. Ginsburg, Trans.). New York: Fordham University Press.

Canguilhem, G. (2011). *Œuvres complètes, tome 1: Ecrits philosophiques et politiques, 1926–1939*. Paris: Vrin.

Castoriadis, C. (1978). Technique. In *Les carrefours du labyrinthe* (pp. 289–224). Paris: Seuil.

Dagognet, F. (1985). *Rematérialiser: matières et atérialismes*. Paris: Vrin.

Dagognet, F. (1989). *Éloge de l'objet. Pour une philosophie de la marchandise*. Paris: Vrin.

Dagognet, F. (1993). *Le musée sans fin*. Seyssel: Champ Vallon.

Daston, L. (Ed.). (2000). *Biographies of scientific objects*. Chicago: University of Chicago Press.

Daston, L. (Ed.). (2004). *Things that talk: Object lessons from art and science*. New York: Zone Books.

Daumas, M. (Ed.). (1962). *Histoire générale des techniques. Tome I: Les origines de la civilisation technique*. Paris: Presses Universitaires de France.

Daumas, M. (Ed.). (1965). *Histoire générale des techniques. Tome II: Les premières étapes du machinisme*. Paris: Presses Universitaires de France.

Daumas, M. (Ed.). (1971). *Histoire générale des techniques. Tome III: L'expansion du machinisme*. Paris: Presses Universitaires de France.

Deforge, Y. (1985). *Technologie et génétique de l'objet industriel*. Paris: Maloine.

Dessauer, F. (1927). *Philosophie der Technik: Das Problem der Realisierung*. Bonn: F. Cohen.

Diesel, E. (1939). *Das Phänomen der Technik, Zeugnisse, Deutung und Wirklichkeit*. Berlin: VDI-Verlag.

Ducassé, P. (1958). *Les techniques et le philosophe*. Paris: Presses Universitaires de France.

Espinas, A. (1897). *Les origines de la technologie*. Paris: Alcan.

Feenberg, A. (2000). From essentialism to constructivism: Philosophy of technology at the crossroads. In E. Higgs, A. Light, & D. Strong (Eds.), *Technology and the good life?* (pp. 294–315). Chicago: University of Chicago Press.

Feenberg, A. (2003). Modernity theory and technology studies: Reflections on bridging the gap. In T. J. Misa, P. Brey, & A. Feenberg (Eds.), *Modernity and technology* (pp. 73–104). Cambridge, MA: MIT Press.

Friedel, R. (1994). The history of the zipper? *American Heritage of Invention & Technology, 10*(1), 8–16.

Gemelli, G. (1987). Communauté intellectuelle et stratégies institutionnelles; Henri Berr et la fondation du Centre International de synthèse. *Revue de Synthèse, 108*(2), 225–259.

Gille, B. (1978). *Histoire des techniques*. Paris: Gallimard.

Gille, B. (1979). La notion de "système technique" (essai d'épistémologie technique). *Culture technique*, 1, Neuilly-sur-Seine: Centre de recherche sur la culture technique.
Gorz, A. (1988). *Métamorphoses du travail, quête du sens*. Paris: Galilée.
Guchet, X. (2005). *Les Sens de l'évolution technique*. Paris: Léo Scheer.
Guillerme, J., & Sebestik, J. (1966). Les commencements de la technologie. *Thalès, 12*, 1–72. Reprinted in: Guillerme J., & Sebestik J. (2007). Les commencements de la technologie. *Documents pour l'histoire des techniques, 14*(2), 49–122.
Hård, M. (1998). German regulation: The integration of modern technology into national culture. In M. Hård & A. Jamison (Eds.), *The intellectual appropriation of technology, 1900–1939* (pp. 33–67). Cambridge, MA: MIT Press.
Harman, G. (2011). *The quadruple object*. Alresford: Zero Books.
Haudricourt, A.-G. (1987). *La technologie, science humaine. Recherche d'histoire et d'ethnologie des techniques*. Paris: Editions de la Maison des sciences de l'homme.
Henri, M. (1987). *La barbarie*. Paris: Grasset.
Hottois, G. (1982). Ethique et technique. *Bulletin de la Société Française de Philosophie Paris, 76*(3), 77–114.
Hottois, G. (2004). *Qu'est-ce que la bioéthique?* Paris: Vrin.
Husserl, E. (1900–1901). *Logische Untersuchungen*. Halle: Niemeyer. 2nd English edition: Husserl, E. (2001). *Logical investigations* (D. Moran, Trans.). London: Routledge.
Idhe, D. (1993). *Philosophy of technology: An introduction*. New York: Parangon.
Ihde, D. (1979). *Technics and praxis*. Dordrecht: Springer.
Ihde, D. (1995). *Postphenomenology: Essays in the postmodern context*. Northwestern: Northwestern University Press.
Ihde, D. (2008). Introduction: Postphenomenological research. *Human Studies, 31*(1), 1–9.
Ihde, D. (2009). *Postphenomenology and technoscience: The Peking University lectures*. Albany: State University of New York Press.
Ingold, T. (2013). *Making: Anthropology, archeology, art and architecture*. London: Routledge.
Janicaud, D. (1985). *La puissance du rationnel*. Paris: Gallimard. English edition: Janicaud, D. (1994). *Powers of the rational science, technology, and the future of thought* (P. Birmingham & E. Birmingham, Trans.). Bloomington: Indiana University Press.
Julien, M.-P., & Rosselin, C. (Eds.). (2009). *Le sujet contre les objets…tout contre*. Paris: Editions du CTHS.
Kapp, E. (1877). *Grundlinien einer Philosophie der Technik. Zur Entstehungsgeschichte des Kultur as neuen Gesichtspunkten*. Brunswick: Westermann.
Kermisch, C. (2011). *Le concept du risque: De l'épistémologie à l'éthique*. Paris: Lavoisier.
Kopytoff, I. (1986). The cultural biography of things: Commoditization as process. In A. Appadurai (Ed.), *The social life of things: Commodities in cultural perspective* (pp. 70–73). New York: Cambridge University Press.
Koyré, A. (1948). Les philosophes et la machine. *Critique, 4*(23), 324–333. Republished in Koyré, A. (1961). *Etudes d'histoire de la pensée philosophique* (pp. 610–630). Paris: Armand Colin.
Kroes, P., & Meijers, A. (2006). The dual nature of technical artefacts. *Studies in History and Philosophy of Science Part A, 37*(1), 1–4.
Krünitz, J. G. (Ed.). (1773–1858). *Oekonomisch-technologische Enzyklopädie oder allgemeines System der Staats- Stadt- Haus- und Landwirthschaft* (Vol. 242). Göttingen and Frankfurt: Joachim Pauli.
Lafitte, J. (1932). *Réflexions sur la science des machines*. Paris: Bloud & Gay.
Lalande, A. (1902–1947). Technologie. In *Vocabulaire technique et critique de la philosophie* (17th ed., 1991, pp. 1106–1107). Paris: Presses Universitaires de France.
Lamard, P., & Lequin, Y.-C. (2005). *La technologie entre à l'université. Compiègne, Sevenans, Belfort-Montbéliard…*. Belfort: Editions de l'Université de technologie de Belfort-Montbéliard.
Latour, B. (1992). *Aramis ou l'amour des techniques*. Paris: La Découverte. English edition: Latour, B. (1996). *Aramis, or, the love of technology* (C. Porter, Trans.). Cambridge: Harvard University Press.

Latour, B. (1994). On technical mediation. *Common knowledge, 3*(2), 29–64.

Latour, B. (2001). A sociology without objects? Remarks to inter-objectivity. *Berliner Journal für Soziologie, 11*(2), 237–252.

Latour, B. (2002). Morality and technology: The end of the means. *Theory, Culture & society, 19*(5–6), 247–260.

Latour, B. (2010). *Cogitamus: six lettres sur les humanités scientifiques*. Paris: La Découverte.

Law, J., Bijker, W. E., Hughes, T. P., & Pinch, T. (1987). *The social construction of technological systems*. Cambridge, MA: MIT Press.

Leca-Tsiomis, M. (1999). *Écrire l'Encyclopédie. Diderot: de l'usage des dictionnaires à la grammaire philosophique*. Oxford: Voltaire Foundation.

Lefebvre, F. (1964). *Vers le cybernanthrope, contre les technocrates*. Paris: Denoël-Gonthier.

Lenay, C. (2013). Minimalist approach for a dialogue between phenomenology, science and technology. *PhiTeCo Lectures (Philosophy, Technology, Cognition): From Perceptual interaction to extended cognition*. https://webtv.utc.fr/watch_video.php?v=17WUR21R2O8O. Accessed 31 July 2016.

Lenay, C., Havelange, V., & Stewart, J. (2002). Les représentations: mémoire externe et objets techniques. *Intellectica, 35*(2), 115–129.

Lenormand, L.-S. (1819). Considérations sur les avantages que présente la Technologie, et sur les moyens propres à faciliter l'étude de cette science. *Revue encyclopédique, 1*, 227–244.

Lenormand, L.-S. (Ed.). (1822–1835). *Dictionnaire technologique, ou Nouveau dictionnaire universel des arts et métiers, et de l'économie industrielle et commerciale* (Vol. 22). Paris: Thomine et Fortic.

Leroi-Gourhan, A. (1943). *Evolution et techniques. I, L'Homme et la matière*. Paris: Albin Michel.

Leroi-Gourhan, A. (1945). *Evolution et techniques. II, Milieu et techniques*. Paris: Albin Michel.

Leroi-Gourhan, A. (1949). Note sur les rapports de la technologie et de la sociologie. *L'année sociologique, 2*, 766–772.

Leroi-Gourhan, A. (1964). *Le geste et la parole. Tome I: Technique et langage*. Paris: Albin Michel.

Leroi-Gourhan, A. (1965). *Le geste et la parole. Tome II: La mémoire et les rythmes*. Paris: Albin Michel. English edition: Leroi-Gourhan, A. (1994). *Gesture and speech* (Vols. I & II, A. B. Berger, Trans.). Cambridge, MA: MIT Press.

Marx, L. (2010). Technology: The emergence of a hazardous concept. *Technology and Culture, 51*(3), 561–577.

Mauss, M. (1927–1928). Technologie. In M. Mauss (Ed.), *Instructions d'ethnographie descriptive* (pp. 27–76). Lectures at the Institut d'ethnologie of the Université de Paris. Reprinted in N. Schlanger (Ed.). (2012). *Marcel Mauss. Techniques, technologie et civilisation* (pp. 295–326). Paris: Presses Universitaires de France.

Mauss, M. (1948). Les techniques et la technologie. *Journal de Psychologie Normale et Pathologique, 41*, 71–78.

Mertens, J. (2010). The mere handicrafts: Ure's dictionary (1839–1853) compared with the Dictionnaire technologique (1822–1835). *Documents pour l'histoire des techniques, 19*, 277–285.

Mertens, J. (2011). The decline of general technology: Léon Lalanne and the ascendancy of applied mechanics. *Documents pour l'histoire des techniques, 20*, 107–117.

Milet, J. P. (2000). *L'absolu technique: Heidegger et la question de la technique*. Paris: Kimé.

Miller, D. (1998). *Material cultures. Why some things matter*. Chicago: The University of Chicago Press.

Montjaret, A. (Ed.). (2014). Le retournement des choses. *Socio-Anthropologie*, 30(2).

Morère, J.-É. (1966). Les Vicissitudes du sens de "Technologie" au début du XIXe siècle. *Thales, 12*, 73–84.

Moscovici, S. (1968). *Essai sur l'histoire humaine de la nature*. Paris: Flammarion.

Parizeau, M. H. (2010). *Biotechnologie, nanotechnologie, écologie: entre science et idéologie*. Versailles: Quæ.

Parrochia, D. (2009). French philosophy of technology. In A. Brenner & J. Gayon (Eds.), *French studies in the philosophy of science* (pp. 51–70). Dordrecht: Springer.
Parrochia, D. (Ed.). (2011). *François Dagognet: un nouvel encyclopédiste ?* Seyssel: Champ Vallon.
Phillips, E. (1706). Technology. In *The new world of words* (6th ed.). Universal dnglish dictionary containing an account of the original or proper sense, and various significations of all hard words. London.
Pinch, T. J., & Bijker, W. E. (1984). The social construction of facts and artefacts: Or how the sociology of science and the sociology of technology might benefit each other. *Social Studies of Science, 14*(3), 399–441.
Pinsart, M. G. (2003). *Genre et bioéthique*. Paris: Vrin
Poirot-Delpech, S. (1988). L'Encyclopédie Française. *Vulgariser les sciences (1919–1939). Acteurs, projets, enjeux. Cahiers d'histoire et de philosophie des sciences, 24*, 7–14.
Puech, M. (2016). *The ethics of ordinary technology*. New York: Routledge.
Reuleaux, F. (1884). *Kultur und Technik*. Vienna: A Kreiss.
Rheinberger, J. (2005). Gaston Bachelard and the notion of 'phenomenotechnique. *Perspectives on Science, 13*(3), 313–328.
Schatzberg, E. (2006). Technik comes to America: Changing meanings of technology before 1930. *Technology and Culture, 47*(3), 486–512.
Schlanger, N. (2012). Une technologie engagée. Marcel Mauss et l'étude des techniques dans les sciences sociales. In N. Schlanger (Ed.), *Marcel Mauss. Techniques, technologie et civilisation* (pp. 17–134). Paris: Presses Universitaires de France.
Schröter, M. (1934). Philosophie der Technik. In A. Baeumler und M. Schröter (Eds.), *Handbuch der Philosophie* (chap. 43). Oldenbourg, München und Berlin: Druck und Verlag von R. Oldenbourg.
Schröter, M. (1935). *Deutscher Geist in der Technik*. Köln: Schaffstein.
Schuhl, P. M. (1938). *Machinisme et philosophie*. Paris: PUF. Revised second ed. 1946.
Sebbah, F. D., & Wilson, D. (2014). The philosopher and (his) technics: From the work of Pierre Ducassé. *Diacritics, 42*(1), 6–21.
Sebestik, J. (2007). Les commencements de la technologie. Postface/préface. *Documents pour l'histoire des techniques, 14*(2), 123–133.
Selinger, E. (Ed.). (2006). *Postphenomenology: A critical companion to Ihde*. Albany: State University of New York Press.
Sigaut, F. (1985). More (and enough) on technology! *History and Technology, 2*(2), 115–132.
Simondon, G. (1958). *Du Mode d'existence des objets techniques*. Paris: Aubier. English edition: Simondon, G. (2016). *On the mode of existence of technical objects* (C. Malaspina, Trans.). Minneapolis: Univocal Publishing.
Sorel, G. (1908). *Les Illusions du progrès*. Paris: Marcel Rivière.
Souriau, E. (1943). *Les différents modes d'existence*. Paris: Presses Universitaires de France.
Souriau, E. (1956). Du mode d'existence de l'œuvre à faire. *Bulletin de la Société française de philosophie, 50*(1), 4–24.
Spengler, O. (1931). *Der Mensch und die Technik. Beitrag zu einer Philosophie des Lebens*. Munich: Beck Verlag. English edition: Spengler, O. (1932). *Man and technics. A contribution to a philosophy of life* (C. F. Atkinson, Trans.). New York: Alfred A. Knopf, Inc.
Stalnaker, J. (2010). *The unfinished Enlightment. Description in the age of encyclopedia*. Ithaca: Cornell University Press.
Steiner, P. (2010). Philosophie, technologie et cognition: état des lieux et perspectives. *Intellectica, 53*(54), 7–40.
Steiner, P., & Stewart, J. (2009). From autonomy to heteronomy (and back): The enaction of social life. *Phenomenology and the Cognitive Sciences, 8*(4), 527–550.
Stiegler, B. (1994). *La technique et le temps, 1. La faute d'Épiméthée*. Paris: Galilée. English edition: Stiegler, B. (1998). *Technics and time, 1. The fault of epimetheus* (G. Collins & R. Beardsworth, Trans.). Cambridge, MA: Stanford University Press.

Tinland, F. (1997a). *La différence anthropologique : essai sur les rapports de la nature et de l'artifice*. Paris: Aubier Montaigne.

Tinland, F. (1997b). *L'homme aléatoire*. Paris: Presses Universitaires de France.

Verbeek, P.-P. (2006). The morality of things: A postphenomenological inquiry. In E. Selinger (Ed.), *Postphenomenology: A critical companion to Ihde* (pp. 117–128). Albany: State University of New York Press.

Verbeek, P.-P. (2008a). Obstetric ultrasound and the technological mediation of morality: A postphenomenological analysis. *Human Studies, 31*(1), 11–26.

Verbeek, P.-P. (2008b). Design ethics and the morality of technological artefacts. In P. E. Vermaas, P. Kroes, A. Light, & S. A. Moore (Eds.), *Philosophy and design. From engineering to architecture* (pp. 91–100). Dordrecht: Springer.

Vioulac, J. (2009). *L'époque de la technique. Marx, Heidegger et l'accomplissement de la métaphysique*. Paris: Presses Universitaires de France.

Vioulac, J. (2016). *Apocalypse de la vérité. Méditations heideggeriennes*. Paris: Ad Solem.

Winner, L. (1986). Do artifacts have politics? In L. Winner (Ed.), *The whale and the reactor. A search for limits in an age of high technology* (pp. 19–39). Chicago: The Chicago University Press.

Wolff, C. F. (1735). *Philosophia rationalis sive Logica*. Francfurt: Officina libraria Rengeriana.

Woolgar, S., & Cooper, G. (1999). Do artefacts have ambivalence? Moses' bridges, Winner's bridges and other urban legends in S&TS. *Social Studies of Science, 29*(03), 433–449.

Zschimmer, E. (1914). *Philosophie der Technik: vom Sinn der Technik und Kritik des Unsinns über die Technik*. Jena: Mittler.

Zschimmer, E. (1920). *Technik und Idealismus*. Jena: Jenaer Volksbuch.

Zschimmer, E. (1937). *Deutsche Philosophen der Technik*. Stuttgart: Enke.

Part I
Negotiating a Cultural Heritage

Chapter 2
Philosophy and Technology in the French Tradition. The Legacy of François Dagognet

Gérard Chazal

Abstract In opposition to philosophers focused on the intimacy of the subject, a number of French philosophers were more concerned with understanding the objective world as it is, and as we built it. In this respect, technology as an historical process, ending up in a set of objects and practices, affords a worthwhile ground for developing such world-oriented philosophical reflections. This paper provides a survey of this philosophical landscape with a special emphasis on the pro-eminent role of François Dagognet who pioneered a material and object philosophy in France.

Keywords Anthropology · Dagognet · History of technology · Technical object · Philosophy · Economic and political context · Technical milieu · Technological risk

Gilbert Simondon, in his 1964–1965 lectures on perception (2006), identified the origin of the broad divide between subject and object-oriented philosophies in Ancient Greece between Ionian philosophers, instantiating the latter category and the legacy of Plato and Pythagoras who initiated the former.

> Neither Pythagoras nor Plato are operators, architects, craftsmen: they contemplate, they isolate themselves in meditative leisure, founding esoteric groups to whom they restrict their teaching; in their doctrines, ethical importance shows that the world is less important than man. (Simondon 2006: 13)
>
> For Pythagoras and Plato, the real stands *a priori*, it is already constituted into its structure prior to taking perceptual information that respects it, whereas for the Ionian physiologists the real takes shape during the manipulative operation that also provides knowledge. (Simondon 2006: 21)

G. Chazal (✉)
Centre Georges Chevrier, Université de Bourgogne, Dijon, France

S. Loeve et al. (eds.), *French Philosophy of Technology*, Philosophy of Engineering and Technology 29,
https://doi.org/10.1007/978-3-319-89518-5_2

Philosophies of Platonic inspiration according to Simondon remain focused on the revelation of the subject and the description of conscious experience. Three major features characterize such philosophical systems: (1) a purely meditative approach, detached from concrete reality, with a strong preference for the intellect rather than the sensory experience of the world; (2) an inclination toward obscure a somewhat esoteric language; (3) anthropocentrism and abstract humanism.

In stark contrast a number of French philosophers defended the philosophical dignity of manufactured objects, and manufacturing processes. They were not interested in the intimacy of the conscious self and concerned themselves with the external rather than the internal world. Convinced that mental representations can be understood only through their external manifestations, such object-oriented philosophies include not only tools, machines, everyday objects, but also fine-arts objects, as well as the world of industrial labor. For them the material and physical world is no less interesting than the mind. Technical objects[1] are integral parts of culture as Simondon claims in the introduction of *Du mode d'existence des objets techniques* (1958). It is clear that for most of these thinkers, human beings, the ultimate object of philosophy, must be defined and understood through their various achievements. They afford a control over nature but also contribute to shape the human mind as François Dagognet emphasized.[2]

> The world of objects, which is immense, is eventually more revealing of the mind than mind itself. To know what we are, it is not necessarily in ourselves that we must look at. Philosophers, throughout history, have remained too exclusively turned towards subjectivity, without understanding that it is rather in things that mind makes itself the more visible. We must therefore make a real revolution, noticing that this is the side of the objects that mind stands, rather than on the side of the subject. (Dagognet 1993a)

Beginning perhaps in the eighteenth century with Diderot's passion for both technology and fine-arts, this tradition of object-oriented philosophy culminated in Gilbert Simondon's works in the mid-twentieth century (2014). Besides Simondon, and in the French-speaking world, I should cite for this type of approach that rejects any disregard for technical practices as well as the clash between two cultures (Snow 1959), François Dagognet, Jean Claude Beaune, Daniel Parrochia, Jean-Pierre Séris, Gilbert Hottois, Bernard Stiegler, Franck Tinland… and the next gen-

---

[1] In French, Gérard Chazal is generally using the term "technique" rather than the more emphatic "technologie." It is translated here as "technology" only to conform to the standard use in English. However, the editors have taken care to maintain this distinction in adjectival form: technical (pertaining to concrete artifacts, processes and practices); technological (pertaining to their abstract rationalization and organization). (see Loeve, Guchet & Bensaude Vincent, General introduction, note 4.

[2] François Dagognet, physician and philosopher, encyclopedic mind, died October 2, 2015. He was both philosopher of life and of techniques and arts. He leaves an immense work with 67 books and numerous papers in collective books or journals. His teaching in Lyon and at the Sorbonne (Paris) trained a whole generation of philosophers. This paper cites some of them in the area of philosophy of technology. The author of this paper was one of his students and acknowledges having received a lot from his teaching.

eration of young philosophers who pursue that tradition in a very inventive way, many of which have contributed to this volume.

Without claiming that there is something like a research school, I would like to stress the features that this group of French philosophers share, each one in his own way and with a specific object. These common features distinguish a French tradition.

Rather than a theoretical corpus, they share an approach of their object that belong neither to "a philosophy of knowledge, of rationality, and of the concept" as Foucault defines it, nor to its opposite, "a philosophy of experience, of meaning, of the subject" (Foucault 1985: 466). The four distinctive features are: (1) the primacy of practice, denial of any contempt of the hand, the will to take the technical object in a philosophical thought, fully; (2) the desire to inscribe any technique in a history; (3) the idea that man constructs himself through its technical artifacts and practices and (4) the connection between Techniques and Arts.

2.1 Practices and Objects

Gilbert Simondon and François Dagognet are the two leading figures who, rather independently,[3] have oriented this tradition. Here I focus on the influence of Dagognet on a number of scholars mainly in Lyons, who together formed a "Lyon school" according to Dagognet's own terms, although it is not technically speaking a "research school."

One major common feature is the focus on the concrete practices for designing and manufacturing technical objects. Dagognet in particular wrote about doctors' and surgeons' practices (1964), industrial production (1989) as well as about contemporary art exhibits and all sorts of objects and materials (1992a). His attention was even drawn to scrap and waste, the most despised things (1997). Jean-Pierre Séris, whose PhD was supervised by Dagognet, was interested in fine-arts and their practices, as much as in machines, expanding his analysis into the concrete figure of *Homo faber* (Séris 1994). A follower of François Dagognet, Jean-Claude Beaune, devoted several volumes to automata (1980a, b, 1983, 1998) whom he depicted as mirror-machines of human beings and analyzers of the human condition; furthermore he elaborated on the notion of "technical milieus" as a material-imaginary complexes that are both a condition and a consequence of our technical activity. Among Dagognet's students, many have pursued philosophical enquiry in the same vein. Daniel Parrochia, an expert in historical studies on aircraft technology (2003) devoted most of his scholarship to mathematical models of graphs and classifications, including the various networks that gradually extend over our planet (1993,

[3] Although the link between the two may be found in Georges Canguilhem: Dagognet was his student, and Simondon wrote his complementary thesis *Du mode d'existence des objets techniques* under his supervision.

1994).[4] Franck Varenne (2007, 2009) has conducted comprehensive studies of computer simulations. Contrasting with the common view of simulations as phenomenal translations of abstract and analytical logical-mathematical calculus, he emphasized their heuristic value as concrete technical artifacts performing synthetic operations.

This short overview should suffice to convey what I mean by a philosophy based on the object rather than on the subject. In all the works above mentioned technology is the central focus of the philosophical approach. From steam engines to computers and video games, from industrial machines to fine arts, we are faced with an attempt to understand human beings through their technical devices and their artefacts. These authors, each in their own way, are concerned with the concrete conditions of our life as it is framed by technical practices, organized by technical products, and conceived through technology. The next generation is currently following research pathways that echo those of Simondon or Dagognet, as this volume testifies.

2.2 A Historical Turn

A second common feature which is also shared by the French philosophers of science is that they take into account the historical dimension of the objects under study. Beaune's reflections on automation, for instance, are developed through a historical analysis of eighteenth and nineteenth centuries' automata. More recently Mathieu Triclot, before writing a philosophy of computer games, wrote a history of the emergence and early development of computer science in *Le moment cybernétique* (2008). When French philosophers of technology quote historians such as Bertrand Gille (1978) and Maurice Daumas (1962, 1965, 1971), it is not simply to provide their topics with an erudite respectability. If Simondon went back to Ionian physiologists in his course on perception, or if Dagognet have conducted extensive studies of Louis Pasteur, Claude Bernard and Etienne-Jules Marey, this is not by intellectual vanity, it is rather to restitute all the human depth of technology.

This historical turn is based on two assumptions.

1. Technologies are always in process. They unfold over time, something that technology books, which deliver the state of the art for a given time, often forget. The state of the art today is closely related to that of yesterday. So Simondon wrote about the genesis of the technical object[5]:

[4] In *Météores* (1997), Parrochia had also shown how the thermodynamic machine model, fluid mechanics, and vacuum pumps, provided the explanatory frameworks for meteors long before computer simulations were available.

[5] It is worth noting that the first chapter of *Du mode d'existence des objets techniques* is entitled "Genèse de l'objet technique: le processus de concrétisation." (Simondon 1958)

> It is from the criteria of genesis that we one can define the individuality or specificity of any technical object. The individual technical object is not such and such a thing, given *hic et nunc*, but that which has a genesis. The unity, individuality, and specificity of a technical object, are the characteristics of consistency and of convergence of its genesis. (Simondon 1958: 20)

Any technology is fundamentally a matter of "technogenesis" which impinges on the future. As a result, thinking about a technology is thinking about its development, as well as about its future. Technology is always considered as an ongoing process of progress and stasis. Therefore, the history of a technology provides insights into its ontological, economic, social and political meanings. They shed light on the present of technology in its most recent forms and matter for the choices the future requires.

2. If the genetic approach is essential for understanding any technology it is because technogenesis plays a crucial part in the anthropogenesis. The interactions of technology with other human activities have to be investigated in the *longue durée*. For instance, humans' working activities and their effectiveness, or their aesthetic preferences themselves, are never absent from technical practices, and this has been the case since the dawn of humanity, as prehistorian Sophie Archambault de Beaune has pointed out. (2008)

The political and economic history of societies is an integral part of the French philosophy of technology. For instance, Dagognet's *Réflexions sur la mesure* (1993b) considers the measurement tools in their technical aspects as well as in their commercial uses, as Dagognet describes a variety of niches where measurement plays a central role in civil society. All fields, including medicine, depend on measuring instruments. The history of medicine is full of instruments and exploration tools. The history of meteorology is closely linked to the improvement of barometers and thermometers. Industries, but also sociology and psychology, rely on the development of measurement instruments. On this broad historical basis Dagognet outlines a philosophy of measurement. In *L'argent, philosophie déroutante de la monnaie* (2011), and in *Philosophie du travail* (2013), he developed a similiar historico-philosophical approach.

As they consider techniques as a historical and temporal phenomenon, French philosophers are concerned with the chronology of inventions. Archambault de Beaune (2008), drawing on Leroi-Gourhan but also on Levi-Strauss, Simondon and Jean-Pierre Séris (1987, 1994), questions the nature of invention and technical innovation since prehistory. Or else, Séris argues that technical convergences (simultaneous inventions) suggest a kind of determinism of the invention and dissemination of technical practices: "the often met cases of simultaneous inventions force to admit, at least as a plausible hypothesis, a quasi-necessity of the invention, once the conditions that make it possible are met." (Séris 1994: 97)

Séris alludes to the impressive number of tools and technical inventions that occurred during the Neolithic period at different times and in different places in absence of diffusion processes. He also mentions that in the pre-industrial era, in

Europe, before Huygens, Watt and others gave their names to well-known machines, many similar anonymous inventions occurred several times.

In addition to the chronology of inventions French philosophers pay attention to the spatial distribution and changes of technical objects. They also seek to understand how the current state of our everyday technological world, of the mundane technological environment, reflects this game of encounters, combinations, accumulations, breaks, crossovers and hybrids of all kinds. There are complex links between technical invention, war, economical requirements, and social demands. For instance, the relationship between the development of aircraft and the First World War is well known. Archambault de Beaune covers these aspects of the technical evolution of societies through the material, social, economic, political of these different aspects of the history of technology. This explains why some philosophers try to unravel this complexity of the history of technology.

2.3 Humans As Technicians

This French tradition considers technologies as an essential dimension of the human being; they define our humanity as well as the language or art. Unless we stick to pure ideas and abstract speculations, anything that is related to humans, exists in a material world that can be shaped, improved, used, and controlled only by technical processes. These processes in turn require tools, machines, devices, and things already developed by human labor.

This applies to all objects of our daily lives, including pieces of fine arts. Even poetry and literature needed the material existence of the voice before marble or paper and digital media today. Science itself cannot be detached from techniques. It is not just that experimental activities require the support of technical instrumentation such as, for example, large particle accelerators which bring together scientists and engineers, as Bachelard emphasized with his concept of "*phenomenotechnique*." (Bachelard 1949) Science is closely related to technology, because we cannot understand the world – which is the purpose of the scientific enterprise – except by transforming it.

As technology is considered a fundamental dimension of human life, the philosophy of technology is a true technical anthropology. Talking about technologies and talking about humans amount to the same thing. In Dagognet's works and in the publications of the so-called "Lyon school," one can find many examples of this strong link between technology and anthropology, among which I select: (1) the philosophy of medical technology, (2) the technological milieu and (3) the existential function of mathematics in technology.

(1) Dagognet, who was both a medical doctor and a philosopher,[6] insisted on the materiality of the human condition. To treat a patient is working on the materiality of the body, with technical tools which become, over time, more and more sophisti-

[6] In this passage, I mainly referred to François Dagognet (1964, 1990, 1992b, 1998a, b, 2003b).

cated. To cope with diseases and death, our fragile physical condition frequently requires technical interventions on the body, so that care has included technological aspects probably since prehistoric times. In this respect, the philosophy of medical technology is necessarily a philosophy of the human—not of an abstract individual, like the Cartesian subject defined by a pure consciousness, but the flesh-and-blood individual, struggling with herself in a constant battle against diseases and death. Dagognet's philosophy of medical technologies is about organic humans with their limbs, viscera, their immune system, hormonal system, nervous system, etc. Does it mean that medical technology breaks up humans, smashes our body into pieces in a reductionist approach that overlooks the individual as a whole to focus on isolated organs? This reproach of reductionism is based on a specific view of technology as dehumanising, while nature alone would have the power to restore health with the assistance of the physician. The doctor's role would only be limited to encourage nature – *sola natura medicatrix*. Against such technophobic assumptions, Dagognet is a staunch advocate of technology as artifice and cunning. For instance, he claimed that industrially produced drugs are purer and safer than the same molecule taken in a "natural" medicine, where it works alongside many other accompanying compounds which can reduce primary effect or induce toxicity. Dagognet welcomed the development of medical techniques from all means that allow the clinician to work within the body, to identify the slightest traces of organ dysfunctions using various medical imaging methods. From stethoscope to Magnetic Resonance Imaging (MRI), medical instruments contribute to put the inside outside, revealing and circumscribing the disease. It is the role of medicine, in Dagognet's view, to develop all kinds of tools to track down abnormalities and dysfunctional systems and to target them with the modern arsenal of chemotherapy and radiotherapy.

The philosophy of medical technology is concerned with sick and healed humans. This conjunction of anthropology and technology leads to a form of humanism that fully assumes the dimension of the human technician and has nothing in common with the abstract humanism developed by idealistic philosophers. In Dagognet's perspective humans are both the subject and the object of technical practices.

(2) A second example of the French anthropological approach to technology is Jean-Claude Beaune's work on technology and technical practices which focus on the concept of "technical milieu," as suggested by the titles of some of his books (1980b, 1983). In addition to the design and use of tools and devices, technological activities involve the creation of the environment in which humans live, that is to say, the world in which we die as well. Our attachment to certain objects that we keep as souvenirs that we want to convey to the future generations testifies for the meaning of artifacts with respect to our mortal condition. Beaune insisted that Vaucanson's automata, as well as most of our modern robots, be they playful or productive, always go beyond their original purpose. As mirror-images of humans they lead us into repetition and mimic our madness while we try to emulate them. Psychiatrists of the nineteenth century used to describe some of their patients as "ambulatory automata." In the technical milieu, the machine exhausts the meaning when it does not reduce life to its share of illusion, as suggested by Vaucanson's Duck that did not really digest, or Kempelen's chess player, a true imposture.

According to Beaune, from Cartesian automata to current industrial robots, machines allow us to develop a philosophy of time that they mechanically measure. Beyond Bergson's intuitionism and Bachelard's "rhythm analysis" (1932, 1950), Beaune's philosophy of technology is a meditation on temporality. As humans are both technical beings and mortal beings, the technical environment may also be fatal. Beaune is perfectly aware of the mortifying aspect of techniques (Beaune 2002). For example in his latest works (2009, 2013), he saw the shadow of "Big Brother" behind the implementation of Big Data.

From end to end, Beaune emphasizes the way we project our fantasies onto our technical accomplishments. It would therefore be irrelevant to separate *Homo sapiens* from *Homo faber*, the human being from his artifacts.

(3) Daniel Parrochia, another Dagognet's former students, insisted on the mathematical structures underlying technology. The idea that the universe can be mathematically described which is at the origin of modern science has been developed prior to Galileo by Renaissance artists and engineers who discovered that realistic representations in the arts required geometrical perspective and that a technical transformation of the world (watches, fortifications, architecture) need mathematical operators and abstract structures. Since then, many authors have endeavoured to show the relationships between technical developments and their underlying mathematical structures – Carnot (1797), Reuleaux (1877), Lafitte (1932) among others. As technologies spread throughout the human sphere, work, leisure, and lifestyle, they also spread these mathematical and scientific structures although they remain invisible. Uncovering such hidden structures is Parrochia's task. He points out the essential role of technology in the human condition. His approach, that I have elsewhere labelled "mathematical existentialism," (Beaune and Chazal 2009) is essential because mathematics underpinning technological achievements enhance the role of humans.

In *Mathématiques et existence* (1991), Parrochia scrutinized the mathematical foundations of any technical activity, as a break with the natural order involved in the process of humanization. This rupture is characterized as a source of anxiety, as clear from ancient mythology. The technique is then both the cause of the break with the natural order and the means to ward it off. Thus understanding the mathematical substrates of technology allows to better control our environment. We basically need, as Parrochia said, "cards," that is to say, sets of mathematical tools that enable us (1) to provide, through the formulation of laws, an explanation – or at least a description – of the world and (2) to provide efficient methods to transform it. In investigating road-building technology, bridges, tunnels, dams, nuclear stations, and aerospace (1997, 2003), Parrochia explores the most concrete aspects as well as the mathematical forms they develop. Mathematics is presented as a major marker and guide for gradually building a technical world from the stone-age industries of prehistory to today's digital technology at work in computer devices and the organization of networks. In *Cosmologie de l'information* (1994) and *Philosophie des réseaux* (1993), Parrochia aims at a global study of technical practice, from Ader to Internet, letting nothing escape this process of "technogenesis," at least in modern times.

There are indeed recent technical developments that the above-mentioned philosophers did not fail to grasp: tools and various devices work less on matter and energy and increasingly on information, that the machines are more and more autonomous. Dagognet, in *Ecriture et iconographie* (1973) and *Mémoire pour le futur, vers une méthodologie de l'informatique* (1979). Beaune in *Philosophie des milieux techniques* (1998) among others, and Parrochia in *Cosmologie de l'information* and *Philosophie des réseaux* (1993, 1994) insist on the process of "dematerialization" of technologies. The word does not imply a lack of material but precisely this increasing importance of information. However, this profound transformation of techniques becomes central for the younger generation of French philosophers, as instantiated in Triclot's (2008, 2011) or Varenne's (2007, 2009) works, and numerous works on computer imaging, as Stéphane Vial's one (2013).

This overview of the French tradition of technical philosophy would not be complete without the presence of both Séris (1987, 1994) and Tinland (1977) who developed an anthropological philosophy of technology. For them, technical artifacts define mankind as much as language and fine arts, and are fully part of culture. On the one hand, Tinland, reflects on the differences technical development induce between man and animal. As the title of his main work *La différence anthropologique* indicates, technicity both links and separates humans from other living forms. On the other hand, in Séris (1987) we can see how our history, first in Europe, is also the development and the progressive theorization of machines. Séris' works emphasize the interaction between mathematics and practice in the production of various "technological" discourses (discourses about "*techniques*" as well as mathematical formalizations of them).

2.4 From Technologies to Fine Arts and Social Risks

In addition to the focus on objects and the prevailing anthropological perspective a third distinctive feature of the French tradition is the disregard for the utilitarian approach to tools and machines, as well as for the economic framework of goods manufacture. As they tend to focus on the genesis of technicity in its historical and anthropological dimension rather than on technological *applications*, French philosophers often include fine arts in the domain of technicity.

We know that Renaissance artists and engineers, mentioned above, did not separate art and *technique*. For a long time the two words were almost synonymous. The tool and the machine always received ornaments that gave them their share of aesthetics. Craftsmen adorned the sleeves of their tools; nineteenth century engineers decorated their steam engines and machine tools with *Art Nouveau* frames. The close association of techniques and aesthetics is still present in the French language since the term "*art*" translates the Greek "*technè*" and refers both to fine arts and to techniques intended as ways of achieving something.

Dagognet instantiates this intimate connection when he moved from his study from medical technology to modern and contemporary art of which he became an ardent defender. In visual arts he praised the technical attention to objects, to their materiality, to the skills embedded in their manufacture. He was one of the first scholars who paid attention to the life-cycle of mundane objects and the transmutation of waste into pieces of fine arts (Dagognet 1997). Literary references are continuously present in the work of Beaune and Parrochia (1996). Parrochia devoted an important book to music: *Philosophie et musique contemporaine* (2006). Among the next generation of philosophers who pursue this tradition of French philosophy of technology, some of them such as Triclot and Vial are especially sensitive to the strange hybrids of art and technology currently met in the digital world (video games, digital interfaces, virtual avatars…).

However, while they combine technologies and fine arts in a single approach, French philosophers cannot ignore the huge amount of human work involved in the manufacture of objects for their needs and pleasure, nor detach objects from their inscription in an economic, political and social environment. They do not overlook that techniques, technologies and skills are developed in an economic, political, and social context which they determine in return. Indeed, since the nineteenth century, whether explicitly in the shadow of Marx, or not, this view hangs over the whole philosophy of technology and over any critical analysis of technology. The socio-economic dimension is constantly present in the work of Beaune, who was familiar with the industries developed in Clermont-Ferrand and Le Creusot, and showed the most acute awareness of the risks attached to techniques when they are exclusively profit-oriented. *In Philosophie des milieux techniques*, he portrayed the role of engineers both as masters of the transformation processes of matter and as mediators the social organization of production. The social dangers of techniques are also important in the work of Hottois (2004). Furthermore Dagognet, who claimed several times to be "socialist" even if he was not part of any political party, did not fail to take an interest in this aspect of technology. At least four books testify to this view: *Philosophie de la propriété, Avoir*, (1992c) *Une nouvelle morale* (1998b), *L'argent, philosophie déroutante de la monnaie*, (2011) and *Philosophie du travail* (2013), this last book written in collaboration with four of his former students.[7] As for Parrochia, extensive analysis in this area can be found in his book *La forme des crises* (2008).

2.5 Conclusion

In conclusion we can retain four key points.

First, the history of technology is an integral part of human history and we cannot understand the latter without the former. Therefore, to understand the process of humanization is also to be interested in technogenesis. Mankind has been shaped by technologies, by their evolution and their increasing empire over the world.

Second, the human phenomenon is characterized by a distanciation from the natural order. The increasing gap between human achievements and natural constraints

[7]This book brings together François Dagognet with Jean-Claude Beaune, Gérard Chazal, Robert Damien and Daniel Parrochia.

is essentially, although not exclusively, the result of technical practices. We cannot therefore understand the human phenomenon if we are not also interested in the techniques and technology and this sole feature justifies the existence of a philosophy of technology. Humanity is constituted as such through technical practices, the use of tools and machines. Man is a being of devices because he is a being of always-renewed desire. Therefore, a philosophy of technology is necessarily an anthropology.

Third, the philosophy of technique must fit into a general philosophy of the social that takes into account the economic and political context in which technologies are developed.

Finally, and this point is of particular importance, this tradition testifies for an immense fondness for technical objects and activities. Far from considering the former as pure means for securing wealth and well-being, and the latter as obscure and messy practices, French philosophers, even among the young generation, praise techniques for themselves. Dagognet's legacy has imprinted among them the respect for technical practices as human achievement. And while they may emphasize the risks generated by technical advances they are neither dogmatic technophobics nor gullible technophilists.

Acknowledgements A preliminary version of this text has been discussed by Mathieu Triclot and Vincent Beaubois during the authors' workshop held in Paris Sorbonne on 22–23 June 2015 for the preparation of the collective book.

References

Bachelard, G. (1932). *L'intuition de l'instant*. Paris: Stock. English Edition: Bachelard, G. (2013). *Intuition of the instant. Studies in phenomenology and existential philosophy* (E. Rizo-Patron, Trans.). Evanston: Northwestern University Press.

Bachelard, G. (1949). *Le rationalisme appliqué*. Paris: Presses Universitaires de France.

Bachelard, G. (1950). *La dialectique de la durée*. Paris: Presses Universitaires de France. English Edition: Bachelard, G. (2016). *The dialectic of duration* (M. McAllester Jones, Trans.). London: Rowman & Littlefield International.

Beaune, J.-C. (1880a). *La technologie introuvable*. Paris: Vrin.

Beaune, J.-C. (1980b). *L'automate et ses mobiles*. Paris: Flammarion.

Beaune, J.-C. (1983). *Le vagabond et la machine*. Seyssel: Champ Vallon.

Beaune, J.-C. (1998). *Philosophie des milieux techniques*. Seyssel: Champ Vallon.

Beaune, J.-C. (2002). *Le balancier du monde. La matière, la machine et le monde. Essai sur le temps des techniques*. Seyssel: Champ Vallon.

Beaune, J.-C. (2009). *Anthropologie des milieux techniques I Engrenages*. Seyssel: Champ Vallon.

Beaune, J.-C. (2013). *Anthropologie des milieux techniques II Machination*. Seyssel: Champ Vallon.

Beaune, J.-C., & Chazal, G. (2009). *Mathématisation du sensible*. Dijon: Éditions Universitaires de Dijon.

Carnot, L. (1797). *Œuvres mathématiques du citoyen Carnot*. Bâle: Dexker.

Dagognet, F. (1964). *La raison et les remèdes*. Paris: Presses Universitaires de France.

Dagognet, F. (1973). *Écriture et iconographie*. Paris: Vrin.

Dagognet, F. (1979). *Mémoire pour l'avenir, vers une méthodologie de l'informatique*. Paris: Vrin.

Dagognet, F. (1989). *Eloge de l'objet*. Paris: Vrin.
Dagognet, F. (1990). *Corps réfléchis*. Paris: Odile Jacob.
Dagognet, F. (1992a). *Pour l'art aujourd'hui*. Paris: DisVoir.
Dagognet, F. (1992b). *Le corps multiple et un*. Paris: Les empêcheurs de penser en rond, Synthélabo.
Dagognet, F. (1992c). *Philosophie de la propriété*. Paris: Presses Universitaires de France.
Dagognet, F. (1993a). Entretien au journal Le Monde. *Le Monde* (reprinted in the obituary of the same newspaper, 4 october 2015).
Dagognet, F. (1993b). *Réflexions sur la mesure*. La Versanne: Encre Marine.
Dagognet, F. (1997). *Des détritus, des déchets, de l'abject, une philosophie écologique*. Paris: Les empêcheurs de penser en rond, Synthélabo.
Dagognet, F. (1998a). *Savoir et pouvoir de la médecine*. Paris: Les empêcheurs de penser en rond, Synthélabo.
Dagognet, F. (1998b). *Une nouvelle morale*. Paris: Les empêcheurs de penser en rond, Synthélabo.
Dagognet, F. (2003a). *100 mots pour comprendre l'art contemporain*. Paris: Les empêcheurs de penser en rond, Synthélabo.
Dagognet, F. (2003b). *Penser le vivant*. Paris: Bordas.
Dagognet, F. (2011). *L'argent, philosophie déroutante de la monnaie*. La Versanne: Encre Marine Les Belles Lettres.
Dagognet, F., Beaune, J.-C., Chazal, G., Damien, R., & Parrochia, D. (2013). *Philosophie du travail*. La Versanne: Encre Marine Les Belles Lettres.
Daumas, M. (Ed.). (1962). *Histoire générale des techniques. Tome I: Les origines de la civilisation technique*. Paris: Presses Universitaires de France.
Daumas, M. (Ed.). (1965). *Histoire générale des techniques. Tome II: Les premières étapes du machinisme*. Paris: Presses Universitaires de France.
Daumas, M. (Ed.). (1971). *Histoire générale des techniques. Tome III: L'expansion du machinisme*. Paris: Presses Universitaires de France.
de Beaune, S. A. (2008). *L'homme et l'outil, l'invention technique dans la préhistoire*. Paris: CNRS éditions.
Foucault, M. (1985). La vie : l'expérience et la science. *Revue de Métaphysique et de Morale, 90*(1), 3–14. English edition: Foucault, M. (2003). Life: Experience and science. In J. D. Faubion, (Ed.) (1998). *Aesthetics, method, and epistemology. Essential works of foucault* (Vol. 2). New York: The New Press.
Gille, B. (1978). *Histoire des techniques*. Paris: Gallimard.
Hottois, G. (2004). *Philosophie des sciences, philosophie des techniques*. Paris: Odile Jacob.
Lafitte, J. (1932). *Réflexions sur la science des machines*. Paris: Bloud et Gay.
Parrochia, D. (1991). *Mathématiques et existence: ordes, fragments, empiètements*. Seyssel: Champ Vallon.
Parrochia, D. (1993). *Philosophie des réseaux*. Paris: Presses Universitaires de France.
Parrochia, D. (1994). *Cosmologie de l'information. Pour une nouvelle modélisation de l'univers informationnel*. Paris: Hermès.
Parrochia, D. (1996). *Ontologie fantôme: essai sur l'œuvre de Patrick Modiano*. Paris: Belles Lettres.
Parrochia, D. (1997). *Météores: essai sur le ciel et la cité*. Seyssel: Champ Vallon.
Parrochia, D. (2003). *L'homme volant: philosophie de l'aéronautique et des techniques de navigation*. Seyssel: Champ Vallon.
Parrochia, D. (2006). *Philosophie et musique contemporaine. Ou, le nouvel esprit musical*. Seyssel: Champ Vallon.
Parrochia, D. (2008). *La forme des crises. Logique et épistémologie*. Seyssel: Champ Vallon.
Reuleaux, F. (1877). *Cinématique, principes fondamentaux d'une théorie générale des machines*. Paris: F. Savy.
Séris, J.-P. (1987). *Machine et communication, du théâtre des machines à la mécanique industrielle*. Paris: Vrin.
Séris, J.-P. (1994). *La technique*. Paris: Presses Universitaires de France.

Simondon, G. (1958). *Du Mode d'existence des objets techniques*. Paris: Aubier. English Edition: Simondon, G. (2016). *On the Mode of Existence of Technical Objects* (trans: Malaspina, C.). Minneapolis, MN: Univocal Publishing.
Simondon, G. (2006). *Cours sur la perception (1964–1965)*. Paris: Presses Universitaires de France.
Simondon, G. (2014). *Sur la technique (1953–1983)*. Paris: Presses Universitaires de France.
Snow, C. P. (1959). *The two cultures and the scientific revolution*. New York: Cambridge University Press.
Tinland, F. (1977). *La différence anthropologique*. Paris: Aubier.
Triclot, M. (2008). *Le moment cybernétique*. Seyssel: Champ Vallon.
Triclot, M. (2011). *Philosophie des jeux vidéo*. Paris: La Découverte.
Varenne, F. (2007). *Du modèle à la simulation informatique*. Paris: Vrin.
Varenne, F. (2009). *Qu'est-ce que l'informatique?* Paris: Vrin.
Vial, S. (2013). *L'être et l'écran, comment le numérique change la perception*. Paris: Presses Universitaires de France.

Chapter 3
On the Current Uses of Simondon's Philosophy of Technology

Vincent Bontems

> *The question of whether machines can think is about as relevant as the question of whether submarines can swim.*
> *Edsger W. Dijkstra.*
> *I imagined many moons in the sky lighting the way to freedom.*
> *Cindi Mayweather.*

Abstract This chapter provides a brief survey of the current uses of Gilbert Simondon's philosophy of technology. The long-lasting lack of translation of his book *Du Mode d'existence des objets techniques* (1958) explains that the English-speaking readership knows little of his rich and deep thinking, but there is a significant trend of simondonian researches in Argentina, France, Germany and Italy. My analysis on the ongoing researches using Simondon's philosophy of technology follows three paths: *techno*-sciences, *techno*-aesthetics and *techno*-politics. The first focuses on the scientific design and technological concretization of research objects and instrumentation: technology is no more the mere outcome of research; it is a full-fledged epistemological element of *techno-science*. Simondon has proposed *techno-aesthetics* as an original theory of an aesthetic *of* technical which inspires many studies about industrial design. His philosophy of technology has also *techno-political* implication since its aim is to reintegrate technology into Culture in order to allow a mutual emancipation of man and machine.

Keywords Gilbert Simondon · Techno-logy · Techno-aesthetics · Techno-politics · Techno-science

V. Bontems (✉)
Laboratoire de Recherche sur les Sciences de la Matière (LARSIM), Commissariat à l'Énergie Atomique et aux Énergies Alternatives (CEA), Gif-sur-Yvette Cedex, France
e-mail: vincent.bontems@cea.fr

S. Loeve et al. (eds.), *French Philosophy of Technology*, Philosophy of Engineering and Technology 29,
https://doi.org/10.1007/978-3-319-89518-5_3

Gilbert Simondon attributed to his research the purpose to reintegrate the significance of the technical reality within Culture.[1] The purpose of this paper is to provide a brief survey on the current uses of his philosophy of technology[2] which try to accomplish this commitment. Therefore I will not summarize Simondon's work,[3] neither many interesting commentaries, nor current uses of his theory of individuation,[4] but only the attempts to apply, to critic and to reactivate his technological concepts in order to understand and transform our Culture.

In France, Simondon earned his reputation as a philosopher of technology in 1958 when the first edition of his PhD *Du mode d'existence des objets techniques* (Simondon 2013) was published, a book which is about to be published in English (Simondon 2017).[5] His other researches on technology remained confidential until recently. His courses about the notion of invention, written around 1970, were finally published in 2005 (Simondon 2005a). All his other papers and works devoted to technology (between 1953 and 1983) have been gathered and republished only in 2014 (Simondon 2014). This delay explains why the core of most current uses of his philosophy of technology is to be found in interpretations of his first book.

The long-lasting lack of translation explains also why, despite the introduction provided by Paul Dumouchel (1992), the English-speaking readership knows little, if not nothing, of his rich and deep thinking. However, this is not the case in the rest of the world, and there is a significant trend of Simondonian researches in Europe, in South America and in Asia.[6]

The aim of this chapter is to provide a brief survey of the current uses of Simondon's philosophy of technology following three paths: *techno*-sciences, *techno*-aesthetics and *techno*-politics. Since these are not self-evident categories, I will briefly define the origins of each field, identify the main contributors and focus on some exemplary contributions from Argentina, France, Germany and Italy. As director of the Atelier Simondon,[7] I had the opportunity to meet and discuss with

---

[1] The very first line of *Du Mode d'existence des objets techniques* is: "The purpose of this study is to attempt to stimulate awareness of the significance of technical objects." (Simondon 2013: 1) In the writings of Simondon "Culture" with a capital C encompasses all human achievements, including technology, in contrast with "culture" with a tiny c, which only means literary and spiritual works.

[2] The translation of "*technique*" is always somehow problematic since "technique" does not have the same meaning. "Technology" is more convenient, but Simondon himself used "*technologie*" in its original French sense of "knowledge of technics." Here, I will translate "*technique*" by "technology" and "*technologie*" by "techno-logy."

[3] For an excellent general introduction, see Barthélémy (2014).

[4] For a complete bibliography, one may of the International Center for Simondonian Studies (CIdES): http://www.mshparisnord.fr/cides/index.php/menu-3.html

[5] There was previously on the Internet an incomplete unofficial translation by Ninian Mellamphy (1980) and Dan and Nandita Mellamphy (2013).

[6] A sign of this rising interest is the 10 days international congress in Cerisy-La-Salle in august 2013 on "Gilbert Simondon and the invention of the future" with researchers from Argentina, Belgium, Canada, Czech Republic, France, Great Britain, Italy, Japan, Peru, Russia, Taipei, Tunisia and United States. See the proceedings (Bontems 2016).

[7] For a description of the activities of the Atelier Simondon from 2008 until 2013, see Bontems (2009a, 2010, 2012), Bontems and Pascal (2011) and Bontems and Beaubois (2013).

many of these (often young) researchers committed to develop new fields of enquiry in Simondon's wake, but I apologize in advance if I have missed other important contributions or if I do not give enough credit to my colleagues. In my view, the specificity of their ongoing researches is that they use Simondon's philosophy of technology not only as a toolbox, neither as a closed paradigm, but as an open system and a methodology of enquiry about the impact of technological progress in our civilization. They do not repeat or try to discard his concepts but to update and improve their operative value.

3.1 Techno-Sciences

The term "techno-science" has been used as a contraction of "technological science" with various meanings since at least 1956 (Raynaud 2015). Still, the Belgian philosopher of science and technology Gilbert Hottois is often credited to have coined it in his PhD (Hottois 1977; this volume). Later he defined the field of 'technoscience' as "the indissolubility of both theoretical and technical-operative poles" (Hottois 1984: 60). In 1993, he was also the first scholar to devote a book to Simondon's philosophy of the 'technological culture' (Hottois 1993). But he did it without establishing any link between the Simondonian concept of technological culture and his own notion of techno-science. Only recently he explicitly identified the Simondonian notion of techno-logy with techno-science: "'Techno-logy' is the technoscientific research." (Hottois 2004: 129) According to him, "many passages and the general scope of Simondon's philosophy of technology light up and acquire another luster if one admits that what matters for him was research, scientific-technological discovery-invention." (Hottois 2004: 129) Although the simondonian notion of technology is obviously broader than just technoscience, to prove his point, Hottois emphasizes this quote of Simondon: "nowadays, the true technical activity is in the field of scientific research, which is oriented, because it is a research, towards objects or properties of objects that are still unknown. The free individuals are the ones who effectuate the research and institute thus a relation with the *non-social* object." (Simondon 2005b: 512) The end of the quote shows without ambiguity that Hottois' notion of technoscience is opposed to the confuse mix of science, technology and society that one may find in Bruno Latour's use of the notion[8] (Latour 1987). As Jean-Hugues Barthélémy has insisted at the same time (Barthélémy 2004, 2005), technoscience should rather be thought as an extension of

[8] In my view, Simondon's philosophy of technology is incompatible with Latour's method: the commitment of the former to explain the functioning of machines contradicts the treatment of scientific instruments as black-boxes; his theory of the cultural phases is much more complex and differentiated than the so-called actor-network theory; his theory of individuation and concretization forbids the use of an anthropocentric category such as "non-human beings" to subsume both living and technical (and fictional) beings.

Gaston Bachelard's notion of "phenomeno-technology," (*phénoménotechnique*) (Bachelard 1985) which means that the contemporary scientific phenomenon are not only observed through technical instruments but even produced by these experimental devices as artificial actualizations of natural potentialities. Therefore, physics and chemistry are developing a technical and rational materialism: "from now on, we must acknowledge that it is only by a misuse of language that we say that the chemical phenomenon is a *natural* phenomenon. The artificial materialism, the scientific chemistry, the rationalism of the inter-materials laws have projected on the 'mineral realm' a network of relations which do not occur *in nature*." (Bachelard 1972) Techno-science is not only applied science but a fundamental technology that creates new artificial possibilities – new technological modes of existence – at all scales.

According to this interpretation of techno-science, Simondon's philosophy of technology is constructively used in the context of the history of scientific instrumentation by Henning Schmidgen. Taking Ian Hacking's phrase, "Experimentation has a life of its own," (Hacking 1983: 150) as a starting point, he investigates how experimental set-ups have emerged and evolved in nineteenth century physiology and psychology: e.g., how Hermann von Helmholtz's path-breaking time measurements of the propagation speed of nervous impulses led to an increasing use of similar precision measurement procedures in physiological and psychological laboratories. Since Simondon has exemplified the difference between technical individuals and technical "settings" (*ensembles*) by the "laboratory for the physiology of sensations" (Simondon 2013: 61), Schmidgen examines the scientific instruments of the nineteenth century which rarely were isolated individuals and worked as components of complex material assemblages, involving energy sources, tools for calibration, etc. (Schmidgen 2005). He also uses Simondon's notion of "technological lineage" in order to reconstruct a series of interconnected experimental set-ups devised by time measuring physiologists between 1861 and 1870 (Schmidgen 2014). He concludes that these instruments are *not* merging in the way Simondon has described "concretization" (the increase of synergy) with respect to combustion engines and vacuum tubes.

However, the Simondonian approaches of techno-sciences do not just extend the analyses of phenomeno-technology. Some of them switch to contemporary laboratory practices that go beyond modern experimental science. While Bachelard was primarily concerned by the epistemological break with the notion of "thing" and the technological constitution of the phenomenon as a scientific object, these analyses of techno-science focus on research objects that are technologically defined as machines and they show how the functioning of the instrument is constitutive of their knowledge. For example, Xavier Guchet refers to Simondon in his studies of nanotechnologies in order to analyse the invention of new ways of technical action on matter at this specific scale (Guchet 2014: 254). Sacha Loeve also transposes Simondon's concepts to nanotechnologies, which he defines as techno-science on account of the functional individualization of scientific objects (such as organic molecules) as technical objects. Some molecules are presented as "nano-machines" in order to emphasize their capacity to individually perform cinematic or electronic

Table 3.1 Matrix of technical values

Type/scale	Simple functioning	Median functioning	Major functioning
	1	2	3
Passive functioning	Solidity	Resistance	Stability
	4	5	6
Active functioning	Mechanical yield	Energetic yield	Transport yield
	7	8	9
Informational functioning	Fidelity	Sensitivity and precision	Complexity

operations (Loeve 2010), and Loeve provides an accurate Simondonian distinction between what are relevant operative analogies and what are only metaphors in the case of these "molecular machines." (Loeve 2015) Both thus provide an "object-oriented" analysis of technoscientific knowledge.

Despite the fact that Bernard Stiegler did not use the term "technoscience," he may be considered as anticipating this second orientation in the first volume of *Technics and Time* (Stiegler 1998), in which he intends to show that the invention, as anticipation of virtual objects, *presupposes* rather than precedes the mode of existence of technical objects. But Stiegler's "organology" remains focused on the human subject while a Simondonian methodology implies to shift the analysis towards the operations of the technical object itself. The need for this theoretical shift is explained by Guchet (2005, see also Guchet, this volume).

Mixing Simondon's "genetic mechanology" with methodologies of conception such as TRIZ (Altshuller 2007) or the C-K Theory (Le Masson et al. 2014), Vincent Bontems has analysed the functioning of scientific instruments and described some of their technical lineages (Bontems 2009b). For example, he analyses the phenomeno-technology of STM in nanotechnologies (Bontems 2011), of the Large Hadron Collider in particle physics (Bontems 2009c) and, in collaboration with Vincent Minier, of space-crafts such as the space observatory Herschel or the Martian-rover Curiosity (Bontems and Minier 2015). He tries also to formalise the genetic mechanology as an operative techno-logy thanks to diagrams (Bontems 2015a). One example of this "diagrammatization" is the following "matrix" (Table 3.1) which summarizes Simondon's final classification of machines.[9]

With this matrix Bontems address the relevant technical value for each aspect of functioning. This allows to assign each object to its paramount functionality or to profile the hierarchy between its functionalities. For example: Herschel's telescope

[9]After reading Jacques Lafitte's book *Réflexions sur la science des machines* (1932) in 1968, Simondon integrated his distinction between passive, active and informational machines (passive machines do not move, active machines are aimed to transmit movement, informational machine are aimed to transmit information) and crossed it with three degrees of functionality, which he defined according to the relation of the functioning with orders of magnitudes (the degree is minor when it is operating only at the scale of the structure of the object itself; it is median when it puts in relation this structure with the micro-structure of the material; and it is major when it puts in relation the structure and micro-structure with other scales of the global environment). See Bontems (2015b).

has an obvious informational functioning, which is to detect and to collimate infrared light, but it is a *simple* informational functionality, while the passive functionality of its structure is a *major* one, since its material (silicon carbide) and external structure are determined by its relation to the cosmic environment (and the transition between the initial environment on Earth and the final environment in outer space).

Analyzing the design and concretization of such big scientific instrument is not only a way to open a black-box in order to better understand the production of knowledge in astrophysics, it is also for Bontems a research-action in knowledge management: what has been learned about innovation with these complex processes may improve the ability of an organization to maintain its inventiveness.

3.2 Techno-Aesthetics

By offering an apprehension of objects aiming to reconcile their technical functioning with their aesthetic and psychosocial integration in the world, Simondon has produced a new perspective that inspires many current researches, especially on design issues. His philosophical reflection about aesthetics in the third part of *Du mode d'existence des objets techniques* was included in a general theory of cultural "phases": according to him, aesthetics thought is an attempt to restore an equilibrium between two opposite phases in the genesis of Culture, technology and religion. Technology is a way to focus on the operative structures in the world while religion expresses the feeling of the background and the value of the unity of the world. Hence, aesthetic thought builds of a specific relation to the world "starting from human effort to rebuilt unity" (Simondon 2013: 265) by creating artificial realities that have also a spiritual value. But it is not limited to a piece of art: Simondon shows how one can also deploy aesthetic thought in relation with a technical object, obtaining an aesthetic experience from its contemplation. A technical object arouses a real aesthetic experience when it is embedded in a context that underlines its own functioning, as in the case of a tractor in a field, during the ploughing.

As a scholar and an artist, Ludovic Duhem has thoroughly worked on this subject, showing how Simondon frees aesthetic thought from the frontiers of fine arts (Duhem 2012). Developing his own theory of techno-aesthetics, he reformulates some classical problems of aesthetics thought (imagination, creation, judgment, interpretation) in Simondonian terms, i.e. as genetic, relational and technical problems (Duhem 2010, 2013). But one may also find in Simondon a more original theory of an aesthetic *of* technical objects – a sort of 'inner aesthetics' of technical objects – i.e. a *techno-aesthetics* as it defined it later on (Simondon 2014: 379–384). As Giovanni Carrozzini argues in his book *Gilbert Simondon filosofo della* 'mentalité technique' (Carrozzini 2011a), techno-aesthetics leads Simondon to develop some suggestions of his friend the phenomenologist Mikel Dufrenne. Simondon's techno-aesthetics is concerned with the inner structure of technical objects, directly linked with their functioning, and techno-aesthetical experience derives from the accord with this functioning structure of technical objects. This accord arouses not from contemplation but throughout the use of this technical object (Carrozzini 2011b). When a techno-aesthetic experience runs out it produces in the user an *orgasmic pleasure*. Furthermore Carrozzini analyses the relation between this way of conceiving aesthetics *of* technical objects and the point of view about technical objects maintained by some of the most famous representatives of Italian industrial design, such as Bruno Munari and his idea of technical beauty as *correctness*: "in the field of design [...], you never give a judgement about the beauty or the ugliness of an object; you just say that it is correct or that it is not correct, and this correctness derives from its functions" (Munari 1981: 320–321). Provided one does not confuses "function" with "utility," the comparison between techno-aesthetics and industrial design may be extended also to the *behavioural design* of the American designer and psychologist Donald A. Norman (2004) because both bring out that our relations with technological objects are not only based on cognitive information but also, and maybe more, on emotions and motivations. The analysis of such psychosocial "halo" combined with the techno-aesthetics point of view, which integrates the synergy between the technological object and its "associated milieu" (*milieu associé*) provides deep insight into current design issues.

As a matter of fact, the application of Simondon's techno-aesthetics to the study of design is one of the strongest and most diversified field of Simondonian scholarship. The Atelier Simondon has organized for 2 years a seminar in the school of industrial design ENSCI (École National Supérieure de Création Industrielle) in which Barthélémy, Bontems, Carrozzini, Duhem, Loeve, Minier, but also Vincent Beaubois, Sebastien Bourbonnais, Frederic Pascal and Victor Petit have presented their contributions. Beaubois is particularly involved in this field, exploring the genesis of western material culture beyond handicraft and industry (Beaubois 2013). Meanwhile Bourbonnais has written about a simondonian approach of digital architecture (Bourbonnais 2009) and Loeve about simondonian techno-aesthetics in nanotechnology (Loeve 2011; Loeve, this volume). All these researchers are not only quoting Simondon but also following his suggestion by elaborating a complete design theory, which avoid both the reductionisms of functionalism and of the "user experience" oriented models.

3.3 Techno-Politics

The third part of *Du mode d'existence des objets techniques* describes Culture has a poly-phased system emerging from an initial "magical phase." There is no such thing as a "political phase" in this theory of cultural phases because "Culture" itself as a whole regulates the functioning of society. And nevertheless, philosophy, the ultimate reflexive phase of Culture, plays a crucial role which has political or, more exactly, techno-political implications: philosophy is supposed to reveal the contribution of each phase to Culture. This is what can justify, despite the absence of a political phase, the techno-political nature of his philosophy of technology. For example, Simondon has deeply modified the problematic of alienation by reintegrating the human-machine and machine-machine relations within Culture. He has thereby re-elaborated the notion of information in a way that revives the epistemology of the critical theory of society, and he has given us a new paradigm to think politics as an open praxis.

Herbert Marcuse has long ago (Marcuse 1964) noticed the relevance of Simondon's philosophy of technology in order to think the cultural limits of the capitalist or socialist productivism, but he did not really base his own thinking on the same ground. Gilles Deleuze has also borrowed some concepts from Simondon but used them in his own (and perhaps loose and confusing) way. More recently, Muriel Combes has been the first scholar to treat Simondon as a political thinker (Combes 1999) but she has limited herself to his theory of psychosocial individuation. Andrew Feenberg has criticized what he perceived as limitations and political ambiguities in Simondon's philosophy of technology (Feenberg 2014) and argued for a mutual enrichment of "mechanology" and science studies. Stiegler's philosophy of "transindividuation" is rooted in Simondon's work and transposes Simondon's analysis of the transfer of the operations from the human operator to the machine to our time when it is the intellectual operations that are transferred (Stiegler 2015). But the field of Simondonian techno-politics has really been opened when both Barthélémy (2008) and Guchet (2010) insisted on the originality of Simondon's ambition to develop a "difficult humanism" which is a "technological humanism." Since the task of humanism, according to Simondon, is that "nothing human remains alien to us"[10] and technology is a concretization of human intelligence and gestures, the most important task for humanism is to reintegrate technology within Culture. Meanwhile Bontems (2013) and Carrozzini (2013) have emphasized the Simondonian analogy between this rehabilitation of the technical objects and the emancipation of slaves.[11] They discuss the issues of this fight against the alienation of machines and they compare it to the Marxist problematic of alienation.

More influenced by Combes, the translator into Spanish of *Du Mode d'existence des objets techniques*, Pablo Esteban Rodríguez, has emphasized the originality of Simondon's critical theory of information in order to elucidate some of the political, scientific and philosophical problems posed by its usual deterministic definition (Rodriguez 2010). Simondon has defined information not only according to the message content (as in Cybernetics) but according to the potential energy contained in the system formed by the transducer, the receiver and the signal propagation medium. Rodriguez relates Simondon's theory and its key notions (individuation, transduction, modulation, metastability, etc.) to a theory of knowledge and power which leads him to show that Deleuze's intention to update Michel Foucault's work in terms of an "episteme of the information" and of "control societies" cannot be

[10] This is a quote of Terence's play *The Self-Tormentor*.

[11] "Recognition of the modes of existence of technical objects must be the result of philosophic consideration; what philosophy has to achieve in this respect is analogous to what the abolition of slavery achieved in affirming the worth of the individual human being." (Simondon 2013: 1).

understood without a close examination of Simondon's philosophy of technology. In order to think the digital culture Yuk Hui has made interesting comments (Hui 2015) on the political implication of a new elaboration of the simondonian concept of information.

But the most impressive work about techno-politics is Andrea Bardin's book about the link between Simondon's epistemology and politics (Bardin 2015). He shows how the theory of individuation processes introduces a new paradigm for the social system beyond the traditional alternative between the organic "body politic" and the mechanical social automaton. Simondon's model is rather inspired by – but not identified to – the model of an "open machine" connecting the metastability of the system (in accord with the epistemology of life sciences and thermodynamics) to the partial indeterminacy of processes (in accord, this time, with the epistemology of quantum physics). By extending the normative schema of the living to the functioning of the machine itself, Simondon does *not* intend to identify the social system to either the organism or the technical object; he is rather questioning any structural classification of them by displaying an operative *analogy between politics and invention*. In fact, he defines all systems based on their different regimes of functioning, i.e. the different (and not *a priori* harmonic) processes of individuation that simultaneously take place within them.

In such a theory the concept of the transindividual plays a demystifying role by demounting the very image of human nature that all philosophical political imagination of the body politic has ever been based on, grounding the expectation of a definitive "neutral" solution to the problem posed by politics. In effect the two classical models for the body politic concurred to the same *neutralization* of political invention (reduced to either a natural – internal – finality or a technical – external – one), and the exclusion of the actual *emergence of finality* from within the social system. In short, conceiving politics itself as a problem to be solved, rather than the field in which collective problems emerge, meant to cancel the field of political invention, namely the process of experimentation in which finality would actually emerge *within* the body politic as the result of political struggles. It is from this perspective that Simondon's model of an "open machine" allows to originally frame the different normativities constituting and crossing the social system, isolating therein the peculiar relation between technical normativity and the properly transindividual function played by normative invention in social systems.

In order to promote such a "Society of Invention," Barthélémy is also currently working on an ambitious "human ecology," which develops and links together his project of a new theory of techno-scientific knowledge and his conception of a "difficult off-centered humanism" (Barthélémy 2015). In general, Simondon's technology appears to many like a new way to empower the ecological and ethical consciousness for breaking with productivism and consumerism in order to initiate transformations of the relationships between human beings and their natural, social and artificial environments (Bontems 2015c, 2016).

3.4 Conclusion

Despite their diversity, all these Simondonian researches on techno-sciences, techno-aesthetics and techno-politics share a common feature: the commitment to understand the role of technicity in Culture. As Bardin and Giovanni Menegalle (2015) emphasize, Simondon's analysis of the mode of existence of technical objects has revealed the inventive and thus anti-conservative power of technology when decoupled from the imperatives of productivity. Although productivity emerges as the single norm organizing the relationship between humans and their techno-symbolic "milieu" and between humans themselves, this norm is not inherent to the technical system itself. The ideology of performance, which rests on the unique criterion of productivity, should not determine our scientific, aesthetic or political thoughts, nor justify a reject of technology itself.

The central aim of Simondon's analysis of technicity is thus to substitute the questionable opposition between technology and culture with the more historical tension between pre-industrial and industrial forms of technical culture (Simondon 2015) that simondonian philosophers have to extend to post-industrial societies. Drawing on a distinction between closed and open social systems, introduced by Henri Bergson, and challenging the homeostatic model put forward by the cyberneticist Norbert Wiener, Simondon used this dichotomy to designate the inner antagonism of the social system in its process of transition across the industrial-technological threshold (when technology has reached a scale that exceeds all particular cultures). Simondon argues that in contemporary societies and at the largest order of magnitude, humans should be considered as "technicians of the human species," because their interventions in the technical system are often returned to them in the form of environmental instabilities that require further techno-symbolic reconfigurations. And this task is even more urgent now that the process of automation affects not only material but also intellectual production. The present danger, according to him, is that if the technicians of humanity are only either technocrats or cultural conservatives, then the political management of technological evolution will take the form of either passive adaptation or active ideological reaction. Central to his program, then, is the possibility of institutionalizing the openness of the technical system from below, starting with the reprogramming of individual cognitive capacities towards collective processes of individuation that do not merely resist but invent and experiment in the human techno-symbolic milieu that he names "Culture", i.e. religious, technical, aesthetical, scientific, moral and philosophical thoughts as different "phases." All cultural phases derive most of their schemes from previous technological operation. As a matter of fact, "Culture" itself comes from an analogy with "agriculture," i.e. with an old stage of technological development. The issue of our time is precisely to resorb the hysteresis between the humanist culture (which is only a part and not the whole of Culture) and the accelerating techno-scientific civilization (which should be considered as a meaningful part of Culture) in order to restore their relative contemporaneity. For this purpose philosophy has to invent a techno-logical encyclopaedism (Bontems 2006; Carrozzini

2006). In this sense, the current uses of Simondon's philosophy of technology may trigger invention and emancipation in our Culture if they amplify the technicity which is only implicit or remains repressed in other cultural phases.

References

Altshuller, G. (2007). *The innovation algorithm. TRIZ, systematic innovation and technical creativity*. Worcester: Technical Innovation Center.

Bachelard, G. (1972). *Le Matérialisme rationnel*. Paris: Presses universitaires de France.

Bachelard, G. (1985). *The new scientific Spirit*. Boston: Beacon.

Bardin, A. (2015). *Epistemology and political philosophy in Gilbert Simondon: Individuation, technics, social system*. Dordrecht: Springer.

Bardin, A., & Menegalle, G. (2015). Introduction to Simondon. *Radical Philosophy, 189*, 15–16.

Barthélémy, J.-H. (2004). Des instruments de connaissance. *Science et Avenir*, Special Issue, 46–50.

Barthélémy, J.-H. (2005). *Penser la connaissance et la technique après Simondon*. Paris: L'Harmattan.

Barthélémy, J.-H. (2008). *Simondon ou l'encyclopédisme génétique*. Paris: Presses universitaires de France.

Barthélémy, J.-H. (2015). La dérivation philosophique de l'homme et les enjeux de l'écologie humaine. *Appareil*. http://appareil.revues.org/2157. Accessed 20 Mar 2015.

Barthélémy, J.-H. (2014). *Simondon*. Paris: Belles Lettres.

Beaubois, V. (2013). Design et technophanie. In J.-H. Barthélémy (Ed.), *Cahiers Simondon. Numéro 5* (pp. 59–74). Paris: L'Harmattan.

Bontems, V. (2006). L'Encyclopédisme de Gilbert Simondon. *Revue philosophique, 2*, 30–49.

Bontems, V. (2009a). Activités de l'Atelier Simondon. In J.-H. Barthélémy (Ed.), *Cahiers Simondon. Numéro 1* (pp. 149–152). Paris: L'Harmattan.

Bontems, V. (2009b). Gilbert Simondon's genetic 'mécanologie' and the understanding of Laws of technical evolution. *Techne, 13*, 1–12.

Bontems, V. (2009c). Actualité d'une philosophie des machines. Gilbert Simondon, les Hadrons et les Nanotechnologies. *Revue de Synthèse, 1*, 37–66.

Bontems, V. (2010). Activités de l'Atelier Simondon. In J.-H. Barthélémy (Ed.), *Cahiers Simondon. Numéro 2* (pp. 147–150). Paris: L'Harmattan.

Bontems, V. (2011). How to accommodate to the invisible? The 'halo' of 'Nano'. *NanoEthics, 5*, 175–183.

Bontems, V. (2012). Activités de l'Atelier Simondon. In J.-H. Barthélémy (Ed.), *Cahiers Simondon. Numéro 4* (pp. 137–141). Paris: L'Harmattan.

Bontems, V. (2015b). Simondons Klassifizierung technischer Objekte. *Jahrbuch Technikphilosophie*, 43–57.

Bontems, V. (2013). Esclaves et machines, même combat. In J.-H. Barthélémy (Ed.), *Cahiers Simondon. Numéro 5* (pp. 9–24). Paris: L'Harmattan.

Bontems, V. (2015a). Le cycle de la technologie chez Simondon. In J.-H. Barthélémy (Ed.), *Cahiers Simondon. Numéro 6* (pp. 7–27). Paris: L'Harmattan.

Bontems, V. (2015c). Simondon. In D. Bourg & A. Papaux (Eds.), *Dictionnaire de la pensée écologique* (pp. 7–27). Paris: Presses universitaires de France.

Bontems, V. (Ed.). (2016). *Gilbert Simondon et l'Invention du futur. Colloque de Cerisy*. Paris: Klincksieck.

Bontems, V., & Beaubois, V. (2013). Activités de l'Atelier Simondon. In J.-H. Barthélémy (Ed.), *Cahiers Simondon. Numéro 5* (pp. 139–144). Paris: L'Harmattan.

Bontems, V., & Minier, V. (2015). Innovation: désir de consommation, d'exploration ou de progrès. In P. Mustière & M. Fabre (Eds.), *La Science: jusqu'où explorer? Actes des rencontres Jules Vernes* (pp. 34–43). Nantes: Coiffard.

Bontems, V., & Pascal, F. (2011). Activités de l'Atelier Simondon. In J.-H. Barthélémy (Ed.), *Cahiers Simondon. Numéro 3* (pp. 149–154). Paris: L'Harmattan.

Bourbonnais, S. (2009). Nature et constructions. Les architectures numériques à la lumière de la philosophie de Simondon. In J.-C. Bignon, G. Halin, & S. Kubicy (Eds.), *Conception architecturale et Approches environnementales* (pp. 97–110). Nancy: Presses universitaires de Nancy.

Carrozzini, G. (2006). *Gilbert Simondon: per un'assiomatica dei saperi Dall'«ontologia dell'individuo» alla filosofiadella tecnologia*. Lecce: Manni.

Carrozzini, G. (2011a). *Gilbert Simondon filosofo delle mentalité technique*. Milano: Mimesis/ Centro Internazionale Insubrico.

Carrozzini, G. (2011b). Esthétique et techno-esthétique chez Simondon. In J.-H. Barthélémy (Ed.), *Cahiers Simondon. Numéro 3* (pp. 51–70). Paris: L'Harmattan.

Carrozzini, G. (2013). Aliénation et travail chez Bataille et Simondon après Hegel et Marx. *Il Protagora, 20*, 343–361.

Combes, M. (1999). *Simondon. Individu et collectivité. Pour une philosophie du transindividuel*. Paris: Presses universitaires de France.

Duhem, L. (2010). Introduction à la techno-esthétique. *Archée*. http://archee.qc.ca/ar.php?page=article&no=343. Accessed 22 Sept 1015.

Duhem, L. (2012). Thinking aesthetic reality (with Simondon). *SubStance, 41*, 16–31.

Duhem, L. (2013). Entrer dans le moule. Poïétique et individuation. *La Part de l'œil, 27*(28), 227–258.

Dumouchel, P. (1992). Gilbert Simondon's plea for a philosophy of technology. *Inquiry, 35*, 407–421.

Feenberg, A. (2014). Simondon and constructivism. *Philosophy & Technology, 28*(2), 302–308.

Guchet, X. (2005). *Les Sens de l'évolution technique*. Paris: Léo Scheer.

Guchet, X. (2010). *Pour un Humanisme technologique*. Paris: Presses universitaires de France.

Guchet, X. (2014). *Philosophie des nanotechnologies*. Paris: Hermann.

Hacking, I. (1983). *Representing and intervening: Introductory topics in the philosophy of natural science*. Cambridge: Cambridge University Press.

Hottois, G. (1977). *L'Inflation du langage dans la philosophie contemporaine*. Bruxelles: Bruxelles University.

Hottois, G. (1984). *Le Signe et la Technique*. Paris: Montaigne.

Hottois, G. (1993). *Simondon et la Philosophie de la 'culture technique'*. Bruxelles: De Boeck University.

Hottois, G. (2004). *Philosophie des sciences, Philosophie des techniques*. Paris: Odile Jacob.

Hui, Y. (2015). Le cycle de la technologie chez Simondon. In J.-H. Barthélémy (Ed.), *Cahiers Simondon. Numéro 6* (pp. 29–46). Paris: L'Harmattan.

Lafitte, J. (1932). *Réflexions sur la science des machines*. Paris: Bloud & Gay.

Latour, B. (1987). *Science in action. How to follow scientists and engineers through society*. Cambridge: Harvard University Press.

Le Masson, P., et al. (2014). *Design theory. Methods and organization for innovation*. Cham: Springer.

Loeve, S. (2010). About a definition of nanotechnology: How to articulate Nano and Technology. *HYLE – International Journal of Philosophy of Chemistry, 16*, 3–18.

Loeve, S. (2011). Sensible atoms: A techno-aesthetic approach to representation. *NanoEthics, 5*, 203–222.

Loeve, S. (2015). La notion d'objet relationnel dans les nanotechnologies (avec et après Simondon). In J.-H. Barthélémy (Ed.), *Cahiers Simondon. Numéro 6* (pp. 59–67). Paris: L'Harmattan.

Marcuse, H. (1964). *One-dimensional man. Studies in the ideology of advanced industrial society*. Boston: Beacon Press.

Munari, B. (1981). La progeattualià techno-artistica. In M. Dufrenne & D. Formaggio (Eds.), *Trattato di estetica. 2. Teorie* (pp. 320–321). Milano: Mondadori.
Norman, D. A. (2004). *Why we love (or hate) everyday thing*. New York: Basic Books.
Raynaud, D. (2015). Note historique sur le mot 'technoscience', *Carnet Zilsel*. http://zilsel.hypotheses.org/1875. Accessed 4 Apr 2015.
Rodriguez, P. (2010). Sur le lien entre humanisme moderne et philosophie de la technique: Martin Heidegger et Gilbert Simondon. In *Revue Ibéroaméricaine de Science, Technologie et Société*. Buenos Aires: Centro Redes.
Schmidgen, H. (2005). Physics, ballistics, psychology: A history of the chronoscope in/as context, 1845–1890. *History of Psychology, 8*, 46–78.
Schmidgen, H. (2014). *Hirn und Zeit: Die Geschichte eines Experiments, 1800–1950*. Berlin: Matthes und Seitz.
Simondon, G. (2005a). *L'Invention dans les techniques. Cours et conférences*. Paris: Seuil.
Simondon, G. (2005b). *L'Individuation à la lumière des notions de forme et d'information*. Paris: Millon.
Simondon, G. (2013). *Du Mode d'existence des objets techniques*. Paris: Aubier.
Simondon, G. (2014). *Sur la Technique*. Paris: Presses universitaires de France.
Simondon, G. (2015). Culture and technics (1965). *Radical Philosophy, 189*, 17–23.
Simondon, G. (2017). *On the mode of existence of technical objects*. Minneapolis: Univocal Publishing.
Stiegler, B. (1998). *Technics and time, 1: The fault of Epimetheus*. Stanford: Stanford University Press.
Stiegler, B. (2015). *La Société automatique, 1: L'Avenir du travail*. Paris: Fayard.

Chapter 4
Christianity and the Philosophy of Technology in France

Daniel Cérézuelle

Abstract In the twentieth century, a number of French Christian philosophers have described technology as a civilization process, each one in his own way. A short survey of Jacques Ellul's, Dominique Chenu's, Emmanuel Mounier's and Jean Brun's major works, shows that their diverging views were rooted in contrasting understandings of the Christian concept of incarnation. The dilemma spiritualization *vs* incarnation which underpinned their view of technology may still be relevant to understand the uncritical sanctification of technology in this century.

Keywords Autonomy · Disincarnation · Incarnation · Spiritualization · Technological spirit

Among the French thinkers who published about "*la technique*" it is not easy to distinguish those who were Christians from those who were not. This chapter will consider only those who have explicitly affirmed their Christian faith and analyzed the role of technology with respect to their understanding of the human condition and vocation as anchored in that faith. Although based on the same sources, their analyses of the role of modern technology diverged widely. Some consider that each human individual is called to incarnate the Spirit in this world through his personal actions. For others, it is through their collective action that humans contribute to the process of spiritualization of the world through a gradual reduction of the otherness of the material world, an act of faith that continues the divine creation. The meaning of incarnation – a central notion in Christianity setting it apart from most other religions – thus appears to have played a key role in the evaluation of technology, and it could still be the case in contemporary debates about the meaning of technological progress.

D. Cérézuelle (✉)
Programme Autoproduction et Développement Social (PADES), Paris, France

S. Loeve et al. (eds.), *French Philosophy of Technology*, Philosophy of Engineering and Technology 29,
https://doi.org/10.1007/978-3-319-89518-5_4

4.1 Jacques Ellul's Theory of the Autonomy of *"la Technique"*

A specialist of the history of law and institutions and a law professor at the University of Bordeaux, Jacques Ellul published a series of books which proved extremely influential. For years Ellul was better known in the United States than in France. A number of activists and protestors, who professed ideals of egalitarianism, liberalism and pioneered the environmental movement in the 1950s found in Ellul's works the confirmation of the growing political power of the military industrial complex and alerts about the impact of technology on the environment and the development. But Ellul's writings also influenced conservative religious groups who liked his caustic comments about false modern idols and Christian-Marxist ideology. The most "activists" among his readers, however, often rebuked Ellul for what was perceived as a pessimistic view of technical progress as an uncontrollable force. On the other hand, conservative theologians, for whom Ellul was a traditionalist, overlooked the progressive and critical dimensions of his theological and social writings. In both cases another reading of Ellul is necessary.

4.1.1 Ellul's Aspirations

When reflecting as a young man on his experience of contemporary social life, Ellul came to believe that growing technological development was not only an impersonal but also a de-personalizing process. It went against the demands of liberty and responsibility that grew out of Christian faith. "We feel it necessary to proclaim certain values and incarnate certain strengths." "[T]he problem facing each one is to know if we can incarnate the necessity that we feel to be inside us," however "nowhere is it a question of living one's thinking and thinking one's acts, but simply just thinking and just earning one's living.[1]" (Charbonneau and Ellul 1937: 55).

In a civilization that has carried to the extreme the divorce between the material and the spiritual, Ellul was concerned with creating living conditions that are compatible, in concrete terms, with the individual's need to be personally responsible for all aspects of their lives, and he took on the task of theorizing this new form of alienation. Just as Marx in the previous century had sought to define the power of capital and the dynamics of its expansion, Ellul considered that in our time technology had become a dominating force.

In his earliest book *La technique ou l'enjeu du siècle* (1954), Ellul contended that modern technology is not neutral and cannot be regulated just by an increase of ethical awareness. As a modern social phenomenon, "*la technique*" (as Ellul called it)

[1] On the relationship of Ellul and Charbonneau with the personalist movement see Christian Roy (1999). [All quotations in this paper have been translated from the original French by Virginia Coulon].

is beyond the control of ethics and politics. Like capital, it tends to develop as an autonomous power that forces individuals and society to comply and conform. For Ellul, technology is more than a mere accumulation of machines and tools ("*les techniques*"). In his view, technology has moved beyond the machine phase and deeply penetrated the social sphere through propaganda, the organization of leisure time, the division of work and the State power. He emphasized that all aspects of human activity were systematically organized according to the principle of maximum efficiency: "The technological phenomenon is the main preoccupation of our time; in every field men seek to find the most efficient method." (Ellul 1954: 21) For Ellul, techno-optimists – whether they be of the left or the right wing – missed the point when they equated technological progress with freedom. Ellul insisted on the fundamental *ambiguity* of technological development. While emancipating mankind from nature, technology imposes on humans a set of abstract, albeit consistent, constraints that become out of control. Technology has become more and more powerful to such a point that it follows its own dynamics, indifferent to the moral, qualitative or aesthetic dimensions of human existence. In *Le système technicien* [*The Technological System*], Ellul insisted on the autonomy of technology resulting from the gradual interaction of its various subsystems. "It is technology that works its own changes [...]. Technology has as a specific given, the feature that it requires its own transformation for itself." (Ellul 1977: 79–80) Furthermore, he continued, technological progress inevitably engenders a more organized and integrated society, taking control of all aspects of individual and collective life. Technological innovations such as railroads, television, or the system of public healthcare, generate a rigid social organization through a process of mutual reinforcement, each technological progress in the organization of one field bringing about growth in others. While emancipating humans from natural constraints, technological progress creates new constraints and dependencies. In short, not only is technique's development autonomous, it is an inevitable phenomenon of today's world that man has no choice but to adapt to. "The technological milieu is no longer a set of resources that we sometimes use (for work or distraction). It is now a coherent ensemble which "corsets" us on all sides, which encroaches upon us, and which we can no longer do without. It is now our one and only living milieu." (Ellul 1977: 42) Self-generating, the technological system follows no preconceived plan. It does not aim to possess or dominate man. It has no intention, no objective. Humans do not interfere in any meaningful way to set it up nor do they manage or design a system generating an autonomous dynamics. Society has become a closed system organized around a homogenous technological structure. "One cannot hope to solve an isolated problem in our society, for our society is an ensemble, whose structure is the technological system. The responses must be all-inclusive, like the society itself." (Ellul 1977: 166) This is essential, for in the modern world the technological system follows its own dynamics, eliminating everything that is non-technical, with a complete technological organization as the result. There are limits, however, to the totalizing power of the technological system. The growth of such systems can only multiply dysfunctions (accidents, dangerous situations, pollution, entropy, etc.) on an ever-global level and lead to consequences regarded as irrational even though they follow the inner logic of the system. Ellul agreed with Bernard Charbonneau that the development of the technological society gives rise to a dialectic between system and chaos that requires a choice between growing social and ecological disorder or total (or totalitarian) organization, if not both at the same time. (Charbonneau 1973)

Contrary to what we often read, Ellul was not a fatalist and did not believe that the coming of a global technological society was inevitable. He assumed that alternative models of social organization were still possible, which would dominate technology instead of being determined by it, and he acted accordingly in many occasions of his life. He was convinced that action was possible because the internal workings of the blind dynamics of contemporary society require human's inner support. But humans can also refuse this support, as Christian revelation encourages them to do. Those who call Ellul a "pessimistic calvinist," and pretend that he sees evil in technology because it is absolutely autonomous and all-powerful, have not read him close enough. In his first book on technology, he insisted that technology's autonomy was not total. And the protestant Ellul, (he was not a Calvinist, but a Barthian[2]), would quote the catholic Charles Péguy who "taught us, not in his writings but in his life, that the whole man was still possible." (Ellul 1954: 223) A few lines further, he added, "There is no such thing as technique by and in itself. In its irresistible forward progress, it is accompanied by man, without whom it is nothing." (Ellul 1954: 223) It seems difficult to be more optimistic!

4.1.2 Technique and Disincarnation

Ellul's views on the technological society are rooted in the notion of incarnation that he considered as a central feature of the Christian revelation. For him, to be incarnated is the condition of striving to express the spiritual truths that are inside us in our daily action within the material world. Freedom of action and truth mean nothing when separated from action. (Ellul 1981) This view is inspired by the example of Jesus whose life and actions in this terrestrial world obeyed the Words of God. In *Présence au monde moderne* (1948), Ellul outlined a Christian ethics for a world dominated by the power of technology and of the State. The volume opened with some reflections about incarnation. "God became incarnate – it is not for us to undo His work." (Ellul 1948: 14) Believers must reconcile their spiritual condition with their material condition. They are responsible for incarnating their spiritual values in the material world, the world, as Ellul says, from which there can be no escape. (Ellul 1948) Above all, Ellul rejected the evolutionary doctrines of salvation like that of Pierre Teilhard de Chardin (1955, 1959), who believed in the "gradual emergence [of the Kingdom], humanity ascending towards God." On the basis of his

[2]Carl Barth's (1886–1968) theology insisted both on the otherness of a transcendant God and on man's freedom. He developped a "dialectical theology" which, together with Kierkegaard's and Marx's writings, was one of Ellul's main sources of inspiration (Rognon 2007).

notion of incarnation Ellul attempted to build up a civilization based on the measure of man. (Ellul 1948) And yet our technological civilization is not on the level of men of flesh. The development of its economic, scientific and technical processes is based on an abstraction that sacrifices concrete man to ideal man. As Ellul claims, the living, concrete man on the street is ruled by instruments that were supposed to bring happiness to abstract man. The man of philosophers and politicians, who does not exist, is the only goal of this prodigious adventure that everywhere has transformed man of flesh and blood into an instrument and given him a life of misery. (Ellul 1948) Ellul's focus on man's incarnate condition led him to refuse the increasing divorce between aims and means that characterizes technological society in his view. For Ellul, the incarnation of the Word in Christ gives should provide a model of action. Human's actions must seek to unite the material and the spiritual. A good action is one whose aims have been incorporated not only into its effects but also into its agents and its means. An efficient act carried out by someone who does not know what he is doing, someone who merely acts as an irresponsible instrument, cannot be a good act. It is not our instruments nor our institutions that count, but ourselves. (Ellul 1948) This is a principle that Ellul repeats over and over in his books: the means must always be compatible with the ends. We cannot bring about good ends by using means which diminish man. It is only through a process by which man becomes disincarnate (which is at the very heart of the West's technological adventure) that it is possible to imagine that the ends can justify the means. There can be no moral justification for resorting to an act that has a negative effect on humanity (for example, assembly line work, bureaucratic depersonalization or political violence). According to Ellul, we reject the idea of instrumentalizing humans' work, the human activity that makes up our field of vision today. (Ellul 1948) It is not enough for a technique or an institution to be impersonally efficient or automatically produce an effect for it to be called good. It can only be good if it also opens up the possibility for everyone to be responsible for one's actions. Whatever the final aim of an act may be, if it diminishes the subject or degrades its object it is a bad act.

It we are to take incarnation seriously, it follows that all our acts, and their effects, must embody our value system. There is nothing new in this, but Ellul was original in perceiving the radical consequences of incarnation and using this as criteria to evaluate depersonalization in today's world and formulate his critical stance about the role of the State and modern technology. He shows how the technological and institutional structures of modern society have become autonomous, a process that is contradictory to the logic of incarnation which places the notion of aims and means on a personal level. In refusing to disassociate aims and means, Ellul points out that this is the consequence that Christians must put into practice, for today it is a question of being, rather than performing. It is important for man who is standing on his two feet out in the world to rediscover the wholeness of personal life. (Ellul 1948) Because he looked at incarnation through Christ and in the life of each one of us in this way, Ellul saw the need to assess and set limits to technology and institutions from the perspective of the role they play in the life of men.

4.2 Technique and Spiritualization: Marie-Dominique Chenu and Emmanuel Mounier

4.2.1 You Said Incarnation?

Ellul's understanding of incarnation was not shared by all Christians.[3] Although some sincere Christians have adopted a critical attitude with respect to the "negative aspects of progress." Catholicism, itself, became ardently progressive in the course of the twentieth century. In the nineteenth century the political conservatism of the French Catholic Church alienated a big percentage of the working class. After Leo XIII's 1891 papal letter *Rerum novarum*, a movement grew within the Catholic world which sought to reconcile the Church with the working class and technical progress. A whole theology of spiritualization, radically different from Ellul's theory of incarnation, started to take form, aimed at marshaling every technological resource in order to "carry out the Work of God." This divine task meant transforming all of nature through technology, an endeavor aimed at overcoming the otherness of the material world considered as ontologically imperfect. This objective led some Christian thinkers to adopt a strong technophile and progressive outlook. Without entering here in a survey of the works of the Jesuit Pierre Teilhard de Chardin, it suffices to mention that his technophile inclinations led him to celebrate the explosion of the atomic bomb in Hiroshima as a manifestation of the growing presence of the divine in matter. Echos of Teilhard's theology can be found in the writings of the Dominican Marie-Dominique Chenu as well as in Emmanuel Mounier's works. Mounier, the founder and executive director of the journal *Esprit* which promoted a version of personalism quite different from Ellul's or Charbonneau's, did not hesitate to write that "nature is not solely the matrix of humanity. It is given to us as an operational field whose boundaries are not fixed but whose general sens is clear. Nature is there to be recreated by man.[4]" (Mounier 1948: 49) For Mounier the importance of technology lies in its ability to allow powerful radical thinking about physical reality that it inspires. He advocated a recreation of nature that would consist in the total domination of physical reality by humans, in other words the *utter and final domination* of matter by the spirit. This explains Mounier's harsh words against the precursors of environmental thinking like Charbonneau and Ellul,[5] who let themselves be caught up, as he puts it in his 1948 book, by their *petty fear of the twentieth century,* raising stumbling blocks on the glorious pathway of mankind's vocation.[6] A collective volume *La technique et*

[3] However, similar views have been developed by Ivan Illich, a Catholic who acknowledged his debt to Ellul.

[4] Mounier's text "*La machine en accusation*" in *La petite peur du vingtième siècle*, became the guidebook of progressive Catholic optimists after World War 2 (Mounier 1948).

[5] Mounier had met Charbonneau and Ellul before War World 2 in the movement *Esprit* and he was already opposed to their environmental approach.

[6] This is also the reason why he separates the individual from the person who for him is an exclusively spiritual principle, separated from the body and its empirical physical, social and psycho-

l'homme published in June 1960 by a group of progressive theologians of the Centre Catholique des Intellectuels français ["Catholic Center of French Intellectuals"], exposed the project to instrumentalize the body and all of nature in the name of a certain definition of incarnation. (Chenu et al. 1960) Such reflections were developed in a context of increasing anxieties concerning technological innovations. *Brave New World* came out in 1932. The atomic bomb was dropped on Hiroshima in 1945 and Ellul's book, *La technique ou l'enjeu du siècle* appeared in 1954. In 1956 Jean Rostand, a biologist and science writer clearly discussed the possibility of genetic manipulation of human beings in *Peut-on modifier l'homme?* (translated as *Can Man be Modified?* in 1959), and Norbert Wiener's 1950 book, *The Human Use of Human Beings: Cybernetics and Society*, was published in French in 1952 as *Cybernétique et société* (Wiener 1950). *La technique et l'homme*, however, did not discuss the concrete problems raised by the technological civilization. It just alluded to work as a form of alienation and to the ugliness of the industrial world, but it was totally blind about environmental issues that already pointed 60 years ago, such as the reduction of biodiversity or the disappearance of local cultures resulting from the industrialization of agriculture.

4.2.2 The Spiritualizing Potential of Techniques

To what extent such a lack of vision, such uncritical and blind optimism can be linked to a specific view of incarnation? To explore this question let us take a close look at Father M.D. Chenu's chapter "Towards a theology of technique" in *La technique et l'homme.* (Chenu et al. 1960) His notion of incarnation seems to be a standard one. He assumed the consubstantiality of spirit and matter and claimed that "[man's] perfection does not consist in overcoming this worldly life, as a fortuitous accident heavy with meaning, but of achieving in this world the full ontological and moral equilibrium of his being." (Chenu et al. 1960: 163) The technological dimension of human life is thus inseparable from its incarnated mode of being. At the same time, however, the technological vocation of humans is described in extremely intellectualistic terms. Man is called to "penetrate the world with his spirit through a process of rationalization, just as he virtuously rationalizes his own body. In this movement the spirit's transcendental nature remains intact [...]." (Chenu et al. 1960: 164) Chenu went further in arguing that the divine truth about humanity "is that the spirit penetrates deeply into the body, into its own body, but also into the body of this world that is accomplished in it; man is the demiurge of this world, answerable to the Creator whose Work he participates in." (Chenu et al. 1960) Nature's matter, including the human body, becomes "the matter of freedom" and through his actions man "finalizes the consecration of the world." (Chenu et al. 1960: 165) Remarkably,

logical characteristics. This would explain why idealistic philosophers and theologians have trouble distinguishing the person from the soul – a distinction that Bernard Charbonneau, for one, categorically refuses.

the notion of spirit is identical to a spirit of reason and the model for spirit's action over matter that Chenu proposes is that of a demiurge, of the God creator of philosophers. Here is the core difficulty. The model could be that of the Father Creator, but it is clear that it cannot be that of the Son made Man. In the last pages of the volume, Chenu replaced the mystery of incarnation with an intellectual model of Creation in which the Father imposes a rational order on the world of matter. But the Son, the incarnated Word, does not work like a creator giving shape to matter. The image of the potter can be that of the Father but not of the Son. Jesus' actions, and his words, were addressed to beings of flesh who were expecting a kingdom of love and freedom and who had the freedom to respond or not to his message. In thus conceptualizing incarnation, from the Father, the Almighty Creator's standpoint rather than from the perspective of the Son who refused the temptation of unlimited power, Chenu had missed the chance to realize that technological power could require some limitation.

Chenu assumed that Christian faith could no longer be circumscribed to the role of guiding inner consciousness because "connected to matter, co-creator of the world, able to invent his own environments, *homo technicus* works to advance history." (Chenu et al. 1960: 164) This broad statement about human's technological vocation is meant to provide us with answers to a whole series of complex questions: How should humans deal with this world that he now has the power to mistreat and to which they still belongs? What ethical criteria should they adopt? At what pace should they proceed? What limits should they assign at each stage of technological development? These questions are of no obvious interest, are even absurd, when incarnation is seen as co-creating the world and when one is convinced that it is thanks to human technology that humans will be able to respond to "*the eager expectation* [of creation], *groaning in the pains of childbirth* […], [in order to be] *brought into the freedom and glory of the children of God*."[7] In such a perspective, the questions about the damage caused by progress, about setting limits to the use of power, and about choosing between several technological models of development matter little. Since there is no unacceptable power level, there are no unacceptable consequences. All these are petty little fears…

For Chenu, the notion of incarnation goes hand in hand with a one-sided understanding of technological action, perceived only, in a Platonic way, as an action of the rational mind over matter that is outside itself. So doing, he does not take into account that because of human's status as incarnated beings, technical action always retroacts on them. Chenu assumes that in the process of technical action "the spirit's transcendental nature remains intact." His fascination for the powerful action of the spirit obliterates its concrete impacts on the bodily condition of humans. In brief the notion of flesh plays no role at all in Chenu's text.

This sample of texts by Catholic intellectuals (Protestants at that time said exactly the same thing), suggests that their love of progress and technology was

---

[7]Here Chenu links three separate verses of Paul's epistle, *Romans* 8–19, *Romans* 8–22, and *Romans* 8–21, inversing somewhat the order – and subtly changing their meaning! [Translator's note: translations taken from the New International [on-line] English Version of the Bible].

based on a very specific, one-sided view of the notion of incarnation. For them, the term referred, in fact, to the spiritualization of nature, meaning the domination of the spirit over the body meant to eventually serve the demands of the mind.[8]

A certain spiritualist view is thus perfectly consistent with a legitimization of the technological instrumentalization of the body and nature, provided that we neglect the ultimate consequences of this attitude. This spiritualist approach to technology is in line with nineteenth century scientism. It is reminiscent of Marcellin Berthelot's and Ernest Renan's views of the glorious future of mankind under the leadership of science. (Berthelot and Renan 1898) Christian techno-optimists condemned spiritualism as outdated, as they considered flesh only as a source of action in the world, and view this action according to the instrumentalist model: the domination of the spirit (the intellect and the will) over matter. In their idealistic and one-sided perspective, they ignored that concrete man of flesh can be as much an object of technology as its subject. They were more concerned about the effects of technology on things than about the power of technology on humans and their lifeworld. Ultimately, under cover of incarnation, the action of the spirit over matter is perceived as domination rather than union, as the act of imposing the spirit's projects and models on the body construed as a passive object.

4.3 Technique and Desire in the Writings of Jean Brun

4.3.1 The Technological Spirit

For Ellul, technics, as a process, refers to any work carried out by applying a certain method in order to obtain a result. It is an anthropological constant that he sets apart from contemporary technology defined, as we saw above, as "the main preoccupation of our time; in every field men seek to find the most efficient method." Ellul's formula is striking. He does not say that the technological phenomenon *requires* humanity's concern, but that it *is* the major matter of concern itself. And this concern is based on the belief that any increase in technique's operating power (it's efficiency) is necessarily beneficial for mankind. If for Ellul this conviction is one of our time, it is obvious that it could also disappear, as would the technological

[8] Their contemporary, Bernard Charbonneau, refused to assimilate spiritualization with incarnation: "Incarnation. Spirit made flesh, not flesh or matter becoming spirit. The inverse of Evolution, whether we talk in lay terms or those of Teilhard de Chardin where Matter becomes Spirit. From the beginning the atom is pregnant with progress towards the noosphere, a fate Matter cannot escape from. Man can intervene but he cannot change the essence of what is happening. There is no more freedom. However, if spirit becomes flesh, matter, reality, it is because it has a different nature. In the beginning there was not one but two terms, Spirit and flesh, as different in their nature as man can conceive them to be, although tightly intermingled and confronted with this individual existence that we partake of. The Spirit becomes flesh in the personal awareness that suffers from the immensity of the distance between them." *Trois pas vers la liberté* (unpublished and undated: 7).

phenomenon. The autonomy of technology is neither an intrinsic nor a permanent feature. It is relative to a certain state of society that tends to worship technology. For Ellul, the determining power of technology over man can only be brought about by man's own subjective dispositions with respect to techniques. In Max Weber terms, the *spirit* of technology would therefore be a *Primum Movens*. Ellul occasionally criticized this technological spirit, in *Exégèse des nouveaux lieux communs* [*A Critique of the New Commonplaces*] (Ellul 1966) as well as *Le bluff technologique* [*The Technological Bluff*]. (Ellul 1988) He was silent, however, about the origins and the source of the technological spirit, letting others take on this task, including Jean Brun, a professor of the history of philosophy at the University of Dijon, who devoted much of his time to the question of the technical imaginary.

According to Brun, if technology is not socially neutral, it is because it is not existentially neutral and several of his works aimed to shed a new light on the fundamental reasons for man's fascination with techniques. Brun wrote a number of books about the philosophy of technology. In *Le rêve et la machine* [*Dream and the machine*], he argues that if technique is problematic, it is not because it lacks a rational basis, but rather because it is hyper rational. (Brun 1992) Technology's potential for dehumanizing man does not come from a narrow understanding of reality, but from the active desire to cut the links to reality that characterizes human existence and that circumscribes its limits. For Brun, this universal aspiration of mankind is particularly manifest in his attempt to control the way reality functions. This explains the specific orientation of technological undertakings in the West, but it had even been the case before the emergence of modern science in the sixteenth and seventeenth centuries. Brun was convinced that the development of technology and science has an ontological basis whose roots go much further back in time. His aim was to uncover its genesis and its profound existential meaning. Philosophers forget all too often that humans first dreamed about their technological inventions before building them and that the history of technique is controlled by a "metaphysical dream-state" ("*onirisme métaphysique*") (Brun 1992: 131) that Brun attempted to describe on four levels.

First, concerning the intellectual and methodological foundations of the world of machines and devices, antiquity had already invented the categories of motor, number, concept, and matter which in turn enabled reality to be seen as an object of operations. In the first part of his book Brun sought to demonstrate that the ability to conceptualize these categories is not so much the result of advances in objective rationality as it is of man's ontological cast of mind that made technical thinking possible.

Secondly, as for practical applications of what began as ideas, Brun endeavored to show that the stuff of technology is made of dreams. The genesis of different techniques reveals a human desire for the transmutation of reality, an urge to go beyond the ontological framework of existence, in particular of its space-time constraints. Technical inventiveness results from a desire to push back all limits, to be ubiquitous, infinite and immortal. To borrow an expression from René Girard, we could say that the idea of technique became associated very early on to human's "deviated aspiration to be transcendent." It is the desire to go beyond our finite boundaries that gives technology its vocation to cure man of the misfortune of existence and to bring about new ways of being. In so doing technique becomes vested with an ontological mission, allowing man to invent new forms of existence.

Third, the desire to go beyond space-time constraints and the experience of reality is never exhausted by the technological devices that have actually been invented. For Brun, nothing expresses the ontological vocation of technology better than the incredible world of imagination that machines have brought forth and whose analysis throws light on our relationships with real techniques. "One of the biggest paradoxes of the modern world is that the machines designed by man as efficient tools for shaping matter in very concrete terms are now seen as great magicians able to shuffle reality's cards in order to re-deal if not re-design them." (Brun 1992: 368) It is in the violence of this desire and in the existential intensity of our dreams about going beyond existence and its finite boundaries that can be found, according to Brun, "the roots of the pulsion to accelerate progress in which the machine is given the task of opening circles, eradicating sameness, killing the subject's very essence and being, as well as that of man himself." (Brun 1992: 328) For Brun, we project our desire to create new forms of being onto technique. This is why technique fascinates us – and makes us forget how important safeguards are.

Fourth, in relation to the question of technological practices, Brun argues that not only does the mystifying power of technique makes us blind to "the damage caused by progress," and that its violence is an active component of technical thinking. The yearning to transmute existence generates an impatience with concrete things and their limits which he experiences as obstacles to be overcome. In *Le retour de Dionysos* ["Dionysos Returns"] Brun gave an inventory of all the pathological forms of exasperation which are engendered by human desire to break out of the self's cage. He showed how the obsession to go beyond ontological limits of existence breeds a culture of cruelty that uses technology's power of transmutation to invent what he calls "technological rituals" [*sabbats techniques*] or "technological orgies" [*orgies techniques*]. (Brun 1969: 49 sq) In *Les masques du désir* [*The Masks of Desire*], Brun insists on "the metaphysical vocation of technique whose utilitarian applications hide its true ontological meaning and soteriological ambition." (Brun 1981: 211) In the chapter entitled "*Les transes techniques*" ["Technical Trances"], he points out how modern humans have used techniques to put an end to their individual and separate selves.

4.3.2 From Fascination to Demythification

Holding out the promise of overcoming existence, technology has a powerful element of myth-making and ritualization which makes us deaf to the alerts of whistle-blowers, indifferent to risks, and blind to the damage caused by progress. In other words this fascination explains why technical thinking and violence so often go hand in hand.[9]

[9] A masterful exploration of this type of thinking can be seen in David Cronenberg's film *The Fly*.

The desire to go beyond the actual existence generates a world of unbridled technophile imagination organized around dreams of absolute power, eternity, and ubiquity exemplified in science fiction as well as in the current movement of "transhumanism." Although most champions of transhumanism claim to be atheist or agnostic, they are close to Teilhard's, Mounier's and Chenu's theology which granted technology a soteriological role, transforming it into a tool of salvation.

From a broader perspective, the technological spirit triggers a vague mentality of scientism in the public at large, accompanied by an often childish faith in progress whose political expression is the multiplication of "great useless projects" that are a challenge not only to economic reasoning but to rationality itself. Since technology is not existentially neutral, it is not socially neutral. The technological spirit, which plays a decisive role in the acceleration and the growing autonomy of technical systems, feeds on technology's myth-building potential. It both prevents us from learning lessons from the past and from anticipating environmental and social risks (risks to liberties for example) that result from technological acceleration.

Brun's philosophy of technology extends Pascal's meditation on man's futile attempts to run away from himself through untertainment ("*divertissement*"). It also updates Chestov and Berdiaev's views of technology and provides a coherent and meaningful vision of the seemingly absurd, perverse or insignificant aspects of our relationship to techniques. First, it cannot be said that technique does not have an existential impact. Second, technique's pragmatic function (that leads us to believe that efficiency is the handmaiden of usefulness) is not always its main function. Third, the irrational and dehumanizing effects are intrinsic features of technology. There is a high risk of failure for any attempt to control technology (or even in more modest terms, to simply accompany it) that forgets to take into account the deep existential roots of machines. Technique can only be controlled when it is demystified. The task is not easy, says Brun, because "the nature of the rational is to lose its reason [*devenir délirant*] to the extent that if it accepted to have its share, it would be in contradiction with itself." (Brun 1992: 312)

4.4 Concluding Remarks

The modernization process is often described as being accompanied by a process of secularization. But humans are religious animals and if, in the West at least, we are witnessing a decline of religious practices it is perhaps because the need for the sacred has mainly been transferred to the State (Ellul 1973) and to technology, to the detriment of concerns about the day-to-day living conditions of real people. The technological spirit breeds an idolatry of techniques whose framework and existential foundations were uncovered by Brun. He thus paved the way to further attempts to uncover the religious dimensions of technology by David Noble, Philippe Breton and Pierre Musso. The transhumanism movement, that so willingly claims its debt to Teilhard de Chardin, is but one recent avatar of the aspiration to go beyond our limits, suggesting that the spiritualization/incarnation divide continues to shape our relationship to technique and contributes to its sanctification.

Acknowledgements A preliminary version of this text has been discussed by Jean-Yves Goffi and Quentin Hardy during the authors' workshop held in Paris Sorbonne on 22–23 June 2015 in view of the preparation of the collective book.

References

Berthelot, M., & Renan, E. (1898). *Correspondance 1847–1892*. Paris: Calman-Lévy.

Brun, J. (1969). *Le retour de Dionysos*. Paris: Desclée.

Brun, J. (1981). *Les masques du désir*. Paris: Buchet Chastel.

Brun, J. (1992). *Le rêve et la machine: technique et existence*. Paris: La Table Ronde.

Charbonneau, B. (1973). *Le système et le chaos. Critique du développement exponentiel* (2014 edition). Paris: Editions Le Sang de la Terre.

Charbonneau, B. & Ellul, J. (1937). Directives pour un manifeste personnaliste (2014 edition). In *Nous sommes des révolutionnaires malgré nous. Textes pionniers de l'écologie politique*. Paris: Editions du Seuil.

Charbonneau, B. (unpublished and undated). *Trois pas vers la liberté*.

Chenu, M. D., et al. (1960). *La technique et l'homme. Recherches et débats du centre catholique des intellectuels français, cahier n° 31*. Paris: Librairie Arthème Fayard.

Ellul, J. (1948). *Présence au monde moderne*. Genève: Roulet. English Edition: Ellul, J. (2016). *Presence in the modern world* (L. Richmond, Trans.). Eugene: Wipf & Stock Publishers.

Ellul, J. (1954). *La technique ou l'enjeu du siècle*. Paris: Armand Colin. English edition: Ellul, J. (1967). *The Technological Society* (John Wilkinson, Trans.). New York: Knopf/Vintage.

Ellul, J. (1966). *Exégèse des nouveaux lieux communs* (2004 edition). Paris: La Table Ronde. English edition; Ellul, J. (2012). *A critique of the new commonplaces* (H. Weaver, Trans.). Eugene: Wipf & Stock Publishers.

Ellul, J. (1973). *Les nouveaux possédés*. Paris: Fayard. English edition: Ellul, J. (1975). *The New Demons* (C. E. Hopkin, Trans.). New York/London: Seabury/Mowbrays.

Ellul, J. (1977). *Le système technicien*. Paris: Calmann Levy. English edition; Ellul, J. (1980). *The technological system* (J. Neugroschel, Trans.). New York: Continuum.

Ellul, J. (1981). *La parole humiliée* (2014 edition). Paris:La Table Ronde. English edition; Ellul, J. (1985). *The humiliation of the word* (J. M. Hanks, Trans.). Grand Rapids: Eerdmans.

Ellul, J. (1988). *Le bluff technologique*. Paris: Hachette. English edition; Ellul, J. (1990). *The technological bluff* (G. W. Bromiley, Trans.). Grand Rapids: Eerdmans.

Mounier, E. (1948). *La petite peur du vingtième siècle*. Neuchâtel: Editions de La Baconnière. English edition; Mounier, E. (1962). *Be not afraid, a denunciation of despair*. New York: Sheed & Ward.

Rognon, F. (2007). *Jacques Ellul, une pensée en dialogue*. Genève: Labor et Fides.

Rostand, J. (1956). *Peut-on modifier l'homme ?*. Paris: Gallimard. English edition; Rostand, J. (1959). *Can man be modified?* (J, Griffin, Trans.). New York: Basic Books.

Roy, C. (1999). Ecological Personalism: The bordeaux school of Bernard Charbonneau and Jacques Ellul. *Ethical Perspectives, 6*(1), 33–44.

Teilhard de Chardin, P. (1955). *Le phénomène humain*. Paris: Seuil. English edition: Teilhard de Chardin, P. (1959). *The phenomenon of man*. New York: Harper Perennial.

Wiener, N. (1950). *The human use of human beings: Cybernetics and society*. Boston: Houghton Mifflin.

Chapter 5
Rise of the Machines: Challenging Comte's Legacy with Mechanology, Cybernetics, and the Heuristic Values of Technology

Ronan Le Roux

Abstract This essay is about the rise of the machines against the legacy of Auguste Comte, or, to be more precise, against the classical picture, that associated with positivism, of modern technology as a mere application of science. I examine how and to what extent three French philosophers pursued non-Comtean insights into the relations between science and technology: some decades before the emerging STS wave that swept in as the 1970s wound down, they lent their attention to original aspects of these relations, acknowledging specific features of technology within a rationalistic view of science. With respect to the legacy of positivism, these three philosophers, Pierre Ducassé, Georges Canguilhem, and Gilbert Simondon, are most significant. To begin, I give an overview of their non-Comtean insights in the institutional context of the Institute for the History of Science and Technology; then, I move focus to their respective interests for cybernetics as an emblematic reference with respect to those insights. The question of why the latter did not evolve into an explicit research agenda, despite their convergence, is raised in the conclusion.

Keywords Antipositivism · Canguilhem · Comte · Cybernetics · Ducassé · Lafitte · Mechanology · Positivism · Simondon

This essay is about the rise of the machines against the legacy of Auguste Comte, or, to be more precise, against the classical picture, that associated with positivism, of modern technology as a mere application of science (Comte 1830: 62–63). This picture has come to be widely dismissed by today's academics in the social and philosophical studies of science and technology: "The order of the world put science before technology. Over the last twenty years, work in the history of science

R. Le Roux (✉)
EA 4395 "Lettres, Idées, Savoirs" (LIS), Université Paris-Est Créteil Val de Marne, Créteil Cedex, France

S. Loeve et al. (eds.), *French Philosophy of Technology*, Philosophy of Engineering and Technology 29,
https://doi.org/10.1007/978-3-319-89518-5_5

and technology has begun to redraw this picture, blurring the boundaries between realms, and insisting on a more reciprocal (and more interesting) relation." (Galison 1997: 185) Consequently, many have promoted a representation characterized by the hybridization, symbiosis, or entanglement of science and technology. Another general aspect of these studies is that they often disregard any idea of positive knowledge, and thus any epistemological concern by the same token. Their subversion of positivism has therefore relied on a deconstruction of the overhanging position of science rather than on the construction of concepts that would give substance to technology by enabling analyses of its specificity.[1] Other studies have not "black-boxed" the relation between science and technology, instead providing reviews and analyses of the different ways in which they are interdependent (e.g. Solla Price 1965; Faulkner 1994; Cuevas 2005).

Postmodern science and technology studies (STS) do not hold the monopoly on antipositivism, and interdependence studies have not accounted for every mode of interdependence possible. In this chapter, I examine how and to what extent three French philosophers pursued non-Comtean insights into the relations between science and technology: some decades before the emerging STS wave that swept in as the 1970s wound down, they lent their attention to original aspects of these relations, acknowledging specific features of technology within a rationalistic view of science. With respect to the legacy of positivism, the writings, activity, and institutional perimeter of these philosophers – Pierre Ducassé (1905–1983), Georges Canguilhem (1904–1995), and Gilbert Simondon (1924–1989) – are most significant. To begin, I give an overview of their non-Comtean insights in the context of the Institute for the History of Science and Technology; then, I move focus to their respective interests for cybernetics as an emblematic reference with respect to those insights. The question of why the latter did not evolve into an explicit research agenda, despite their convergence, is raised in the conclusion.

5.1 Non-Comtean Insights at the IHST

The Sorbonne's Institute for the History of Science and Technology (now the IHPST following the introduction of "Philosophy") is a most interesting entity to scrutinize: while representing the best institutional embodiment of the history of science *à la* Comte, its foundation in 1932 took place at a time when intense concern surrounded technology. The Great War had confirmed the grave doubts raised by the industrial revolution as to the general benefits of the machine age for mankind.

[1] Even though these studies have significantly contributed to focus on the practical and material aspects of scientific work, they have at the same time claimed a non-theoretical stance, whether by staying at a level presumed purely empirical and descriptive, or by reaching a level of generality that offers no analytical operationality, such as the expression "non-human." Also, the extent to which the social-constructivist "turn to technology" actually subverts or repeats the traditional hierarchy between science and technology is questionable.

"Techniques" were identified as a key topic for public debate, for the agendas of both the new and revived social sciences (Espinas, Mauss, Febvre, Friedmann…), as well as for philosophy, with Bergson, Simone Weil, and Pierre-Maxime Schuhl (Schuhl 1938) contributing to various aspects of the topic. In all cases, techniques and machines came to be seen as a challenge to rationality rather than its outcome.

The foundation of the IHST was not isolated from such views. The Institute's first director, Abel Rey (1873–1940), was a friend of Mauss and Febvre, both members of the board (Braunstein 2006; Chimisso 2008). Rey opened the first volume of Febvre's *Encyclopédie Française*, entitled "L'Outillage mental," (Rey 1937) by depicting the historical emergence of rationality as a differentiation passing from primitive to more logical and sophisticated "mental toolsets."[2] Although he claimed a positivist stance, his notion of mental toolset is seen to be of Bergsonian inspiration (Braunstein 2006: 187), since history of science would mirror the dependence of *homo sapiens* on *homo faber* rather than the contrary (Bergson 1934: 67).

Gaston Bachelard (1884–1962), in his PhD dissertation supervised by Rey (Bachelard 1927), sharply distinguished a technical rationality, autonomous from, analogous to, and likely to inspire, scientific rationality.[3] This early idea of his would soon after be steamrollered by the concept of *phénomenotechnique*, a shift that received little if any attention. Henceforth, Bachelardian technology would oscillate between two conditions: either split into two opposites, with phenomenotechnique to one side (the only truly rational artifacts[4]), and everything else to the other side (now devoid of any specific rationality and confined to aesthetic emotions – cf. Bachelard 1970: 60); or else with all technological artifacts subject to the same axiomatic rationality propagating from theory to technology (that is to say, the contemporary meaning of "technology" as applied science was, for Bachelard, already the paradigm for technology in general, i.e. techniques and artifacts). In the first case, phenomenotechnique is a normative concept complying with "regional epistemologies": disciplines with weaker theoretical frameworks than Physics and Chemistry, and whatever mode of existence technology may have there, are simply disregarded by Bachelard. In the second case, the ontology of technology is no longer subordinated to a regional clause and falls entirely under a single model: the

[2] "Every [mental toolset] pushes knowledge further, giving it more amplitude and enabling it to better assess previous efforts. Hence, the chronological order somewhat express the successive stratifications of intellectual technology and gradual improvement in the classification and subordination of the various families of instruments we have at our disposal today." (Rey 1937: 1°14–1)

[3] Bachelard refers to Franz Reuleaux's theory of machines: "Kinematics is a formal science of undeniable purity. Its technical realizations thus offer examples paradigms likely to arouse and adjust speculative thinking. Language earned its richness and precision through the hand rather than through the brain"; "In order to acquire knowledge about a machine or an artifact, one must follow a different method to the pure scientific method." (Bachelard 1927: 156–164)

[4] "...an experiment is not 'careful' unless it is complete, that is, unless it is conducted according to a well-conceived plan based on a mature theory. (…) Instruments are nothing but theories materialized" (Bachelard 1934: 9, 13); "A measuring instrument always ends up as a theory: the microscope has to be understood as extending the mind rather than the eye." (Bachelard 1938: 24)

propagation of an axiomatic correction from scientific theory to technical artifact.[5] In any case, Bachelard can no longer conceive of technology in any other terms than those of phenomenotechnique, which, by its complete submission to scientific theory, comes close to Comte's hierarchy.[6]

Pierre Ducassé (1905–1983), professor, director of study, and secretary general for the scientific works of the Institute and its journal *Thalès*, was the third pillar of the IHST from its inception until his departure in 1955 (Le Roux, 2011b). Unlike Rey and Bachelard, Ducassé was deliberately involved in subverting Comte's philosophy of technology, although he was initially a positivist himself. He was also a collaborator of Henri Berr's "Centre international de synthèse." In 1933, Berr's *Revue de synthèse* published a short paper by an engineer and architect, Jacques Lafitte (1933), who, the previous year, had authored a book with the title *Réflexions sur la science des machines* (1932). Lafitte called for a science of machines, for which he proposed the name *mécanologie* (Guillerme 1971; Hart 1976; Mitcham 1979; Le Roux 2009). It should study the world of machines as it is, describe and explain how and why machines are as they are. Borrowing inspiration from morphology and natural history, mechanology would be simultaneously a social science, for it should deal with human invention and its social conditions. Influenced by Lafitte and Bergson, Ducassé distanced himself from positivism, acknowledging that the *Système* was not only overwhelmed by the machine age it was supposed to frame, but also that it resisted "opportunities of fresh intellectual insights" brought about by machinism (Ducassé 1957: 15). Lafitte's mechanology responded to that very concern. In 1945, Ducassé published the first French general history of technology (Ducassé 1945), a small book that ends on an explicitly mechanological tone,[7] echoed in 1948 by the creation of a small journal, *Structure et évolution des techniques* (SET). A very significant step toward the "new type of investigations" Ducassé was crying out for (Ducassé 1958: 1), this journal welcomed short papers by authors of very diverse domains. He also spread his concerns over time measurement into history of science.[8]

[5] "The dialectical movement that begins with the dialectics of the axiomatic systems is thus prolonged by the formation of axiomatic systems in Physics, and eventually by the formation of axiomatic systems in technology. (…) Technical change is often determined by ground-level revolutions. We used to stress this essential discontinuity. We gave the basic example of the *sewing machine*, which found its rational form when breaking with the attempts to imitate the gestures of the sewer, thereby laying a new foundation for sewing. (Bachelard 1949: 133)

[6] My interpretation thus goes against that of Rheinberger (2005).

[7] "We are just beginning to suspect that the machines around us live a sort of autonomous existence [footnote referring to Lafitte's book (1932)]. As any other production of life, they provide us with problems of structure and problems of evolution. Their families might offer even more complexity than those of the living beings, and their classification is not fixed. Machines produce and reproduce themselves according to general laws of which we are not aware, and whose logic may overcome our old intellectual habits." (Ducassé 1945: 131)

[8] Ducassé (1948, 1949). While still at the IHST, Ducassé was appointed to the university of Besançon, neighbouring Switzerland, where he took interest in the local clock-making ecosystem.

Georges Canguilhem (1904–1995) took charge of the IHST in 1955, winning out over Ducassé to succeed Bachelard under circumstances that remain to be documented. The recent publication of Canguilhem's early works has made it plain that he was a philosopher of technology (Braunstein 2000, 2011), before and beside working on the history and philosophy of medicine and biology. Canguilhem argued that technical creation is how man opposes the determinism of his milieu by organizing his environment, hence his general investigations concerning the concept of technique (Canguilhem 1937; 1938a; 1938b) which carry a clearcut dismissal of positivism.[9] Showing that Descartes, although an ancestor of positivism concerning the relations between science and technology,[10] also had more subtle ideas on the subject, Canguilhem infers one of his lifelong mottos: "The technical initiative lies within the requirements of life."[11] Science depends on technology (and not the contrary), but not in the way that pragmatists claim.[12] Canguilhem explicitly refers to Bergson (Canguilhem 1938a: 506), though with a slight departure as he believes that machines imitate life in a non superficial way. The famous conference "Machine et organisme" (Canguilhem 1961) would thus expand the topic with two novelties. First, another original form of dependence of science upon technology, by which technical artifacts inspire scientists with analogies for explaining organisms[13]; we could speak of a "heuristic value" of machines. The second novelty is subtle: again addressing the question of "the originality of the technical phenomenon with respect to the scientific phenomenon" (ibid.: 102), Canguilhem criticizes those who have seen in machines mere "solidified theorems" (ibid.). Besides the "Cartesian" target, how can one not also recognize Bachelard's phenomenotechnique between the lines? Canguilhem thereby calls for a stop to subordinating the ontology of technol-

[9] "Is technical activity a mere prolongation of objective knowledge, as is commonly thought in echo of positivist philosophy, or is it the expression of an original 'power', profoundly creative, and for which science, sometimes after the fact, elaborates a development program or a code of precaution?" (Canguilhem 1937: 490)

[10] "Awareness of what is possible technically comes to us from knowledge of what is necessary theoretically. Up to this point, nothing in the Cartesian philosophy bearing on technology seems anything other than obvious, if by obvious we mean the long standing familiarity modern thinking has with a subject of reflection which, from Da Vinci to Marx via the Encyclopedists and Comte, has advanced a development now become standard." (Canguilhem 1937: 494)

[11] "Thus, the ultimate irreducibility of technology to science, of making to knowing, the impossibility of a continuous and total transformation of science into action, would equate to affirming the originality of some 'power.' It would seem that considering technology to be an always somewhat synthetic action, thus impossible to analyze in and of itself, does not, even from a Cartesian point of view, imply denying it any value, since it can still be seen as a mode of creation, however inferior." (Canguilhem 1937: 497)

[12] "...this dependence is not the prolongation of an *élan* at all, but rather its arrest. The emergence of scientific thought is conditioned upon the failure of technical thought. (...) Science then appears as a reflection over failures and obstacles." (Canguilhem 1938a: 504)

[13] "When Descartes looks to machines for analogies to explain the organism, he refers to spring and hydraulic automata. And so, he becomes a tributary, intellectually speaking, of the technical forms of his time, of the existence of clockworks and watches, of watermills, artificial fountains, pipe organs, etc." (Canguilhem 1961: 106)

ogy to regional epistemologies: phenomenotechnique cannot even withdraw into the solid shell of Physics, since even its superior theory-ladenness could not account for the entire existence and properties of instruments. He had the intuition of an alternative logic behind their genealogy, calling for a "technological phylogenesis" to account for their development.[14] Technical borrowings, which play a key role in the "evolution of instrumental techniques," are indeed beyond the scope of phenomenotechnique. Like Ducassé with time measurement, Canguilhem projects an original insight stemming from his non-Comtean philosophy of technology into his history of science.

Even though Gilbert Simondon (1924–1989) never formally studied or worked at the IHST, he certainly dwells within the perimeter covered by the topic. He took the argument of the dependence of science upon technology to a deeper philosophical level, confronting Comte's law of three stages: instead of a succession of a theological, a metaphysical, and a positive (i.e. science-applicative) stage, technical thinking and religious thinking permanently combined with each other into theoretical thinking; science is a "system of compatibility" between them (Simondon 1958: 206). Furthermore, science cannot account entirely for the existence and organization of technical artifacts, which evolve as problems of internal compatibility between structures and functions are gradually overcome.[15] Technical artifacts are then definitely not a mere application of science, even in the industrial age: Simondon indeed agrees that the industrial age is that wherein technical species most embed science, *but* only to the extent that scientific knowledge of internal effects is paralleled by a mature technical knowledge of the internal organization of

[14] "... the history of Physiology (...) overtly shows how one good turn deserves another in the borrowing of instrumental techniques. For example, (...) it is the famous *kymograph* (1846) who made C. Ludwig famous. Now, according to the technological phylogenesis, there is no doubt the ancestor of this instrument is J.-L.-M. Poisefeuille's *hemodynamometer*. Ludwig's proper ingenuity consisted in combining Poisefeuille's arterial manometer with a graphic recorder. So that when E. J. Marey got involved in the development and improvement of the graphic method in France, he found himself indirectly indebted to Poisefeuille as he was directly indebted to Ludwig." (Canguilhem 1963a: 232)

[15] "The difference between the technical object and the physicochemical system taken as an object of study lies in the sole imperfection of the sciences; the pieces of scientific knowledge used as a guide to predict the universality of the mutual actions that occur within the technical system are limited by a certain amount of imperfection; they do not enable prediction of every effect with rigorous accuracy; that is why a certain gap remains between the system of technical intentions corresponding to a defined purpose and the scientific system of knowledge about the causal interactions aiming at that very purpose; the technical object is never entirely known (...). The ultimate distribution of functions to the structures and the exact computation thereof could only be achieved provided that the scientific knowledge of every single phenomenon likely to occur in the technical object was completely acquired; since this is not the case, a certain difference remains between the technical scheme of the object (which includes the representation of some human ends) and the scientific picture of the phenomena taking place within it (which only includes schemata of efficient, mutual or recurrent causality). (...) the mode of existence of the concretized technical object being analogous to that of spontaneously produced natural beings, they [technical objects] can legitimately be considered natural objects, that is to say submitted to an inductive study. They are not just an application of existing scientific principles." (Simondon 1958: 35–36, 47–48)

the artifact, that is, an empirical knowledge of the relationships between structures and functions. This specific technical knowledge inhabits the very space of the "concrete sciences" that Comte decided to delete from his system.[16] Not only are technical objects not just an application of science, but they can be an empirical source of original knowledge for science, an eventuality that was absolutely excluded by Comte's doctrine.[17] Finally, another of Simondon's oppositions to Comte lies in the heuristic value of technical artifacts, through which they inspire theoretical models and concepts. Simondon, however, explores the idea in a more systematic way than Canguilhem, with examples such as: Cicero's metaphors taken from agriculture, boat sailing, and war (Simondon 1961: 132); hylomorphism as a cognitive habit deeply rooted in the ancient practice of brick molding, whose long-term influence Simondon criticizes as an ontological obstacle (Simondon 1964); chaining technology as an underlying model inspiring Descartes' rules for the direction of the mind, and, of course, cybernetics (Simondon 2009).

5.2 The Inductive Value of Cybernetics in Postwar France

One major international postwar intellectual event was Norbert Wiener's book on cybernetics (Wiener 1948). First published by a French publishing house, right next to the Sorbonne, it caught immediate attention from academic circles and beyond. The topic soon became a fashionable subject for a noticeable spectrum of French philosophical debates. It fostered enthusiasm among our three philosophers, as it

[16] "In principle, the category of technique is the background to every abstract construction of positivism, but never for its own sake, never, so to speak, as an autonomous agent. (...) Concerning technicality, (...) the effective content of the concrete references finds itself transformed over the course of the Comtean enterprise, beginning to end: a transformation acknowledged and desired by Comte, in deep compliance with the logic of his work (...). Indeed, as Comte turns to the sociological, moral and religious effectuation of the concrete variables his theoretical philosophy entails, nothing seems to change in the order of knowledge and its relation to action; everything changes, in fact, because the methods for action are not exactly linked to abstract thinking as they were before. Without being altered in its general systematic function, the relation between science and technology is 'reduced' by the suppression of one intermediary, though previously conceived of as necessary: 'concrete' science. Renouncing the concrete sciences is, for Comte, (...) renouncing any basic diversity, any special originality of the disciplines of action, conceived of as relatively autonomous systems of technico-scientific complexes." (Ducassé 1957: 13–14)

[17] "Studying the functioning schemes of concrete technical objects entails a scientific value, for these objects are not deduced from a single principle; they testify to a certain mode of functioning and compatibility that exists factually and was constructed before it was predicted: this compatibility was not included in each of the separate scientific principles used to construct the object; it was discovered empirically; from the observation of this compatibility, one can work backwards toward the separate sciences to raise the problem of the correlation between their principles and found a science of correlations and transformations that would be a general technology or mechanology." (Simondon 1958: 48) Simondon later claimed that he was not aware of Lafitte's book at the time; it does not seem possible to verify whether he did indeed coin the same word independently.

significantly echoed and amplified their non-Comtean insights, and even had them reciprocally involved in the life of French cybernetics.

Cybernetics directly conflated two key points of Comte's system: clear-cut divisions of labor between disciplines[18] as well as between theoretical men and practical men, and unidirectionality of knowledge flow between science and technology.[19] Wiener's book would have been unto satanic verses for the father of positivism: first, cybernetics made and promoted heuristic use of machines, either as models for biological, social and cognitive processes, or as sources of unpredictable artificial behaviors demanding empirical study (such were the machines constructed by cyberneticians, giving birth to experimental robotics). Second, cybernetics relied on interdisciplinary, boundary-crossing collaborations between engineers, mathematicians, biologists, social scientists, physicians, and involved the circulation of formalized concepts and methods, whereas Comte was openly hostile, in particular, to the use of mathematics in sciences such as Biology. Finally, whereas Comte notoriously pretended to assess which sciences were likely to hold their promises and thus legitimately deserved a future, cybernetics would emerge as one of the most impressive postwar scientific and technical endeavors, in the minds of both scientists and engineers and the general public. Not only was reverse knowledge flow from machines to scientific theories and interdisciplinary collaborations possible, but it was fertile, producing a counter-specialization not in the monopoly of the positivist philosopher.

Implications of cybernetics for classification and synthesis of knowledge did not remain unnoticed in the French intellectual landscape (Le Roux, 2018: Chap. 10). Henri Berr's *Centre de synthèse*, founded on a similar idea of fighting excessive specialization (but wagering that specialization and generalization are not mutually exclusive), was friendly towards cybernetics: besides Ducassé, Suzanne Colnort, a historian, philosopher, and assistant to Maurice Daumas, wrote a few papers on cybernetics. In 1958, the 21st *Semaine de synthèse*, a more or less annual high-level interdisciplinary conference, featured a talk by mathematician Benoît Mandelbrot

[18] Comte's objective was to fight the *effects* of excessive specialization, not specialization itself: boundaries were to remain solid between the disciplines. Philosophers alone were the caste both permitted and supposedly able to proceed to the combination of knowledge; mathematicians should certainly not attempt to unify separate fields (Pickering 2007: 443). In Comte's classification, mathematics deals with simple objects and is useless when it comes to complex objects such as organisms or societies (Comte 1830: 148–153).

[19] Comte had little consideration for the machine theories taught at Polytechnique, because of their intermediary position between pure theory and pure practice (Vatin 2003). Purity mattered: the priority task was to reorganize the pure sciences, and *only then* "deduce" the conditions of their application. Therefore, machines could never claim any heuristic value: their theory could attain legitimacy only *after* the pure sciences of which it would be an application were *complete*, and hence beyond the reach of any modification. Also, a reverse flow of knowledge from machine theories to fundamental science would imply a move from a more specialized to a more general knowledge, inconceivable for Comte, for "high capacities in theoretical sciences and high capacities in applicative sciences are essentially distinct to the point that they exclude each other and could never be found in one and the same head." Comte, quoted by Ducassé (1939: 230). "Capacity for philosophy and capacity for details are *mutually exclusive*." Comte, quoted by Petit (1994: 64)

on cybernetics. Besides the *Centre de synthèse*, a figure of embodied encyclopaedism was François Le Lionnais: a chemical engineer and mathematician, famous for editing *Les Grands courants de la pensée mathématique* in 1948 and the foundation of the artistic movement Oulipo in 1960 (Motte 1986), he wrote papers on cybernetics and chaired at the first international congress of cybernetics in Belgium, since he worked for Unesco on the teaching and diffusion of science. Another book of encyclopedic spirit he edited included papers from Wiener and others involved in cybernetics (Le Lionnais 1959), such as François Russo, who stresses that the concept of information "points toward redistributions and clusterings likely to modify significantly the classic conception of the system of the sciences" (Russo 1959: 22). It is of course Comte that the clerk has in his sights here. Russo earns his place next to our three philosophers in a bigger picture (Le Roux 2013a, b) as he repeatedly discusses the concept of machine in relation with cybernetics (Russo 1955: 44, 45). Such discussions were not rare among cyberneticians, in France or abroad; but it was probably a characteristic feature of the French to try to link cybernetics to mechanology, as opposed to more abstract, mathematical theories of machines (Le Roux 2009a).

Ducassé played a key role in the life of French cybernetics. At the IHST, he welcomed the seminar of a temporary "Cercle d'études cybernétiques" (Le Roux, 2018: Chap. 2). Among its forty members were mathematicians, engineers, biologists, physicians, philosophers, historians, including many of the scholars mentioned in this paper: Ducassé, Dubarle, Russo, Colnort, Le Lionnais, Mandelbrot, David, Wiener himself, and… Lafitte, who gave four of the fourteen talks at the seminar. Cybernetics almost revived his project, but he stepped back due to fears of seeming outdated. Ducassé did not just offer physical but also editorial space to the CECyb. The entire seventh issue of *Thalès*, the IHST journal of which Ducassé was the secretary, was curated by the CECyb and stressed non-Comtean approaches to machines: Ashby underlined the heuristic status of one of his experimental prototypes (Ashby 1951); Russo, expressing dissatisfaction with the theories of Reuleaux and Lafitte, proposed to situate cybernetics within a "general phenomenology of machines" (Russo 1951); and Louis Couffignal, who had presented himself since the late 1930s as the heir apparent of the *Cours de machine* at Polytechnique, which he envisioned updating, gave the outline of his theoretical project (Couffignal 1951), trying to compete with "American" cybernetics. In parallel, Ducassé's journal *S.E.T.* became a (non-technical) tribune for cybernetics in France (Le Roux 2011b). Ducassé drew closer to Couffignal, in whose theoretical project he saw the best version of the "concrete sciences" abandoned by Comte (Ducassé 1957). After 1954, Couffignal wrote many papers in *S.E.T.*

Simondon was also involved in the life of French cybernetics, although to a much lesser extent. He seems to have remained outside the network of Ducassé, *S.E.T.*, and the CECyb (Le Roux 2011a), even though he knew Ducassé's journal (quoted in the bibliography of his *Mode of existence of technical objects*). Just before his *agrégation* in Philosophy and the publication of Wiener's book, he obtained a certificate in Psychophysiology under the direction of Alfred Fessard, the leader of the renewal of French neurophysiology and soon to be member of the

CECyb. Psychophysiology was one of the scientific hotspots of cybernetics. Simondon began his doctoral research circa 1952, at a time when cybernetics was attracting much attention; it is therefore not surprising that cybernetics played a key role in shaping his nascent philosophical system, and probably inspired him to reflect deeply on technology, leading to the birth of his philosophy of technology. Major aspects of his system derive from a critical dialogue with cybernetic ideas. Two early manuscripts (Simondon 1953a, b) confirm the paradigmatic value of cybernetics in Simondon's philosophy and show that he had grasped its encyclopedic implications: "Wiener's book can be considered a new *Discourse on the Method* (…) But why is this aspect of the technological relation[20] between the sciences called cybernetics? Auguste Comte would have called it philosophy." (Simondon 1953a: 38, 42) This comparison with Descartes and Comte is corrected in the second text: first, Simondon no longer equates cybernetics to Comtean reflexivity. Like mathematics and positivist epistemology, cybernetics is not a science like others, defined by an object, but a transversal method; but, unlike mathematics and epistemology, it does not separate teleology from objectivity. The comparison between Wiener and Descartes is also revisited.[21] These corrections reveal Simondon's first critical step toward cybernetics, with just a little paradox as the departure from Comte on the one hand is compensated on the other hand by a Cartesian stance urging cybernetics to find its general "logic" from which to apply its principles instead of proceeding empirically. Still, this "universal cybernetics" is a universal technology, and even a "pure technology"[22] – a nonsense for Comte! The early manuscripts show that Simondon had a good knowledge of the cybernetic literature. At the beginning of his PhD, he returned to the *École Normale Supérieure* trying in vain to

[20] "What is the nature of the relations between the sciences? They are of two profoundly different kinds: a science can meet another by encroaching progressively upon its object (…) but a science can meet another one because it needs it *as a technique* within a domain from which it is not intending to escape. (…) In the operative interscientific (if not suprascientific) compatibility, one discovers a mode of relation to the object that is not merely scientific, but also technical. (…) The no man's land between the specific sciences is not another specific science, it is a universal technological knowledge, an interscientific technology that does not aim at a theoretical *object* selected in the world, but a *situation*. (…) In order to define the nature and the value of cybernetics, this interscientific technology, it suffices to understand that it does not aim at identifying a complicated process with a simpler process – as commonly believed – (for instance the human thinking with the functioning of a mechanical system), but rather to establish equivalences between various situations faced by the scientist when dealing with such or such an object" (Simondon 1953a: 40–42).

[21] "...even if, from a historical point of view, Wiener's book may be as unto a modern *Discourse on the Method*, from a doctrinal point of view it lacks the latter's internal unity; on the basis of brilliant examples, one grasps the concerns to which cybernetics responds, though no cybernetic method is anywhere defined. In *Discourse on the Method*, on the other hand, or more precisely in the fragments that appendix it, there are examples of application of the Cartesian method, within *Dioptrics* and the *Meteors*; but prior to this demonstration through examples, one finds a definition of the rules of the method and an affirmation of their universal validity. Thus, we believe that the most urgent contribution needed by the [present] work is the elaboration of a cybernetic logic." (Simondon 1953b: 197)

[22] "...there can be a pure technology, as there is a pure science, and this pure technology is cybernetics." (Simondon 1953a: 43)

set up a reflection group on cybernetics with alumni and teachers, such as Foucault, Althusser, and others (N. Simondon 2015). In 1962, he was the secretary of the fourth international colloquium of Royaumont, choosing the speakers, among whom were Wiener, Mandelbrot, Couffignal, and others.

Canguilhem had a deep interest in cybernetics too, but never sought to get involved in actual cybernetic circles; his information about cybernetics was more indirect, based on second-hand sources. Another difference between him and Ducassé and Simondon is that his use of cybernetics, although non-Comtean, was less connected to his explicit criticism of positivism: his deep interest in technology, and the arguments he found there against positivism, received little input, if any, from cybernetics. Canguilhem referred to cybernetics in respect to the epistemology of the life sciences, his interest was the heuristic value of machines as conceptual models and the methodological value of mathematics. His famous paper on models and analogies in biology (Canguilhem 1963b) draws heavily from cybernetic references. His archives confirm the topicality of cybernetics. By contrast, in his grand synthesis "Machine et organisme" (Canguilhem 1961), cybernetics is remarkably absent.

In 1971 and 1976, two conferences on mechanology were organized in Paris by a group of Canadians. The proceedings (Le Moyne and Hart 1971, 1976) significantly echo the story told in this paper, as Lafitte (who died in 1966) was the tutelary figure of the conference, Simondon its guest star, Canguilhem a chairman and also responsible for the synthesis of the first conference, several members of the CECyb were speakers or attendants, along with leading figures in history of technology, sociology of work, and computer science. The *Centre de synthèse* supported the conference, and Suzanne Delorme, the successor of Henri Berr, chaired one of the sessions. These conferences raise questions about continuity and the cumulative nature of machine theories and their relation to other scientific disciplines. The proceedings had a very confidential diffusion; so that if they are to be taken as symbols embedding non-Comtean insights, they are far from equating the latter's widespread diffusion.[23]

5.3 Conclusion

The insights of our three philosophers concerning the relations between science and technology show remarkable convergence; why have they have remained scattered and embryonic, scarcely visible, instead of translating into an explicit research agenda? Why, as of today, have these insights had no impact on historical epistemology?

Ducassé abandoned historical epistemology when he left the IHST; Simondon was scarcely involved in it. Canguilhem's case is the most enigmatic: deeply

[23] Thibault & Hayward (2014); Iliadis (2015).

interested in technology until at least the 1970s,[24] claiming explicit antipositivism,[25] he was interested in how machines seem to follow specific laws,[26] possibly selective (Canguilhem 1961: 123–124), and coined the expression "technical phylogenesis." But in "Machine et organisme" (which synthesizes many of these earlier arguments), after calling for an "organology" as a biologically-inspired theory of machines (see Hoquet, Chap. 16, this volume) in opposition to Descartes (and implicitly, I believe, to Bachelard too), Canguilhem finally backtracked, concluding that while certain machines are the product of a rational calculus, others have an "irrational origin" (Canguilhem 1961: 125). The specific rationality implied by the organologic project of analyzing the patterns of machine genesis is abandoned in favor of "irrationality." Two early clues might help in understanding this switch to irrationality: that the technical artifact, because it is a synthesis, is not subject to analysis (Canguilhem 1937: 497; 1942: 115); and, in line with the explicit influence of the French anthropological school he had carefully read, that tools were primitively linked to magic (Canguilhem 1944: 8). It is striking that in the 1965 update of "Machine et organisme," Canguilhem discusses neither cybernetics nor Simondon's genetic technology. Canguilhem the technologist leaves the last word to a romantic Canguilhem (poetic even, surprisingly referring to Heidegger in his 1942 course on Bergson), for whom technical creation belongs more to art than to science.

In the end, it seems that all three philosophers seemingly avoided the direct confrontation with Bachelard's phenomenotechnique implied by their non-Comtean insights. Barely crossing that line, how could their converging insights have survived their diverging trajectories in an institutional space where positivism was so deeply entrenched?

The development of heterodox insights thus suffered from the institutional weakness of philosophy of technology. Our philosophers had scant ties with each other. Ducassé and Simondon were at the margins of philosophical institutions, while Canguilhem was at the center, but mainly viewed as a philosopher of medicine. There was no other figure to represent or unify the field, no common space where insights could be openly discussed, and integrated. Such weakness makes for more uncertain inter-generational transmission and renders the field more dependent on individuals, that is, on more scattered topics of interests, and on the contingency of individual acquaintances. This institutional weakness could barely provide relays to put heterodox insights at work – J.-C. Beaune was one of them –, or compete with post-structuralism: machines and cybernetics finally found a place in the emerging

[24] Box GC16.1, courses, notes and seminars on history of technology, XIXth century technological thinking. Boxes GC10.4 and GC11.3.4: courses on science, technology and their relations.

[25] For instance, he praised Daumas in his stance against positivism for having "protested against any scienticist conception of the history of technology," in particular in his historical study of scientific instruments (Canguilhem 1969).

[26] "*Tools derive from tools*. The major obstacle is to account for the very first ones" (Canguilhem 1944: 8). Canguilhem underlines the ontological independence of technology with respect to science: "The chronological precedence [of technique over science] reflects a priority of *essence*"; "The concept of machine includes that of *trick* (*mekanè*), of *trap*, of *fooling*. This concept is incompatible with the concept of science" (ibid.: 4, 16).

postmodern philosophical landscape – trading off epistemological rationality. Machines have undoubtedly conquered much discursive space in French philosophy, and have become more fashionable than Auguste Comte; yet their logic of development was debarred from social constructivism and postmodern metaphorics (including Deleuze and Guattari's reference to a "machinic phyllum" – see De Landa 1997).

Countering this hiatus, I wished to show that there have been significant research insights which combine both the in-subordination of machines *and* a rationalistic frame. These provide inspiration for updating rationalism along lines sensitive to the technical dimensions of science. One task is the further construction of mechanology and its application to the study of scientific instruments (non-Bachelardian phenomenotechnique). Another task, for which Beaune was a historical relay (Beaune 1979), is a methodical study of the heuristic value of machines as models for analogies: from the rationalist perspective of 'French epistemology', contrasting with cultural studies of science, it should consider the scientific value and performance of analogical reasoning. Non-Comtean insights are thus likely to contribute to the cross-fertilization of philosophy of science and philosophy of technology (Ihde 1991; Hottois 2004).

Acknowledgments Most of the material for this paper was taken from my 2010 PhD dissertation and briefly presented at the 2014 HOPOS Conference in Ghent, Belgium. I am grateful to helpful assistance and comments from N. Queyroux, G. Thibaut, N. Simondon, A. Iliadis, and Ch. Chimisso; and to linguistic correction (C. J. Robertson) thanks to financial support from Université Paris-Est team EA 4395 Lettres, Idées, Savoirs.

References

Ashby, W. R. (1951). Statistical machinery. *Thalès, 7*, 1–7.

Bachelard, G. (1927). *Essai sur la connaissance approchée*. Paris: Vrin.

Bachelard, G. (1934). *Le Nouvel Esprit scientifique*. Paris: Presses Universitaires de France. English Edition: Bachelard, G. (1984). *The New Scientific Spirit* (Trans. Goldhammer, A.). Boston, MA: Beacon Press.

Bachelard, G. (1938). *La Formation de l'esprit scientifique. Contribution à une psychanalyse de la connaissance objective*. Paris: Vrin. English Edition: Bachelard, G. (2002). *The Formation of the Scientific Mind. A Contribution to a Psychoanalysis of Objective Knowledge* (Trans. McAllester Jones, M.). Manchester: Clinamen Press.

Bachelard, G. (1949). *Le Rationalisme appliqué*. Paris: Presses Universitaires de France.

Bachelard, G. (1970). *Le Droit de rêver*. Paris: Presses Universitaires de France.

Beaune, J.-C. (1979). Les rapports entre la technolcgie et la biologie, du XVIIe au XIXe s.: L'automate, modèle du vivant. In P. Delattre & M. Tellier (Eds.), *Élaboration et justification des modèles: applications en biologie* (Vol. 1, pp. 199–214). Paris: Maloine.

Bergson, H. (1934). *La pensée et le mouvant. Essais et conférences*. Paris: Félix Alcan. English Edition: Bergson, H. (2012). *The Creative Mind: An Introduction to Metaphysics* (Trans. M. L. Andison). Mineola, NY: Dover.

Braunstein, J.-F. (2000). Canguilhem avant Canguilhem. *Revue d'histoire des sciences, 53*(1), 9–26.

Braunstein, J.-F. (2006). Abel Rey et les débuts de l'Institut d'histoire des sciences et des techniques. In M. Bitbol & J. Gayon (Eds.), *Épistémologie française* (pp. 173–191). Paris: Presses Universitaires de France.

Braunstein, J.-F. (2011). À la découverte d'un "Canguilhem perdu". In G. Canguilhem (Ed.), *Oeuvres complètes, tome. 1: Écrits philosophiques et politiques (1926–1939)* (pp. 101–137). Paris: Vrin.

Canguilhem, G. (1937). Descartes et la technique. In G. Canguilhem (2011), *Oeuvres complètes, tome. 1: Écrits philosophiques et politiques (1926–1939)* (pp. 490–498), 2011. Paris: Vrin.

Canguilhem, G. (1938a). Activité technique et création. In G. Canguilhem (2011), *Oeuvres complètes, tome. 1: Écrits philosophiques et politiques (1926–1939)* (pp. 499–509), 2011. Paris: Vrin.

Canguilhem, G. (1938b). Review of Schuhl (1938). In G. Canguilhem (2011), *Oeuvres complètes, tome. 1: Écrits philosophiques et politiques (1926–1939)* (pp. 510–511), 2011. Paris: Vrin.

Canguilhem, G. (1942). Cours et notes sur Bergson, 1941–42. Archives of G. Canguilhem, CAPHES, Paris, File GC 11.1.6.

Canguilhem, G. (1944). Technique et science. Archives of G. Canguilhem, CAPHES, Paris, File GC 11.3.4.

Canguilhem, G. (1961). Machine et organisme. In G. Canguilhem (1998). *La Connaissance du vivant* (pp. 101–127). Paris: Vrin. English Edition: Canguilhem, G. (2008). *The knowledge of life* (S. Geroulanos & D. Ginzburg, Trans.). New York: Fordham University Press.

Canguilhem, G. (1963a). La Constitution de la physiologie comme science. In G. Canguilhem (1994) *Études d'histoire et de philosophie des sciences* (pp. 226–273). Paris: Vrin.

Canguilhem, G. (1963b). Modèles et analogies dans la découverte en biologie. In G. Canguilhem (1994). *Études d'histoire et de philosophie des sciences* (pp. 305–318). Paris: Vrin. English Edition: The role of analogies and models in biological discovery. In A. C. Crombie (Ed.), *Scientific change* (pp. 507–520). London: Heinemann.

Canguilhem, G. (1969). Unpublished untitled manuscript written for M. Daumas's inaugural lesson at the Conservatoire national des Arts et Métiers. Archives of G. Canguilhem, CAPHES, Paris, File GC16.1.9.

Chimisso, C. (2008). *Writing the history of the mind: Philosophy and science in France, 1900 to 1960s*. Aldershot: Ashgate.

Comte, A. (1830). *Cours de Philosophie positive* (Vol. 1). Paris: Rouen Frères.

Couffignal, L. (1951). La Mécanique comparée. *Thalès, 7*, 9–36.

Cuevas, A. (2005). The many faces of science and technology relationships. *Essays in Philosophy, 6*(1), article 3. http://commons.pacificu.edu/cgi/viewcontent.cgi?article=1173&context=eip. Accessed March 2018.

De Landa, M. (1997). The machinic phylum. *TechnoMorphica*. http://v2.nl/archive/articles/the-machinic-phylum. Accessed March 2018.

De Solla Price, D. (1965). The science/technology relationship, the craft of experimental science and policy for the improvement of high technology innovation. *Research Policy, 13*(1), 3–20.

Ducassé, P. (1939). *Essai sur les origines intuitives du positivisme*. Paris: Alcan.

Ducassé, P. (1945). *Histoire des techniques*. Paris: Presses Universitaires de France.

Ducassé, P. (1948). Pour l'histoire de la mesure du temps. *Thalès, 5*, 113–127.

Ducassé, P. (1949). Relations entre sciences et techniques, à propos de la mesure du temps. In *Les journées d'information pour la coordination des enseignements de la philosophie et des sciences physiques dans les classes de philosophie et de sciences expérimentales, Strasbourg, novembre 1949* (pp. 11–25). Paris: Publication du Centre national de documentation pédagogique.

Ducassé, P. (1957). Auguste Comte et la philosophie des techniques. *Structure et évolution des techniques, 55–56*, 11–15.

Ducassé, P. (1958). *Les Techniques et le philosophe*. Paris: Presses Universitaires de France.

Faulkner, W. (1994). Conceptualizing knowledge used in innovation: A second look at the science-technology distinction. *Science, Technology and Human Values, 19*(4), 425–458.

Galison, P. (1997). Three laboratories. *Social Research, 64*(3), 1127–1155.

Guillerme, J. (1971). À propos de "mécanologie" de Jacques Lafitte. In J. Guillerme (2008) *L'art du projet: histoire, technique, architecture* (pp. 47–50). Wavre: Mardaga.
Hart, J. (1976). Introduction. In Le Moyne, J. & Hart, J. (1976). *Mécanologie 2. Les Cahiers du Centre culturel canadien*, 4 (pp. vii–xii).
Hottois, G. (2004). *Philosophie des sciences, philosophie des techniques*. Paris: Odile Jacob.
Ihde, D. (1991). *Instrumental Realism*. Bloomington/Indianapolis: The Indiana University Press.
Iliadis, A. (2015). Mechanology: Machine typologies and the birth of philosophy of Technology in France (1932–1958). *Systema, 3*(1), 131–144.
Lafitte, J. (1932). *Réflexions sur la science des machines*. Paris: Bloud & Gay.
Lafitte, J. (1933). Sur la science des machines. *Revue de Synthèse, 6*(2), 143–158.
Le Lionnais, F. (1959). *La Méthode dans les sciences modernes*. Paris: Éditions Science et industrie.
Le Moyne, J., & Hart, J. (1971). Colloque La Mécanologie, 18-19-20 mars 1971. *Les Cahiers du Centre culturel canadien*, 2.
Le Moyne, J., & Hart, J. (1976). Mécanologie 2. *Les Cahiers du Centre culturel canadien, 4*.
Le Roux, R. (2009). L'Impossible constitution d'une théorie générale des machines ? La cybernétique dans la France des années cinquante. *Revue de Synthèse, 130*(1), 5–36.
Le Roux, R. (2011a). Simondon technologue, une individuation inachevée. Dialogues (et silences) avec les théories des machines dans la France d'après-Guerre. In J.-H. Barthélémy (Ed.), *Cahiers Simondon* (Vol. 3, pp. 93–108). Paris: L'Harmattan.
Le Roux, R. (2011b). Pierre Ducassé et la revue Structure et évolution des techniques (1948–1964). *Documents pour l'Histoire des techniques, 20*, 119–134.
Le Roux, R. (2013a). Influence des projets de théories des machines sur l'histoire des techniques chez P. Ducassé et Fr. Russo. In A.-L. Rey (Ed.), *Histoire et Méthode. Quelle histoire font les historiens des sciences et des techniques?* (pp. 337–348). Paris: Classiques Garnier.
Le Roux, R. (2013b). Un Jésuite, des machines, une histoire teilhardienne des techniques ? Russo dans le contexte de la pensée catholique française d'après-Guerre. *Bulletin de la Sabix, 53*, 17–25.
Le Roux, R. (2018). *Une Histoire de la cybernétique en France, 1948–1975*. Paris: Classiques Garnier.
Mitcham, C. (1979). Documentation: Analysis of Machines in the French Intellectual Tradition (Espinas, Lafitte, Weil). *Research in Philosophy and Technology, 2*, 189–234.
Motte, W. (Ed.). (1986). *Oulipo: A primer of potential literature*. Chicago: Dalkey Archive Press.
Petit, A. (1994). L'Impérialisme des géomètres à l'École polytechnique. Les critiques d'Auguste Comte. In B. Belhoste, A. Dahan-Dalmedico, & A. Picon (Eds.), *La formation polytechnicienne 1794–1994* (pp. 59–75). Paris: Dunod.
Pickering, M. (2007). Auguste Comte and the Académie des sciences. *Revue philosophique de la France et de l'étranger, 132*(4), 437–450.
Rey, A. (1937). Évolution de la pensée. In A. Rey, A. Meillet, & P. Montel (Eds.), *L'Encyclopédie française, t.1: L'Outillage mental, pensée, langage, mathématiques*. Paris: Société de gestion de l'Encyclopédie française.
Rheinberger, H. (2005). Gaston Bachelard and the notion of "Phenomenotechnique". *Perspectives on Science, 13*(3), 313–328.
Russo, F. (1951). La Cybernétique située dans une phénoménologie générale des machines. *Thalès, 7*, 69–75.
Russo, F. (1955, January). Fondements de la théorie des machines. *Revue des questions scientifiques*, 43–74.
Russo, F. (1959). Le Système des sciences et des techniques et l'évolution de la classification des sciences. In F. Le Lionnais (Ed.), *La Méthode dans les sciences modernes* (pp. 17–23). Paris: Editions Science & Industrie.
Schuhl. P.-M. (1938). *Machinisme et philosophie*. Paris: Presses Universitaires de France.
Simondon, G. (1953a). Cybernétique et philosophie. In G. Simondon (2016) *Sur la philosophie (1950–1980)* (pp. 35–68). Paris: Presses universitaires de France.

Simondon, G. (1953b). Épistémologie de la cybernétique. In G. Simondon (2016) *Sur la philosophie (1950–1980)* (pp. 177–199). Paris: Presses universitaires de France.

Simondon, G. (1958). *Du Mode d'existence des objets techniques*. Paris: Aubier.

Simondon, G. (1961). Psycho-sociologie de la technicité. In G. Simondon (2014) *Sur la technique (1953–1983)* (pp. 27–129). Paris: Presses Universitaires de France.

Simondon, G. (1964). *L'Individu et sa genèse physico-biologique*. Paris: Presses Universitaires de France.

Simondon, G. (2009). Technical mentality (A. De Boever, Trans.). *Parrhesia, 7*, 7–27.

Simondon, N. (2015). Gilbert Simondon and the 1962 Royaumont Colloquium: Information, Cybernetics, and Philosophy (A. Iliadis, Trans.). *Philosophy & Technology, 29*(3), 284–291.

Thibault, G., & Hayward, M. (2014). Jean Le Moyne's Itinéraire mécanologique: Machine Poetics, Reverie, and Technological Humanism. *Canadian Literature, 221*, 56–72.

Vatin, F. (2003). Comte et Cournot. *Une mise en regard biographique et épistémologique. Revue d'histoire des sciences humaines, 8*, 9–40.

Wiener, N. (1948). *Cybernetics, or control and communication in the animal and the machine*. Paris: Hermann.

Chapter 6
We Have Never Been Wild: Towards an Ecology of the Technical Milieu

Victor Petit and Bertrand Guillaume

Abstract Oddly, many of those who have pioneered the philosophy of political ecology in France (André Gorz, Cornélius Castoriadis, Félix Guattari, Serge Moscovici in particular) have usually dismissed both expected terms "Nature" and even "environment." This distinctive feature is due to the choice French scholars have made to closely intertwine ecology and the questioning of technology, leaving aside concerns for environmental ethics. This chapter aims to clarify the distinction between ecology of nature and what we call "ecology of technology," and to better grasp the idea of an ecology against Nature. Then it turns to the key distinction between the concept of environment and that of milieu in order to reconstruct the long and non-unequivocal historical path which led a number of French philosophers from the philosophy of the technical milieu to a political ecology.

This chapter has been written in French for this volume and translated by John Stewart, University of Technology Compiègne.

Keywords Castoriadis · Ecology of nature · Ecology of technology · Environment/milieu · Guattari · Gorz · Friedmann · Machinic ecology · Moscovici · Technical milieu

This chapter argues that one major specific feature of French philosophy of technology is the promotion of an ecological approach to technology characterized by the emphasis on the concept of "technical milieu." Although none of the philosophers

---

V. Petit
Connaissance, Organisation et Systèmes TECHniques (COSTECH), Université de technologie de Compiègne (UTC), Compiègne, France

B. Guillaume (✉)
Centre de recherches et d'études interdisciplinaires sur le développement durable (CREIDD), Université de technologie de Troyes (UTT), Troyes, France
e-mail: bertrand.guillaume@utt.fr

S. Loeve et al. (eds.), *French Philosophy of Technology*, Philosophy of Engineering and Technology 29,
https://doi.org/10.1007/978-3-319-89518-5_6

mentioned in this chapter uses the phrase "ecology of technologies" they all share the conviction that the question of ecology and the question of technology are one and the same.

As we use it, the term "ecology" refers neither to the science of the same name, nor to the concept of "nature" at least in the sense of the "wilderness." We deliberately use "ecology" in a broader sense (Ingold 2001) to point out that French philosophers, as far as they questioned technological development as such, less focused on deleterious impacts of technology on nature, than they highlighted how modern technology ("industrial machinism") shapes the dynamic intercourse between humans and their environment in a new and questionable way. The "technical milieu" is the new milieu of human beings. In this respect, it is undoubtedly a relevant issue for ecology defined as the science of the *oikos*, i.e. of the way humans have built their artificial "house" on earth.

The term "technical milieu" was first coined in the mid-twentieth century and to some extent, it can be considered a specifically French contribution to the philosophy technology – even though not all French scholars have adopted it. However, the aim of this chapter is less to identify national characteristics, than to emphasize the strong link between political ecology and technology, a feature that was missing in Daniel Parrochia's survey of French philosophy of technology (Parrochia 2009). Incidentally most authors mentioned in this chapter do not portray themselves as philosophers of technology, and it is perhaps a French trademark never to make technology the object of a separate, cloistered interrogation.

6.1 Ecology of Nature, Ecology of Technology

6.1.1 Two Traditions?

When environmental ethicist and philosopher of ecology Hicham-Stéphane Afeissa asks, in an essay addressed to a French-speaking audience (Afeissa 2009), what is to be understood by the "philosophy of ecology," he distinguishes two approaches: on the one hand an interrogation about how the technologically-shaped relationships between human beings and their environment have deeply changed after the nuclear bomb (Anders, Jaspers, Arendt, Jonas); on the other hand, the debates about the value of nature and of ecological and ethical communities (Aldo Leopold, Lynn White Jr., Arne Naess, John Baird Callicott). Thus his definition suggests that the philosophy of ecology is split between ecology of technology and ecology of nature. This divide Afeissa rightly highlights is a central distinction to understand the French tradition, which is clearly situated on the former side. Here, the "ecology of technology" refers neither to a research program studying technology as an ecosystem in its own right, nor to anything like environmental ethics's concern with the conservation and protection of nature. The French ecology of technology carries no opposition between nature and technology. As a consequence, the concept of wilderness understood as a "pure" nature untouched by technology doesn't make any sense in it. No *paysages* without *paysans*! In other words, it is possible to talk about ecology with no reference, neither to "nature" nor to "environment." The French ecology of technology has rather great concern with "cultures" and "*milieux*."

In his study of the French tradition of political ecology illustrated by Jacques Ellul, Bertrand de Jouvenel, René Dumont, Serge Moscovici, André Gorz, Cornelius Castoriadis, and Felix Guattari, Kerry Whiteside points that the French debate on ecology is not framed as a strong divide between ecocentric and anthropocentric approaches. It is rather formulated as a critique of the invasion of the technosciences in our daily lives (Whiteside 2002). Despite the variety of their respective approaches, none of the French political ecology philosophers separates the question of ecology from the question of technology. The philosophies of technology and the philosophies of ecology seem to go hand in hand. Significantly, philosopher Dominique Bourg – an important contributor to the research area on sustainable development – came to ecology through the philosophy of technology. In particular, he primarily focused on how human beings are from the beginning "exosomatized," a term he coined to point out that humans have always exteriorized their biological possibilities into artifacts. Being human has always meant being artificial – "*L'Homme artifice*," as Bourg titled a stimulating book that to a certain extent inherited from paleoanthropologist and prehistorian André Leroi-Gourhan's concept of "exteriorization" (Bourg 1996). Significantly, Bourg's philosophy of ecology, initially apprehensive of *deep ecology*, led him to an *industrial ecology* (Bourg 2003).[1]

It is worth noting however that understanding French ecology in the light of these anthropological statements related to the co-shaping of humans and technology – a focus that may end up in diminishing the importance of nature as such in this research area – is a questionable and potentially misleading viewpoint. In particular, Michael Bess (2003) has described French ecological thought as that of a *light green society*, torn between the attractions of modernity and technological progress on the one hand, and the fear of losing contact with countryside nature on the other. So nature seems with no doubt a great concern of French scholars. But as Geneviève Massard-Guilbaud (2012) has perspicuously remarked, Bess's interpretation is distorted by his identification of "environment" with "nature." In our view the singularity of the French approach, if there is one, is rather to be sought in the concept of "milieu," which precisely denotes neither nature (unsullied by any technology), nor environment (unsullied by any subject); and if there is any nostalgia, it is that of the *technical culture* of the peasantry, i.e. a form of know-how. The revolts against mechanization and technologies do not mean that "nature" would be sanctuarized – not even by Rousseau, who features as the advocate of nature *par excellence* (Deneys-Tunney 2000). However this opposition between ecology of nature and ecology of technology has a relevance which is philosophical rather than historical or cultural. To be sure, this divide between two ecological traditions is not the apanage of France,

[1] His current position, with Kerry Whiteside (Bourg and Whiteside 2010), is nevertheless to take his distance from industrial ecology, and to develop reflections about ecological democracy with a decrease rather than a growth of overall human activity.

and the same could be found on both sides of the Atlantic.[2] The original contribution of French scholars is elsewhere, namely in the aim of overcoming the divide itself – a challenge that gave rise to the concept of "technical milieu."

6.1.2 Centrality of the Milieu

The term "technical milieu" does neither separate *technology* and *life*, nor the technologies of *matter* from the technologies of *living organisms*; nor the *ethics* of the living from industrial *politics*; nor the philosophy of the *environment* from the philosophy of *technology*. The "ecology of technology" is closely dependent on the concept of "milieu" rather than concerned with the notion of "environment." Contrary to the latter, the former emphasizes the co-shaping of the individual and his/her milieu. Indeed, thinking the human-technology relationships as the shaping of a "technical milieu" allowed to underline how both humans and technology are co-constituted – a very common statement today, but not at that time. In this respect, the importance of French philosopher Gilbert Simondon (2005) cannot be exaggerated. Even before English translations be available, Simondon's philosophy has prompted increasing interest worldwide in the last decade (see Bontems, Chap. 3 this volume). We suggest that this interest could be relevantly extended to the political ecology landscape, insofar as Simondon precisely introduced an original and insightful way of thinking technology in terms of human-technology co-constitution (Simondon says "individuation"). In this respect, he intended to overcome any instrumentalist definition of technology, and argued that far from being a set of means at the disposal of humans, technology and especially industrial machines reshape the humans-world relationships. And despite the fact that Simondon's concept of "associated milieu" cannot be equated to the concept of "technical milieu" (the "associated milieu" is the milieu of the machine, not of human beings themselves), he undoubtedly provides relevant perspectives for questioning the relationships between humans, their technology and their milieu in the industrial era. Significantly, for those who know how to read between the lines, he has also proposed an interpretation of the ecological crisis, which origin is to be sought not in anthropocentrism, but in hylemorphism.[3] To be sure, both Simondon's view that the individual and the milieu are co-constituted, and his clear distinction between environment and milieu, are echoed in the environmental philosophy of

[2] It is quite possible, for example, to read Lewis Mumford as an actor of the ecological movement.

[3] The hylemorphic conception relies on the view of brute "primary" *matter*, informed by technical activity, a to construct an economy. Hylemorphism, as Simondon argued (2005), is the scheme that enables technologies to function as black boxes. The "milieu" is not a neutral "matter" to be in-formed or de-formed, it is the place situated *between* the input and the output of the mould, where matter takes form, and where individuation operates.

Naess or Callicott. Thus, the concept of milieu could be considered a bridge between the philosophy of technology and the philosophy of nature.

6.1.3 Ecology Against Nature?

French political ecology seems to specifically rely on the rejection of nature as a leading concept, and to rather focus on the concept of "milieu."

This statement seems however challenged by another French contributor to the philosophy of technology, Bernard Charbonneau, who also pioneered political ecology in France (Roy 1992). In the 1930s, he and Ellul wrote what they called a "personalist manifesto" (1935) – a plea against the empire of technology and technocracy over our civilization. A few years later, they transformed it into a manifesto for nature ("A feeling for nature, a revolutionary force" 1936): the struggle for nature and the struggle for freedom, they argued from now on, should be considered inseparable (Charbonneau and Ellul 2014). Most of Charbonneau's later works developed this revolutionary feeling for nature as an alternative to the *diktat* of accumulation (productivism and consumerism) and technocratic rationality (territorial planning) (Charbonneau 1969). So contrary to our previous claim, nature definitely seems to have been a major concern in the emerging landscape of political ecology in France. However "nature" in Charbonneau's and Ellul's views was not, for all that, something like the wilderness, a virgin nature exempt from any technology. Charbonneau unequivocally cautioned against an "integrism of nature" or the latent "eco-fascism" of ecology (Charbonneau 1980).[4] For Ellul as well, "nature" was nothing but an image of technology (Ellul 1977) and his ecology was in no way concerned with nature, but rather with the human milieu (Ellul 1982).[5]

Charbonneau was an exception among French philosophers who, whatever their political inclinations and their religious convictions, were generally hostile to environmentalist movements based on a feeling for nature. Psycho-sociologist Serge Moscovici (1925–2014), another important contributor to the political

[4] Charbonneau assumed that total emancipation of mankind with respect to nature would be paid for at the price of total social control, and therefore a loss of liberty. But the safeguard of nature would lead, in his view, to a sort of eco-fascism: "Ecologism is the ideology tailor-made by and for a handful of experts and civil servants charged with managing the tiny sector of a chemically pure nature from which man – even the most vetted naturalist – is excluded. (…) Since the safeguard of nature can only be ensured by mankind, ecological science only provides a part of the hand. Ecologist risk losing sight of the realities which are cultural, psychological, economic and social – and hence political." (Charbonneau 1980: 92, 95)

[5] "Ecology cannot make of the 'natural milieu' a system to which one must either come back or keep to. And even if one agrees to interpret 'nature' (which is already an artificial concept created by man) as a system, it cannot be question of coming back purely and simply to nature to give free sway to natural laws, nor of having recourse to nature to know what 'is to be done.' (…) Bernard Charbonneau proposed that ecologists should rather use the term 'countryside' than that of Nature, because what is important is precisely this association, this correspondence between man and his milieu." (Ellul 1982: 13, 14)

ecology "*à la française*," exemplifies this general attitude. His book titled *Essai sur l'histoire humaine de la nature* ("Essay on the human history of nature") was a plea against naturalistic and technophobic viewpoints, and against any dualism of nature *versus* society as well (Moscovici 1968). Moscovici broke with the ahistorical concept of nature and he carefully defined it in a close relationship to technology, arts or society. There is no "nature" independent from artifacts. Nature should always be considered in terms of "states of nature" that combine material forces and human activity in various historical ways – thus Moscovici historicized the old political-philosophy concept of "state of nature." Moscovici's political ecology, defined as the study of nature as an historical production involving humans and technologies, emphasizes the co-evolution of material forces and the human species, considered as an aspect among others of nature's own history. Moscovici considered that technologies intervening on nature are political practices in themselves. He never militated for any "going back" to nature, or for reconciling technology and nature. He rather considered that today's task is to struggle for "chosing our state of nature." (Moscovici 2002) Moscovici introduced a clear-cut distinction between "natural ecology" and "historical ecology"; only the latter is political,[6] for it understands that mankind does not so much "adapt" to pre-existing environments, but rather creates its own milieu, and thus, that – by so doing – it has to choose between different states of nature (Moscovici 1972). The link between Moscovici's ecological engagement and his social psychology – a discipline he contributed to develop in France – is profound: both engagements assume that "reality" and "nature" are social and political constructions, and that change is to be sought on the side of the active minority rather than the academic majority (Moscovici 2002).

In order to better understand the contrast between a nature-centered and a technology-centered ecology, it is useful to refer to a lecture delivered by Georges Canguilhem in 1973, when the concept of "environment" was introduced in France. Canguilhem is well-known for his works on the philosophy of biology and of medicine, but he also prompted concerns for ecology. He argued that the milieu of a biological organism is different from the human environment, which is in reality a technically, socially and historically constructed "milieu." For Canguilhem, the science named "ecology" deals exclusively with the former; the environmental movements have nothing to say about it, but only about the latter (Canguilhem 1990). When environmental movements pretend to ground their claims on ecological statements considered as scientific ones, they do nothing else than ideology: they keep their values veiled under the curtain of science. To some extent, Canguilhem anticipated the pitfall of movements such as "Green growth": "protecting" the environment can indeed be the means for transforming it into a market commodity and selling it; and conversely, it is important to beware of "islands of anti-technological purity." Canguilhem dismissed both neoliberal discourses and anti-capitalist

[6] In his 1968 essay, Moscovici even names this form of ecology "political *technology*," "technology" here meaning the normative study and organization of techniques, as he uses "*techniques*" in the rest of the book (Moscovici 1968).

counter-discourses, putting them back-to-back because none of them were emancipated from the standard view of technology as applied science. In his view technology was rather to be conceived as a tactic for living. Canguilhem's critical views about environmentalism were based on his conviction that technology is not so much an application *of science*, but a *fact of life*; a conviction that undermines any "back-to-nature" movement. Thus Canguilhem's biological philosophy of technology condemns *both* "technicism" and "naturalism." (see Hoquet, Chap. 16 in this volume).

To be sure, Canguilhem's search for an ecology of technology did not remain unanswered. In many respects, another contributor to this political ecology "*à la française*," François Dagognet (1924–2014), may be considered Canguilhem's heir insofar as he also maintained that technology and life could not be dissociated in an ecological perspective (Dagognet 2000).[7] To be sure, Dagognet is not the champion of nature but rather of artifacts that shape the "milieu" of humans (see Chazal, Chap. 2 in this volume). Struggling against any naturalism and correlated essentialist philosophical attitudes, Dagognet conferred great value to materials transformed by humans, to artifacts that continuously blur boundaries – in a nutshell, he prized the world as the result of human activities. Definitely, in this philosophical landscape from Charbonneau and Ellul to Dagognet, there is no place for any sanctification of nature, nor for the concept of "wilderness."

The aversion for all sorts of claim to nature, and the associated fear of eco-fascism, are a constant feature of the French intellectual landscape. Many French "ecologists" refuse to confer any moral status to nature. They do not only refuse any appeal to nature, or any return to it, but they also oppose to a technocratic management of the environment. They see the scientific alerts about environmental damages as risks of technocratic domination by experts and as a threat to democracy. With regard to risks, it is worth noting, as philosopher Floran Augagneur argues, that this French tradition is very different from the German one, in particular as it is instantiated by Hans Jonas and Ulrich Beck: "The deviation of political ecology, its original sin, would be to have believed in the necessity of a political management of risk, and to the fact that this motivation is linked to an insufficiency of knowledge." (Augagneur 2015: 335)

It would be completely misleading however to conclude that an ecology "against nature" would mean an ecology "for technology"; it rather means first and foremost that "nature" and "technology" have always been closely intertwined in history, and that they articulate each other to shape a "milieu." As Gorz tirelessly argues, militant ecology is not interested in "pure nature"; it rather aims at defending the world we live in, the *milieu of our life*, in order to protect the culture of daily life and its

[7] In an article entitled "Nature denatured, nature naturing," (1976) largely devoted to the philosophy of François Dagognet, who was his student, Canguilhem makes a violent attack on the "naturism" which was rampant at that time. The work of François Dagognet, a philosopher of science and technology, comprises a whole part which one could call a philosophy of "objects and waste," a philosophy which has never doubted the compatibility of ecological thought with industrial thought (Dagognet 1989, 1997, 2000, 2002).

savoir-vivre (Gorz 2008). "The defense of the *milieu of life* in the ecological sense, and the reconstitution of a *lived world*, condition each other and support each other."

6.2 From Technical Milieu to Political Ecology (1945–1970)

6.2.1 *Milieu vs. Environment*

Political ecology "*à la française*" thus developed against any kind of divide between nature and technology. In this respect, the distinction French contributors often made between both terms "environment" and "milieu" gained high relevance. What does this mean exactly? How to define a "milieu" by contrast with an "environment"?

The French term "milieu" designates (i) the middle or center and its surroundings; (ii) the "in-between two places" (mi-lieu); (iii) the ambient atmosphere; and (iv) the *medium* (middle-term, intermediate or mediator). In French language the term "milieu" was first used to refer to physical and moral surroundings in the eighteenth century, then it was adopted by biologists (see in particular Claude Bernard's concept of "*milieu intérieur*"), by sociologists and geographers in the nineteenth century. The term "milieu" says both more and less than the term "environment." It says more, because it is not on the outside, but *between* the inside and the outside. It says less, because it refers to the unique experience of a living organism in a place, whereas the "environment" is identical for all beings which find themselves in a place and stays outside the living beings. While the environment is objective, the milieu is "trajective." (Berque 2014)[8]

6.2.2 *Various Conceptualizations of the Technical Milieu*

The notion of "technical milieu" was introduced in the twentieth century, first and more or less at the same time, at the end of World War II, by Leroi-Gourhan and philosopher and sociologist Georges Friedmann, then by Ellul and Simondon.[9] Following Leroi-Gourhan, the "technical milieu" involves the assumption that technologies have both an internal and an external milieu. The former one, Leroi-Gourhan argues, is the ensemble of ideas and knowledges that individualize each ethnic group. The "milieu technique" is part of this "*milieu intérieur*." It can be defined as the set of knowledges oriented towards material activities within the group. The "*milieu extérieur*" gathers physical features of the external world and also ideas and knowledges hold by others groups. While "environment" designates

[8]According to the "mesology" of Augustin Berque (2014), the term "trajective" applies to that which is situated in the "mi-lieu" between the objective and the subjective.

[9]The birth of the term, and the link between the concepts of "technical milieu" and "technical culture," have already been mentioned in this volume (Beaubois and Petit, Chap. 21 this volume).

objective features that belong to the surroundings of the group, "milieu technique" involves both external (in particular material) constraints and the activities of the group. Far from being a set of fixed properties, the "milieu technique" is the result of dynamic mediations between the group and its "environment."

It is worth noting that in French philosophy, from Auguste Comte to Canguilhem, the concept of "milieu" has been the inseparable corollary of organisms. Both concepts of "social milieu" and "technical milieu" have inherited the intertwining of biological and cultural aspects of human activities, and are still anchored in life. To a large extent, the distinction between "environment" and "milieu technique" refers to what Canguilhem himself called a "biological philosophy of technology" – a trend that has been especially developed in Germany and France. Henri Bergson of course, but also Canguilhem, Leroi-Gourhan, Simondon, Dagognet militated for a non-intellectualist view of technology: far from being applied science, technology is above all the continuation of life by means of non-organic instruments. Note that these names are those we already mentioned above: this is not a surprise, insofar as, with no doubt, the political ecology "*à la française*" bumped this biological trend in the philosophy of technology. The concept of technical milieu phrases this junction.

6.2.3 *The Technical Milieu from the East to the West (Friedmann)*

With Friedmann and Ellul however, the technical milieu gained new and sometimes more critical significations – a shift that can be also noticed in Leroi-Gourhan's later works, by the way. Friedmann began his intellectual career with Marxism and he drew a boundary within the Marxist movement between those who did or did not take into consideration the technical milieu as such. He introduced the concept of "technical milieu" in opposition to the "natural milieu." According to him, the divide between both concepts of "milieu" gradually displaced the prevailing conflict between communism and capitalism: indeed for him, as well as for Ellul, the "technical milieu" overcomes the misleading East/West political opposition. Friedmann who initially defended "machinism" on the basis of his Marxist inclinations ended up advocating humanism, which left him uneasy about the domination of the "technical milieu."

In the post-war period Georges Gurvitch, a sociologist and professor at the University of Strasbourg, organized a symposium to discuss the concept of "technocracy." (Gurvitch 1949) On this occasion, Friedmann expressed his fears about "technicism" (consisting in looking at all human problems from the sole technical viewpoint), and he encouraged a "technological humanism," a culture of the technical milieu, in order to avoid "the anarchic conditioning of modern man by the technical milieu which grows ever more dense around him." (Friedmann 1949) All disciplines contribute to question technology, and its progressive image loses some of its shine. As a matter of fact, the period known as the "Glorious Thirties" (1945–1975) gave rise to more ecological and technological debates than is generally thought (Pessis et al. 2013). Despite the grand technocratic programs launched by the political regime of De Gaulle, which aimed at rebuilding the power of the French nation, there was a strong movement of "techno-critique," (Jarrige 2014) which paved the way to political ecology prior to 1968.

Between the 1940's and the 1960's, Friedmann thus shifted "from a Marxist critique of the 'capitalist illusion' to a moral critique of the 'technicist illusion' and 'disenchanted work'." (Bidet 2011: 52) In *La Puissance et la Sagesse* (1970), he came to understand that the technical milieu does not stop at the frontiers of the capitalist world[10] and this statement led him to criticize his own (Marxist) book published three decades before, *La Crise du progrès* (1936). According to Friedmann Marx misunderstood the power of the technical milieu, and overlooked that individuals are free to resist this power (Friedmann 1970). While Friedman's analysis of "machinism" had been previously directed against economic heteronomy, his analysis of the "technical milieu" was redirected against technical heteronomy (Vatin 2004; Bidet 2011).

Friedmann's famous description of "work fragmented into bits and pieces" (Friedmann 1956) was based on his broad view of the human milieu as biological, psychological, sociological as well as technical, with no possibility to isolate one dimension from others – so he maintained the strong link between technology and life, but he shed light on the increasing distance between the technical milieu and the natural one. He criticized the technicist (or mechanist) vision of workers and their relation to the technical milieu (Friedmann 1946), a critique that is also to be found in Canguilhem (1947, 1955). It is worth noticing that many sociologists of labor and ergonomists subsequently based their analysis of the workers' condition on the dichotomy between organism and milieu (Schwartz 1988; Clot 1995), and claimed that emancipating labor means above all that workers be capable of "individuation processes," i.e. of rearticulating all aspects of human existence, including biological ones, within a unique experience (Bidet 2011).

The concept of "technical milieu" is also associated to Friedman's shift from the sociological studies of labor to the study of leisure when Friedmann founded the "Centre for the study of mass communication" in 1961. He never stopped claiming that the notion of "technical milieu" does not suggest, neither an external determinism, nor extrinsic constraints, and could be viewed as a kind of screen between man and nature. The technical milieu is above all a "human milieu." (Friedmann 1970)

So the concept of technical milieu was loaded with critical insights on modern societies. However, in the second part of the past century, French political ecology emphasized a new concept that was supposed to intensify this criticism, and consequently pushed the "milieu" to some extent in the background: the concept of "technical system."

[10] It is indeed because the technical milieu is homogeneous that "the dichotomy between the milieu of the workers (or those who execute tasks) and that of the organizers (where the decisions are taken) is, in the end, reproduced everywhere." (Friedmann and Naville 1961: 377)

6.3 From Technical Milieu to Technical System

6.3.1 Ellul's Approach to Political Ecology

Indeed in the second part of the twentieth century, the concept of "technical milieu" has been gradually eclipsed by the concept of "technical system." (Triclot 2012) Undoubtedly, Ellul has been a major contributor to this shift. In particular, from *La Technique ou l'enjeu du siècle* (1954) to *Le système technicien* (1977), he no longer focused on the coupling between the individual and the technical milieu, and he consequently ceased to question the continuity between the biological, the technical, the psychological and the social. The concepts of "technicist society" and "technical milieu" lose part of their relevance, in the benefit of this new concept: the "technical system." A high degree of autonomy is the major characteristic of the technical system: "This system thus seems to be largely independent of man (just as the natural milieu was also independent)." (Ellul 1977) To be sure, in arguing that the technical milieu is no longer *our* milieu, Ellul broke up with the meaning that this concept previously got. In a way, he misinterpreted it right from the start: "Technology has already penetrated deeply into man. Not only does the machine tend to create a new environment for man, but it furthermore modifies man in his very being. The milieu in which this man lives is no longer his own milieu." (Ellul 1954) Thus he divorced humans from the technical milieu, just as the ecologists criticized by Canguilhem divorced the latter from nature. Consequently, as far as he misunderstood the concept of technology as a "milieu" that gathers all aspects of human existence and collective activities – including culture as a whole – he adopted an anti-technology position. How could the technical milieu be considered the milieu of human beings if art, ethics as well as culture are absent from it? In the aftermath of 1968, Ellul wondered why the workers were unable to make a revolution, and claimed that the proletarian class had been absorbed, devoured by the technicist society to the point of sharing its objectives and aspirations (Ellul 1969). For the rest of his life he would develop his critical reflections on the "technological bluff," and gave up all hope of escaping from it.

Ellul was a lonely figure in the French landscape of the 1950s, as he confessed. This situation evolved in the 1980s, when technology became a fully-fledged philosophical topic. At this moment Ellul joined a growing community of philosophers concerned with contemporary issues related to technology, including among others Jean Chesneaux (1983), Dominique Janicaud (1994) and Gilbert Hottois (1984). Due to the large spectrum of his concerns, his legacy is extremely diverse (Rognon 2012). Considered a radical critic of the technical system, Ellul is also viewed in opposition to "environmental protection" or to "ecological modernization" (Lamaud 2013), by those who refer to his "Plea against the defense of the environment." (Ellul 1972)

6.3.2 Political Ecology as Techno-critique? (Jouvenel, Castoriadis)

With no doubt, in case this tension between both concepts of "technical milieu" and of "technical system" would be underestimated, the history of French political ecology would remain misunderstood.

The phrase "political ecology," in its current use, dates back from 1957. It was introduced by the political scientist, economist and futurologist Bertrand de Jouvenel.[11] De Jouvenel's aim was to overcome the shortcomings of political economy by reintegrating into the fluxes measured by economic science the exchanges of matter and energy between human society and the planet. Jouvenel (1976) criticized the cult of the "civilization of power," and the reversal of values whereby the means justify the ends. By contrast, he strove for a democratic determination of ends (for example by political discussion of public expenditure in Research & Development). He did not criticize the power of technology as such, but he rather claimed for the reorientation of research towards renewable energies and cyclical technical ecosystem.[12]

René Dumont, an agricultural scientist who was the first "ecological" candidate in a Presidential election (1973), exemplifies the same ambivalence regarding technology. Like Friedmann, Dumont shifted his intellectual positions. After the Liberation (1945) he actively supported the modernizing and productivist dynamic of French agriculture; but in the 1960s he radically changed his mind and condemned agricultural technologies based on chemistry and fossil fuels. He doubted that the so-called "Third World" countries could develop on the basis of our industrial model. However, although Dumont condemned productivist technologies in the short term, he cared to examine each technology through case-by-case studies and fostered "intermediary technologies." (Dumont 1973)

To better understand how technology prompted great interest in social sciences in France during the "Glorious Thirties" of 1945–1975 let us mention two journals from the New Left and critical Marxism: *Arguments* (1956–1962) and *Socialisme ou Barbarie* (1949–1967). The latter was leaded by Cornelius Castoriadis and developed strong arguments against the "axiological neutrality of technology" thesis: far from being neutral, modern technology is above all an instrument of class exploitation. To Marx, Castoriadis replied: "in fact, there is no autonomy of technology, nor there is an immanent tendency of technology towards an autonomous development." (Castoriadis 1975) In his article "Technique" for the *Encyclopædia Universalis*, Castoriadis clearly recognized in Marx the father of a non-instrumental

[11] Since 1957, Jouvenel has published two articles on the subject of political ecology: "De l'économie politique à l'écologie politique," republished in Jouvenel (1976); and "L'économie politique de la gratuité," republished in Jouvenel (2002).

[12] He defends what would be called today "industrial ecology" or "circular economy": "This is what makes us attentive about closing the circuits," he wrote in 1965: "either by substituting our materials by others which can be digested by natural agents, or by supplementing the action of the latter by artificial agents." (Jouvenel 2002: 246)

and socio-historical conception of technology; but at the same time he criticized Marx for going not far enough in his critique, and in the end for failing in challenging Plato. In order to overcome the separation of means and ends, of the social and the technological, the notion of technology must be reconsidered (Castoriadis 1978). "Neither ideally, nor in reality, can one separate the technological system of a society from what that society is." (Castoriadis 1986) However, Castoriadis worried about the tremendous "autonomization" of technoscience (Castoriadis 1990: 87–124). To be sure, there is no contradiction in Castoriadis's claims insofar as the "autonomy of technoscience" is recognized for what it is, i.e. a social imaginary construction that can be deconstructed. As a philosopher of democratic autonomy, Castoriadis thus challenged the "pseudo-autonomy" of technoscience and economy and became engaged in political ecology (Castoriadis and Cohn-Bendit 1981). In his view, both the ecological and the democratic projects go hand in hand; and they can only be realized by means of a novel "imaginary institution" – a term Castoriadis coined – of the role of technology in society.

So it is worth noticing that in the French context, but also abroad, there is no clear boundary between political ecology and techno-critique movements. The aforementioned authors, all involved in the birth of French political ecology, pointed out a close link between the ecological perspective and a an alternative "technical project" contrasting with the industrial one, which still remains to be imagined: "If particular technologies are 'rational activities', technology itself (we use the term here with its common restricted meaning) is absolutely not. In its historical reality technology is a project whose meaning is uncertain, whose future is obscure, and whose finality is indeterminate, obviously given that the idea of making ourselves 'masters and possessors of nature' is strictly meaningless." (Castoriadis 1975: 111)

6.4 Toward a Political Ecology of Technology (1970–1990)

6.4.1 Design and Environment: Two Contemporaneous Terms in France (Baudrillard)

The involvement of a new actor, *design*, makes even more complex the birth of a political ecology "*à la française*." Both terms "environment" and *design* were imported in France at the same time (Beaubois and Petit, Chap. 21 this volume) In the 1970 Aspen symposium, devoted to *Environment by design,* Jean Baudrillard commented: "the theoretical concept of 'environment' is related to the practical concept of 'design' – a concept that can be analyzed as the production of consumption (the production of signs by human beings)." (Baudrillard 1972) "The social control of the air, of water, etc. under the sign of protection of the environment," he wrote, "is obviously a case where humans themselves enter more deeply into the field of social calculation. When nature, air, water, after having been simple productive forces, become rare commodities and enter into the field of value, humans themselves enter deeply into the field of political economy." (Baudrillard 1972) Baudrillard later took his distance from an "ecology of media," and considered technology the "accursed moiety," the impossibility of back-tracking: "if the special attribute of mankind consists in never exhausting its possibilities, the very essence of technology is to develop its own possibilities until the point it even goes beyond them." (Baudrillard 2007)

The marriage of design and environmental thinking is just another way of naming the "technological turn" of ecology, exemplified by Buckminster Fuller on the other side of the Atlantic. In France, this turn led to a split in political ecology between, on the one hand, those like André Gorz who advocated a felicitous sobriety, a voluntary simplicity and a deliberate reduction in the scale of human activity, and on the other hand, those like Félix Guattari who developed an ecology involving machines, in a more techno-optimistic tone.

6.4.2 Technology, Autonomy, Ecology (Gorz)

Gorz, just like Friedmann, Dumont, etc., has changed his view with respect to technology and its evolution.[13] He started from a critique of consumerist society,[14] in a Marxist tone close to Jean-Paul Sartre's views, then he reoriented his critique of capitalism towards a critique of labor and its technical milieu. For him as well as for Castoriadis, ecology and autonomy are the same combat; however, for Gorz, such issues must be addressed through careful examination of labor, salaried employment, incomes etc. From Friedmann to the present days, French political ecology tied close links with the study of industrial labor and workers' condition.

In *Ecologie et liberté* (1977), Gorz writes that the difference between ecologism and socialism bears precisely on the technological presuppositions of economic

[13] "At the time I thought, rather like Radovan Richta, that technological evolution would gradually eliminate repetitive, non-qualified work in favor of work which would become increasingly intellectual, technically advanced, and thus potentially favorable to the flowering of autonomous capacities. But from 1969 onwards, I realized that this would be to expect from technological evolution a political effectiveness that it obviously does not possess. I began to thematize this aspect following a stay in the United States during which I had long discussions with Stephen Marglin, with Herbert Marcuse – I had been the first, in 1964, to give an account in *The Nation* of his *One-dimensional man* – and with critical engineers and doctors." (Gorz 2000: 222–223)

[14] "It is by way of a critique of the model of opulent consumption that I became an 'ecologist' before the term was invented. My point of departure was an article which appeared in an American weekly around 1954. This article explained that for the American capacities of production to be exploited to the full, consumption would have to increase by at least 50% over the next 8 years; but that people were quite unable to define what their 50% additional consumption would be consist of. (…) My interest for techno-criticism owes much to my reading, in 1960, of *Critique de la raison dialectique* by Sartre; to 10 days passed in East Germany, at the same period, visiting factories and looking in vain for the germs of workers' power; then, starting in 1971 or 1972, to my discovery of Illich who had entitled *Retooling Society* a preliminary version of *La Convivialité*." (Gorz 2008: 14–16)

questions. Gorz, an heir to Ivan Illich, shared his conviction that societal choices are always imposed on us by way of technological choices; without a struggle for different technologies, the struggle for a different society is in vain. To move from a capitalist hetero-regulation to a convivial auto-regulation one has to change tools (Gorz 1977). "Socialism is not immunized against techno-fascism" and "it is no better than capitalism if it uses the same tools." (Gorz 1977: 35) Gorz's political ecology is first and foremost a discussion about the tools that need to be adopted in order to favor an autonomous relation to our sociotechnical milieu; and this is why, by the end of his life, he came to be interested in the hackers movement, and "fabbers." (Gorz 2008)

Gorz, just as Friedmann, made a move towards a new technical culture based on the unity of production and consumption. Throughout his work, he expressed his fear of an authoritarian ecology dominated by engineering, of a techno-fascism, of a geo-power. When he wrote "technology is to be understood as nature creating itself by means of humans," (Gorz 2003: 131) he was not supporting the claim, rather he was summarizing the ideology that he condemned, that of a transhumanism. Thus Gorz initiated a bifurcation between two ecologies: one is the accomplishment of capitalism, the other still seeks to avoid that fate.

6.4.3 Machinic Ecology (Guattari)

On the other side of the spectrum of French political ecology, Félix Guattari (1930–1992) coupled ecology and machine and redefined both in an "*ecosophie*," "which has the perspective of never considering as separate the material and the value-laden dimensions of the problems under consideration." (Guattari 2013: 408) Far from reducing ecology to environment, Guattari assumed that there are "three ecologies": environmental, mental and social (Guattari 1989). The machine is no longer a mere technical object, but a new organization productive of subjectivity. Conversely, "Subjectivity is not only human. It is also machinic. For me, there is no frontier between man, society, and *tekhnè*, the appropriation of the environment, the constitution of existential territories." (Guattari 2013: 332) In "Ecology and the workers' movement. Towards an ecosophic recomposition," Guattari attempted to reconcile the social struggles of yesterday with the machinic struggles of tomorrow: "On all these fronts [like that of the ecology of work in a digital milieu], the dialogue between ecology and the workers' movement seems to me essential. There is always a risk of seeing ecology go over to conservatism, a return to the previous status quo, with reductionist systemic conceptions, incapable of being articulated with the rhyzomatic mutations of the evolving phyla of machinism." (Guattari 2013: 419)

Guattari first joined the "Green" party but he soon left it to invest in the ecology of media (Goddard 2011) and the contemporary movement for "accelerationism"[15]

[15]This recent intellectual movement of the Left, which advocates accelerating the contradictions of capitalism and those of technological disruptions, is open to various interpretations, but in any case

(Mackay and Avanessian 2014). Among the ecological activists in France, Guattari was probably the least techno-skeptical. He believed to the end in the machinic possibility of a post-media society; and all the questions he addressed turned around "the old problem of technology": "I would characterize the object of politics as ecosophic. [It is] an object with four dimensions: the traditional material and economic fluxes; machines and the associated eco-systems; the universes of political and moral values; and existential territories (…). The proper intellectual challenge is to rethink the old problem of technology." (Guattari 2013: 337–338)

6.4.4 Tentative Reformulations of the Technical Milieu (Beaune, Stiegler)

What about the concept of "technical milieu" in these new orientations of French political ecology? Is it definitely dismissed? On the contrary, as philosopher Jean-Claude Beaune (one of the major representative of the so-called "Lyon school" of the philosophy of technology, see the introduction, Chap. 1 of this volume) puts it, "It is only by having recourse to the notion of 'technical milieu' that it possible to overcome the exteriority of the technical object." (Beaune 1980) Despite their individual independent views, most of the authors mentioned above share a strong attachment to the notion of technical milieu and the assumption that technology and life are not dissociable.[16]

In our view, it is in the prolific work of Bernard Stiegler (1994; Chap. 18 this volume) that one finds the most complete and the most accomplished legacy of French philosophy of the technical milieu. His interpretation of the technical milieu is indeed reminiscent of Leroi-Gourhan's description of the evolution of our exosomatic being; of Friedmann's analysis of the psychopower; of Simondon's description of the "associated milieu" of individuation; of Guattari's attempt at a general ecology[17]; and finally of Gorz's unabated claims for a non-salaried working milieu. Stiegler's works can be summarized as an active meditation over the question raised by the development of digital technologies. His non-profit organization *Ars Industrialis* militates for a new political economy of digital technologies, for "digital studies" aimed at serving a new industrial policy of the digital, and for "an economy of contribution" which calls for the creation of an associated

involves admitting the failure of eco-socialism, that of Gorz for example. Alex Williams and Nick Smicek have published in 2013 a manifesto of this movement, essentially Promothean, which has been the subject of debate. See: http://criticallegalthinking.com/2013/05/14/accelerate-manifesto-for-an-accelerationist-politics/

[16]Whereas the "point of view of the cultural milieu always tends to reconsider the reference to nature," "the point of view of the technical milieu deepens the concept of nature and surpasses it in the direction of life." (Beaune 1999: 18)

[17]Which in Stiegler's works takes the form of a "general organology." (Stiegler 2004, 2005; Hoquet, Chap. 16 this volume)

technical milieu. With no doubt, the concept of technical milieu, whatever its conflicting relationships with rival concepts such as technical system, has been of major importance in the shaping of political ecology in France, and contributed to confer it a particular coloration.

6.5 Conclusion

From this short survey of the history of political ecology in France, it is clear that in the French context the term "ecology" was not associated to the intrinsic value of nature, and rather raised the issue of the status of technology. This non-naturalist and non-environmental ecology is rather an ecological philosophy of the "milieu" (*meson* and *oïkos*) as opposed to the environment, which rejects the alternatives of "technology *or* nature," as well as anthropocentrism *or* ecocentrism.

It is nevertheless difficult to provide a homogeneous definition of this ecology of technology, which oscillates between "ecological history and economics" (Jouvenel), "ecological technologies" (Gorz) and a "machinic ecology" (Guattari). Each author opens up his/her own research pathway: Jouvenel points in the direction of a "circular economy," Gorz towards the *makers* and their "*open technology*," and Guattari towards the "ecology of the media" and accelerationism. Moreover, our aim was not to provide an exhaustive survey of the field. We omitted for example the phenomenological approach instantiated in Jean-Luc Nancy's "*ecotechnie*," (Nancy 2000)[18] as well as the perspective of an ecology of the subject rather than one of the environment (Conley 1997a, b), and Erich Hörl's attempt to extend what he describes as the fourth period of Simondon's encyclopedic enterprise (Hörl and Plas 2012; Hörl 2013). Yet the best way to summarize the major presupposition of the French movement of general ecology of technology, is by using Bernard Stiegler's own words (2004, 2005): a general ecology is an "organology," in other words a philosophy of the technical milieu, a mi-lieu of the biological, the psychical and the social. The ecology of technology and the concept of "milieu" continue to nourish each other.

References

Afeissa, H.-S. (2009). *Qu'est-ce que l'écologie?* Paris: Vrin.

Augagneur, F. (2015). Ecologie Politique (Idées). In D. Bourg & A. Papaux (Eds.), *Dictionnaire de la pensée écologique* (pp. 332–335). Paris: Presses Universitaires de France.

Baudrillard, J. (1972). *Pour une critique de l'économie politique du signe*. Paris: Gallimard.

Baudrillard, J. (2007). *Pourquoi tout n'a-t-il pas déjà disparu?* Paris: Éditions de L'Herne.

[18] For Derrida, this "ecotechnie" constitutes the singularity of Nancy, "taking into account technology and technical ex-appropriation right from the "phenomenological" threshold of the lived body." (Derrida 2000, p. 70)

Beaune, J.-C. (1980). *La technologie introuvable*. Paris: Vrin.
Beaune, J.-C. (1999). *Philosophie des milieux techniques*. Seyssel: Champ Vallon.
Berque, A. (2014). *La mésologie, pourquoi et pour quoi faire?* Nanterre La Défense: Presses universitaires de Paris Ouest.
Bess, M. (2003). *The light-green society: Ecology and technological modernity in France, 1960–2000*. Chicago: University of Chicago Press.
Bidet, A. (2011). *L'engagement dans le travail. Qu'est-ce que le vrai boulot?* Paris: PUF.
Bourg, D. (1996). *L'Homme-artifice*. Paris: Gallimard.
Bourg, D. (2003). *Le nouvel âge de l'écologie*. Paris: Éditions Charles Léopold Meyer.
Bourg, D., & Whiteside, K. (2010). *Vers une démocratie écologique*. Paris: Le Seuil.
Canguilhem, G. (1947). Milieu et normes de l'Homme au travail. *Cahiers Internationaux de Sociologie, 3*, 120–135.
Canguilhem, G. (1955). *La formation du concept de réflexe aux XVIIe et XVIIIe siècles*. Paris: Vrin.
Canguilhem, G. (1990). La question de l'écologie. In F. Dagognet (Ed.), *Considérations sur l'idée de nature*. Paris: Vrin.
Castoriadis, C. (1975). *L'institution imaginaire de la société*. Paris: Seuil.
Castoriadis, C. (1978). *Les Carrefours du labyrinthe* (Vol. 1). Paris: Seuil.
Castoriadis, C. (1986). *Domaines de l'homme. Les carrefours du labyrinthe* (Vol. 2). Paris: Seuil.
Castoriadis, C. (1990). *Le Monde morcelé. Les carrefours du labyrinthe* (Vol. 3). Paris: Seuil.
Castoriadis, C., & Cohn-Bendit, D. (1981). *De l'écologie à l'autonomie*. Paris: Seuil.
Charbonneau, B. (1969). *Le Jardin de Babylone*. Paris: Galimard.
Charbonneau, B. (1980). *Le Feu vert*. Lyon: Parangon.
Charbonneau, B., & Ellul, J. (2014). *Nous sommes révolutionnaires malgré nous. Textes pionniers de l'écologie politique (1935–1945)*. Paris: Le Seuil.
Chesneaux, J. (1983). *De la modernité*. Paris: La Découverte.
Clot, Y. (1995). *Le travail sans l'homme? Pour une psychologie des milieux de travail et de vie*. Paris: La Découverte.
Conley, V. A. (Ed.). (1997a). *Rethinking technologies*. Minneapolis: University of Minnesota Press.
Conley, V. A. (1997b). *Ecopolitics. The environment in postructuralist thought*. London: Routledge.
Dagognet, F. (1989). *Éloge de l'objet : Pour une philosophie de la marchandise*. Paris: Vrin.
Dagognet, F. (1997). *Des détritus, des déchets, de l'abject : une philosophie écologique*. Paris: Les Empêcheurs de penser en rond.
Dagognet, F. (2000). *L'invention de notre monde: L'industrie, pourquoi et comment?* Paris: Encre marine.
Dagognet, F. (2002). *Les dieux sont dans la cuisine*. Paris: Les Empêcheurs de penser en rond.
de Jouvenel, B. (1976). *La civilisation de la Puissance*. Paris: Fayard.
de Jouvenel, B. (2002). *Arcadie. Essais sur le mieux-vivre*. Paris: Gallimard.
Deneys-Tunney, A. (2000). *Un autre Jean-Jacques Rousseau. Le paradoxe de la technique*. Paris: Presses Universitaires de France.
Derrida, J. (2000). *Le toucher, Jean-Luc Nancy*. Paris: Galilée.
Dumont, R. (1973). *L'utopie ou la mort*. Paris: Le Seuil.
Ellul, J. (1954). *La Technique ou l'Enjeu du siècle*. Paris: Armand Colin. English edition: Ellul, J. (1967). *The technological society* (J. Wilkinson, Trans.). New York: Knopf/Vintage.
Ellul, J. (1969). *Autopsie de la révolution*. Paris: Calmann-Lévy.
Ellul, J. (1972). Plaidoyer contre la "défense de l'environnement". *France Catholique, 1309*, 20.
Ellul, J. (1977). *Le Système technicien*. Paris: Calmann-Lévy. English edition: Ellul, J. (1980). *The technological system* (J. Neugroschel, Trans.). New York: Continuum.
Ellul, J. (1982). Écologie, technique et société. *Cahiers du CPO (Centre protestant de l'ouest), 46*(47), 13–20.
Friedmann, G. (1936). *La Crise du progrès : Esquisse d'une histoire des idées (1895–1935)*. Paris: Gallimard.
Friedmann, G. (1946). *Problèmes humains du machinisme industriel*. Paris: Gallimard.

Friedmann, G. (1949). Les technocrates et la civilisation technicienne. In G. Gurvitch (Ed.), *Industrialisation et Technocratie* (pp. 43–60). Paris: Armand Colin.
Friedmann, G. (1956). *Le Travail en miettes : Spécialisation et loisirs*. Paris: Gallimard.
Friedmann, G. (1970). *La Puissance et la Sagesse*. Paris: Gallimard.
Friedmann, G., & Naville, P. (Eds.). (1961). *Traité de sociologie du travail*. Paris: Armand Collin.
Goddard, M. (2011). Towards an archaeology of media ecologies: 'Media ecology', political Subjectivation and free radios. *The Fibreculture Journal, 17*(5), 6–17.
Gorz, A. (1977). *Écologie et Liberté*. Paris: Galilée.
Gorz, A. (2000). Entretien. In F. Gollain (Ed.), *Une critique du travail, entre écologie et socialisme* (pp 219–242). Paris: La Découverte.
Gorz, A. (2003). *L'immatériel. Connaissance, valeur et capital*. Paris: Galilée.
Gorz, A. (2008). *Ecologica*. Paris: Galilée.
Guattari, F. (1989). *Les trois écologies*. Paris: Galilée. English Edition: Guattari, F. (2000). *The three ecologies* (I. Pindar, & P. Sutton, Trans.). London: Continuum.
Guattari, F. (2013). *Qu'est-ce que l'écosophie?* Paris: Editions Lignes.
Gurvitch, G. (Ed.). (1949). *Industrialisation et technocratie*. Paris: Armand Colin.
Hörl, E. (2013). A thousand ecologies: The process of cyberneticization and general ecology. In D. Diederichsen & A. Franke (Eds.), *The whole earth: California and the disappearance of the outside* (pp. 121–130). Berlin: Sternberg Press.
Hörl, E., & Plas, G. (2012). Le nouveau paradigme écologique. Pour une écologie générale des médias et des techniques. *Multitudes, 51*(4), 74–85.
Hottois, G. (1984). *Le Signe et la technique. La philosophie à l'épreuve de la technique*. Paris: Aubier.
Ingold, T. (2001). *The perception of environment: Essays in livelihood, dwelling and skill*. London: Routledge.
Janicaud, D. (1994). *Power of the rational*. Bloomington: Indiana University Press.
Jarrige, F. (2014). *Techno-critiques. Du refus des machines à la contestation des technosciences*. Paris: La Découverte.
Lamaud, I. (2013). Against enviromental protection? Ecological modernization as "technician ecology". In H. M. Jerónimo, J. L. Garcia, & C. Mitcham (Eds.), *Jacques Ellul and the technological society in the 21st century* (pp. 83–96). Dordrecht: Springer.
Mackay, R., & Avanessian, A. (Eds.). (2014). *Accelerationist reader*. London: Urbanomic.
Massard-Guilbaud, G. (2012). La France, Une "société vert clair?" Enjeu – Retour sur *The light green society: Ecology and technological modernity in France, 1960–2000*. Vingtième Siècle. Revue d'histoire, *113*(1), 205–210.
Moscovici, S. (1968). *Essai sur l'histoire humaine de la nature*. Paris: Flammarion.
Moscovici, S. (1972). *La Société contre nature*. Paris: Union générale d'éditions.
Moscovici, S. (2002). *De la nature. Pour penser l'écologie*. Paris: Éditions Métaillé.
Nancy, J.-L. (2000). *Corpus*. Paris: Éditions Métaillé.
Parrochia, D. (2009). French philosophy of technology. In A. Brenner & J. Gayon (Eds.), *French studies in the philosophy of science* (pp. 51–70). Dordrecht: Springer.
Pessis, C., Topçu, S., & Bonneuil, C. (Eds.). (2013). *Une autre histoire des « Trente Glorieuses »*. Paris: La Découverte.
Rognon, F. (2012). *Générations Ellul. Soixante héritiers de la génération de la pensée de Jacques Ellul*. Genève: Labor et Fides.
Roy, C. (1992). Aux sources de l'écologie politique : le personnalisme gascon de Bernard Charbonneau et Jacques Ellul. *Revue des annales canadiennes d'histoire*/Canadian Journal of History, *27*(1), 67–100.
Schwartz, Y. (1988). *Expérience et connaissance du travail*. Paris: Messido.
Simondon, G. (2005). *L'individuation à la lumière des notions de formes et d'information*. Paris: Jérôme Millon.

Stiegler, B. (1994). La technique et le temps, 1. *La faute d'Épiméthée*. Paris: Galilée. English Edition: Stiegler, B. (1998). *Technics and time, 1. The fault of epimetheus* (G. Collins, & R. Beardsworth, Trans.). Cambridge, MA: Stanford University Press.

Stiegler, B. (2004). De la misère symbolique: Tome 1. *L'époque hyperindustrielle*. Paris: Galilée. English Edition : Stiegler, B. (2014). *Symbolic misery, vol. 1: The hyperindustrial epoch* (B. Norman, Trans.). Cambridge, MA: Polity Press.

Stiegler, B. (2005). De la misère symbolique: Tome 2. *La catastrophe du sensible*. Paris: Galilée. English Edition : Stiegler, B. (2014). *Symbolic misery, vol. 2: The catastrophe of the sensible* (B. Norman, Trans.). Cambridge, MA: Polity Press.

Triclot, M. (2012). "Milieu technique" : Généalogie d'un concept. In D. Parrochia & V. Tirloni (Eds.), *Formes, systèmes et milieu techniques après Simondon* (pp. 15–28). Lyon: Jacques André éditeur.

Vatin, F. (2004). Machinisme, marxisme, humanisme: Georges Friedmann avant et après-guerre. *Sociologie du travail, 46*(2), 205–223.

Whiteside, K. H. (2002). *Divided natures, French contributions to political ecology*. Cambridge, MA: The MIT Press.

Chapter 7
The Role of the Philosophy of Technology in French-Language Studies of Video Games

Mathieu Triclot

Abstract This chapter analyzes the role played by the philosophies of technology in the development of French-language studies devoted to video games. This intellectual tradition is well represented within the field, in a variety of forms which are characteristic of the main trends in the French-language philosophy of technology. The field of games studies allows comparing and contrasting these different approaches to a single object. I analyze the specificities of the French-language field, its theoretical options, and the role played by the philosophy of technology in its points of divergence from the dominant trends in game studies. Four studies located at the intersection between philosophy of technology and video games studies are specifically discussed: (1) the reference to Simondon in Etienne Perény's works; (2) the framework of "technological macro-systems" (Gras) according to Raphaël Koster; (3) the alliance between philosophy of technology and phenomenology in the work of Elsa Boyer (Husserl, Derrida and Stiegler); and lastly, (4) the utilization of the "philosophy of technical milieus" (Beaune) in our own Philosophy of video games.

Keywords Game studies · Play studies · Video games · Philosophy of technology · Phenomenology · Artificial perception · Philosophy of technical environments · Epistemology · French philosophy

---

M. Triclot (✉)
FEMTO-ST Institute, Besançon, France

Recherches et Études sur le Changement Industriel, Technologique et Sociétal (RÉCITS), Université de Bourgogne Franche Comte, Besançon, France

Université de Technologie de Belfort Montbéliard (UTBM), Belfort, France
e-mail: mathieu.triclot@utbm.fr

S. Loeve et al. (eds.), *French Philosophy of Technology*, Philosophy of Engineering and Technology 29,
https://doi.org/10.1007/978-3-319-89518-5_7

This paper analyzes the role played by the philosophies of technology in the development of French-language studies devoted to video games. Since about 2005, numerous studies in this field have been published in French. These studies show considerable differences compared with the tradition of English-language game studies and the way it has developed, in particular in the United States and northern Europe (Rueff 2008, Zabban 2012).

Philosophy of technology has played an important role in the constitution of this French-language field of study. Not only is this particular intellectual tradition well represented within the field, but it is present in a variety of forms that are characteristic of the main trends in the French-language philosophy of technology. The field of games studies is therefore of particular interest as it allows comparing and contrasting these different approaches to a single object.

As it will be shown, the debates taking place within French-language video game studies echo older quarrels concerning the different ways of apprehending the technical domain. However, it will also be considered to what extent video games help to bring about a renewal of these intellectual traditions and the fostering of new questions, methods, or investigative practices.

The existence of a relatively autonomous field of "game studies *à la française*" provides a particularly interesting basis for investigating the contemporary dynamics in the humanities, between the internationalization of research practices and the shaping of knowledge by local intellectual traditions and institutions. Philosophy of technology has thus contributed to the opening of new paths for games studies, based on intellectual traditions that were rarely solicited or simply unknown in the English-language field. This contribution has the particularity of being written from the point of view of an actor of the field. It proposes a combination of this actor's account, with the limits such a point of view implies, and an epistemological study of the way this field of knowledge has developed.

It begins by examining the characteristics which serve as the foundation for video games as an object of research and which explain the diversity of available approaches. Secondly, the specificities of the French-language field, its theoretical options, and its points of divergence from the dominant tendencies in game studies are analyzed. Thirdly, four studies located at the intersection between philosophy of technology and video games studies are specifically discussed: the reference to Simondon in Etienne Perény's works; the framework of « technological macrosystems » (Gras) according to Raphaël Koster; the alliance between philosophy of technology and phenomenology as demonstrated by the question of artificial perception in the work of Elsa Boyer (Husserl, Derrida and Stiegler); and lastly, the utilization of the "philosophy of technical milieus" (Beaune) in our study *Philosophie des jeux video* (Philosophy of video games).

7.1 Theorizing Video Games

That a French-language field of video games studies should have seen the light of day, with a certain autonomy in relation to the English-language field, is due in part to the difficulties that affect the development of video games as an object of research.

The first of these difficulties comes from the blurring of categories under which video games can be apprehended. Indeed, video games are distinguished by a constitutive heterogeneity, straddling the fields of technology, art and culture, sports and competition, pictures and fiction, the world of toys, etc. (Sidre 2014). The video game thus offers a multiplicity of "theoretical angles," and the disciplines that can lay claim to it are numerous.

Moreover, this effect of categorial heterogeneity is reinforced by the scale and range of the video games domain. Video games appeared in the early 1960s in the first laboratories dedicated to the nascent field of computer science in the United States (Donovan 2010; Triclot 2012a). This history of more than 50 years has aggregated a considerable quantity of material, including the games themselves, a large variety of graphic forms, peripherals, and software and hardware solutions, as well as secondary sources (magazines, commercial documentation, advertising, etc). Today, video games represent one of the main cultural practices in France, so much so that it is difficult to define the whole range and diversity of ways of playing (Boutet 2012; Berry et al. 2013; Coavoux et al. 2014). Moreover, video games cannot be artificially separated from other cultural practices or from other recreational activities (Berry 2012). The field is thus immense.

Lastly, a final element weighs upon the development of video games as an object of research: the fact that video games are a part of popular culture as a product of cultural industries as well as of consumer culture (Kline et al. 2003). The strategies necessary to negotiate this deficit of legitimacy represent an important aspect of the epistemological trajectory of the field and offset the risk, admittedly symbolic but always present, that the seriousness of the research might be contaminated by the futile character of its object (Rufat and Ter Minassian 2012). Furthermore, such academic research does not arrive on virgin territory. It has to contend with a mass of already available knowledge, from the industry itself, the players, technical training programs, etc. The way in which the division between academic knowledge and lay knowledge is negotiated is thus one of the essential coordinates of the field. Indeed, Anglo-Saxon game studies have evolved in close conjunction with the professional knowledge of game design (Juul 2005, Bogost 2010).

7.2 Constitution of the French-Language Field

These characteristics allow taking stock of the diversity of current epistemological strategies to make video games an object of knowledge. In the case of the French-language field, several other dimensions must be taken into consideration. The first is the fact of "lateness." Indeed, "French research has lagged one or two decades behind Anglo-Saxon scientific production in the field." (Amato and Perény 2008: 4) "In France, research in game studies is both more recent and less structured than in Anglo-Saxon countries." (Rufat and Ter Minassian 2012: 8). This time-lag effect has encouraged a reflective and critical approach to the Anglo-Saxon corpus.

If the publication of the online review "game studies" in 2001 is a good indicator of the structuring of the English-language field (Aarseth 2001), then it must be admitted that French-language studies have appeared with a time-lag of several years. But beyond the factor of mere lateness, French studies do not benefit from the same level of organization, with a space of shared discussion. The first studies in France were divided between information and communication science (SIC), with three Ph.D. dissertations by Vincent Mabillot (2000), Sébastien Genvo (2006) and Etienne Armand Amato (2008); computer science, with the work of Stéphane Natkin (2004); and sociology, with the pioneering work of Pierre Bruno (1993), the dissertation by Laurent Tremel on role-playing games (2001), the book edited by Nicolas Auray and Sylvie Craipeau (2003), and the dissertation by Manuel Boutet (2006).

The case of information and communication science is particularly representative, insofar as the video game object was regularly investigated in this domain, starting in 1983 (Querzola and Verebelyi 1983), but always on the margins, as an example to illustrate other dynamics, without being elaborated for itself. Consequently, when referring to the period between 1999 and 2004 when many video games were studied in seminars, Perény and Amato emphasize that "it is surprising that despite its public recognition, the video game was not at that time able to constitute an object of research as such. Rather, it allowed researchers to investigate and explore the significant variations in interactivity, to illustrate the paradoxes of interactive texts and to test the validity of a vocabulary centered on interactivity and participation." (Amato and Perény 2011)

The French-language field experienced a breakthrough around 2010, as indicated by an increase in the number of publications and the appearance of new disciplines and new forms of exchanges in the field. The organizing of day-long seminars "game studies? à la française!" starting in 2011, is an indication of the recognition of this turnaround as the moment when "shirts were tucked in," when "the 'casual style' of the earlier studies would soon no longer prevail in disciplines where video games research are now recognized." (Amato et al. 2011)

This period corresponded to a number of publications: of dissertations by Alexis Blanchet (2010) and Vinciane Zabban (2011), of a collective work edited by Samuel Rufat and Hovig Ter Minassian under the performative title *Les jeux video comme objet de recherche (Video games as an object of research)* (2011), and of *Philosophie des jeux video (The Philosophy of video games)* (Triclot 2011). A particularity of this French editorial context is that in addition to the academic production, a type of "fan culture" publication also appeared, notably with the work of Pix'n Love editions, who published journalistic monographs which served as materials for research and occasionally academic studies (Blanchet 2010, Audureau 2014).

7.3 Game Studies or Play Studies

The late development of the French-language field tended to foster critical attitudes to Anglo-Saxon studies. The article by Vinciane Zabban devoted to the structuring of Anglo-Saxon game studies helps to shed light, by contrast, on the differences in positioning characteristic of the French-language field (Zabban 2012). According to Zabban, game studies are based on two tightly linked commitments: the primacy given to the game itself (rules and settings) over ways of playing (recreational practices and attitudes), and the choice of the development of a new discipline (ludology) supported by an argument concerning the singularity of games.

It seems to us that the French-language field is characterized by diametrically opposed theoretical commitments: the maintaining of the context of existing disciplines rather than the demand for a "new" ludology, and studies of *play* rather than game studies.

The most obvious point of divergence and the most prevalent in French-language studies is the primacy given to the study of play as experience and practice in contrast to an internalist analysis of games centered on design and systems. This attention to the game as the activity of players reflects the importance of the socio-anthropological context (Berry, Boutet, Coavoux, Koster, Rufat, Ter Minassian, etc). It also coincides with historical studies which examine the material, technical and economic dimensions of gaming practices (Blanchet, Sidre, Perény). This orientation brings the French-language studies close to the positions presented by the journal *Play and Culture* or to the anthropological studies by Taylor (2006), Boellstorff (2008) and Servais (2013).

It is interesting to note that the primacy given to play is rooted in the French-language tradition. While the work of Roger Caillois (1958) represents a major reference in international studies, benefiting from translations from 1961 onward, the writings of philosopher Jacques Henriot (1969, 1989) or psychologist Jean Chateau (1946), for example, are unknown in the field of international game studies. Yet these writers all move towards a definition of play as an activity, which gives priority to the habits as well as to the attitudes of the players. The Caillois, Henriot and Chateau definition minimizes the role of rules, in a break with the formalistic definition of games widely shared in contemporary game studies, but which can also be found within the French-language structuralist tradition, notably in Lévi-Strauss (Wendling 2010).

The effects of tradition and transmission thus come fully into play. Not only is this corpus accessible to the French-language reader, but its authors played a role in the structuring of the research. Jacques Henriot in particular was the founder of the "sciences du jeu" (play sciences) department at the University of Paris 13, a department still active today through its laboratory Experice, working in the domains of education science and sociology of games (Brougère 2003).

On the other hand, the question of the uniqueness of the ludic domain, and its corollary, the demand for an independent discipline, is open to debate. While the passage through existing disciplines is clearly affirmed by some researchers (Rufat and Ter Minassian 2012), other actors in the French-language field are more sensitive to arguments of singularity. This is notably the case among researchers in the domain of information and communications science. Amato's research, for example, aims at characterizing the uniqueness of video games as a "cybermedium," refusing to dilute the question of apparatus in the potentially infinite number of possible uses. But this reevaluation of the technical apparatus does not in any way eliminate the reference to play and to forms of experience, especially concerning the question of avatars (Amato and Perény 2010).

Genvo's work stems from an analogous position: although he is certainly one of the researchers closest to Anglo-Saxon-style game studies through the importance given to game design and the demand for a ludological perspective, his ludology is nevertheless based on the French tradition focused on play, particularly in Henriot (Genvo 2013).

7.4 Philosophy and Philosophy of Technology

The studies undertaken from the point of view of the academic discipline of philosophy first appeared during the second wave of French-language game studies, around 2010. My hypothesis is that the lack of legitimacy of the video game played a crucial role in this slow recognition, the gap between legitimate culture and popular culture being undoubtedly at its maximum.

The emergence of these philosophical studies occurred in the absence of a unified framework. The approaches are characterized by their diversity, representative of the French-language philosophical domain: aesthetics (Olivier Robert, Thomas Morisset, Dork Zabunyan); phenomenology (Elsa Boyer, Brice Roy); epistemology (Martine Robert), and philosophy of technology (Elie During, Mathieu Triclot).

Philosophical references, however, extend beyond the philosophical field to inform other disciplines. Thus the "gameifying" (ludicization) model is presented in Genvo's work in reference to Deleuzian "assemblage," the description of the concretization process of interactive images as found in Perény is based on Simondon's categories, the socio-anthropology of recreational pleasures presented by Koster is founded on a reference to the "technological macro-systems" of Alain Gras, and the observing of game situations in the anthropological studies of Manuel Boutet is inspired by pragmatist philosophy.

Among these studies of a philosophical orientation, I have chosen to present a limited corpus based on the double criteria of the importance given to the philosophy of technology and the state of achievement (publication or dissertation defense). The following four approaches have thus been selected: (1) the work accomplished by Perény with reference to Simondon; (2) Koster's socio-anthropological study of recreational pleasures, included in the work of the CETCOPRA laboratory; (3) the phenomenology of artificial perception developed by Boyer, informed by a critical re-reading of Husserl, Derrida and Stiegler; and lastly, (4) the analysis of "systems of experience" in video games in Triclot's work which continues the tradition of "the philosophy of technical environments." Perény and Koster's work is of particular interest because it derives from associated fields – science of information and communication, sociology, anthropology – while attributing a central position to the philosophy of technology.

Table 7.1 Corpus of French video game studies

	Perény	Koster	Boyer	Triclot
Concepts	Concretization	Technical macro-systems	Artificial perception	Systems of experience
Paradigms	Aspen movie map	Network games, museology	Serious games, Farocki's films	Historical shift from computer gaming in the universities to the arcades
Terrain	History	Interviews	Phenomenology	History
Number of games	< 5	Around 30	< 5	Around 100
Criticism of game studies	Ignoring of the technical making of images	Ignoring of experience and social context	Not mentioned	Ignoring of play
Blind spots	Uses	Techniques	Diversity of situations	Semiology and game design

These works, while illustrating the diversity of available options in the French-language field, also allows comparing the productivity of these different theoretical frameworks. In order to facilitate discussion of these studies, I will analyze each of them through the prism of a common reading grid: key concepts and theoretical innovations, paradigmatic examples, relationship to empirical materials, number of video games mentioned, criticism of Anglo-Saxon game studies, and finally, blind spots inevitably produced by the conceptual framework. The purpose will be to question and highlight the theoretical styles, whose differences are not limited solely to the notions activated, but also depend on writing practices, types of depiction (diagrams, tables, iconography), and the terrain and objects which have been given priority.

Beyond the diversity of approaches and the theoretical effects thus produced, the involvement of philosophy of technology on the video game terrain also raises the question, I believe, of the relationship between philosophical work and empirical data. Indeed, the works mentioned above enter fully into the scope of French-language studies of games, where we have already seen that they favor the notions of play, forms of experience, and the historicity of phenomena in its technical and material aspects. Within this field, philosophy of technology occupies a position of exchange with the other disciplines in the humanities. Once the philosophy of technology structurally defined by a domain of objects issuing from concrete human activities, the question of the articulation between philosophy and other knowledge approaches arises. What is the philosophy of technology? Is its role solely to provide an epistemological framework for historiography? What space, more or less autonomous, can it preserve in relationship to the perspective of socio-anthropology? How can the role of philosophy of technology be negotiated within the approaches to knowledge aimed at an empirical content? (Table 7.1)

7.5 The Concretization of Interactive Images

Images interactives et jeu video: de l'interface conique à l'avatar numérique (Interactive images and video games: from iconic interface to digital avatar) by Etienne Perény appeared in 2013. This study reviews a thirty-year research trajectory. As a member of the Paragraphe laboratory (Paris 8), a pioneering research center for the computerization of texts and writing, in 1983 Perény founded the Atelier de Vidéomatique, specialized in the study of interactions with images. The central thesis of the book is expressed in a direct reference to Simondon (1958): video games constitute the "concretization" of the interactive image, "this interactive evolution of the image only being fully achieved via the generalization of digital avatars." (Perény 2013: 7)

The debt to Simondon throughout the book concentrates on three main points. The first consists in supporting the demand for a return to the technical object itself, in the form of a nuanced and well-informed description of its genesis and of its "concretization" process. This genetic approach calls upon Simondon's ontology of technology: "technical element, object, individual, and ensemble." This perspective introduces a critical distance from the ludological tendency in Anglo-Saxon game studies as well as from the socio-anthropological tendency in French-language game studies. Perény thus proposes a model which distinguishes, in several concentric circles, "pure technicity" from secondary dimensions which, according to Simondon's theoretical lexicon, are a part of "cultural over-determination." "The playability specific to the interactive (video) image" is placed at the center of the model.

The second major debt to Simondon concerns the notion of "technical culture." It is not only a question of "refusing the facile opposition between culture and technology" but also of bringing about, in an optimistic inflection of Simondon's thinking, a transformation in the relationship to technicity, due to the computerization of societies. Perény defends the idea of a spreading of technological culture via popular culture, thanks partly to hackers who "give voice to the code" but also, on a much wider level, to players who "experience coupling with the electronic signal." Games reintroduce into the general culture "the awareness of the nature of machines, of their mutual relationship, of their relationship with man and with the values implied by these relationships," according to the wish expressed by Simondon himself.

The cultural valorization of these new forms of coupling with computers leads us to the notion of "associated milieus," the third concept in Simondon's theories on technology which inform Perény's reflections. Perény sees computer networking and the generalized use of avatars as the basis of a new "technological environment" which insures and intensifies man-machine couplings.

This is a crucial point in that it illuminates the manner in which Simondon's categories are rethought in Perény's essay. "Concretization," especially, appears less as a process of functional over-determination than as an elaboration of an associated milieu. Accordingly, Perény does not offer many analyses of concretization, along the lines of the paradigmatic study of the motor cooling fins presented by Simondon.

Yet descriptions of this sort would be possible, by analyzing, for example, the manner in which the first programmers were able to push the capacities of their machines to the maximum, while expertly taking advantage of their technical limitations (Montfort and Bogost 2009). In this way, "concretization" as seen in Perény is less about technological individuals in and of themselves than about the operator-machine relationship.

This first transformation can be understood in conjunction with the fundamental question of the essay, that of the effects of the apparatus on the subjects, analyzed under the category of "coupling." In a departure from Simondon's thesis, Perény introduces a question typical of a Leroi-Gourhan type of artificialist anthropology concerning the transformations of subjects linked to the technological milieu. As Perény recognizes, this question necessitates the contribution of other references and theoretical contexts. To this end, as a kind of logical "supplement" to Simondon's theories, Perény refers to the work of Akrich, Latour, Deleuze and Guattari. One can nevertheless wonder to what extent these additions lead to a transformation of the Simondonian framework. Perény modestly presents his essay as a simple "application" of Simondon's concepts, but its trajectory raises questions about, and touches on, research areas that invite not only to "complete" the initial framework, but also to introduce conceptual innovations that would inflect it from within.

One of Perény's most remarkable concepts, therefore, is the idea of "impure synthesis." The impurity of synthesis can be understood, in relation to the interactive image, as a polemic against the idea of the purely textual or numerical character of the image, and as a decisive reevaluation of the numerical-analogical coupling in the interactive image. That such a synthesis requires being described as "impure" at the end of the concretization process raises doubts concerning the strict application of Simondon's theoretical framework.

Moreover, it is impossible not to wonder to what extent the process of concretization complies with elements like playing and hacking, which concern re-appropriations, more or less gratuitous, of the technical object. Considering the video game as the product of a Simondonian concretization process results in a paradoxical thesis, since video games can be seen rather as an outstanding example of cultural over-determination, submitting computers to an unexpected use.

The Simondon framework produces a number of blind spots: the links between video games, counter-culture and hacker culture, and the fact that this same hacker culture encounters a certain resistance on the part of laboratories, are all left out of the discussion (Weizenbaum 1976; Markoff 2006). Yet we must ask ourselves if these historical elements are only a marginal record (which can be treated as cultural over-determination) or if they play a determining role. The fact that French computer science, for example, has not developed a game culture, with the exception of a handful of literary programs, even while American games were known, raises questions.

These blind spots are no doubt the price to pay for the constitution of a new area of intelligibility. The turn back to the apparatus and to its fundamental effects thus allows for a clarification of the "holding power" of the interactive image (Turkle 1984). The description of a new relationship of play to machines, thanks to the incorporation of body patterns, as an element of impure synthesis, seems highly pertinent, and in keeping with the descriptions we can find in Boyer's writings. Perény describes a "new commerce" with images, based on "body memo inscriptions of sensory-motor and kinesthetic origins, produced by the very act of interaction, and which are imprinted on the memory of the subject."

Video games are based on these highly powerful interactive effects which Pereny's studies illuminate in a remarkable manner. But the relationship between concretization, play, appropriation, and usage undoubtedly pushes the Simondon framework "to the wall," to use Elie During's expression quoted by Perény (During 2006), and encourages the development of a form of conceptual innovation within Simondon's system.

7.6 Socio-anthropology of the Video-Gaming Experience

Raphael Koster's dissertation "*Le jeu video comme manière d'être au monde: socio-anthropologie de l'expérience vidéoludique*" *(The video game as a way of being in the world: the socio-anthropology of the video-ludological experience)*, defended in March 2013, offers a counterpoint to Perény's work. Like Perény's studies, it provides a form of alliance between philosophy of technology and a second discipline, that of sociology. But while the philosophical references are less developed, both numerically and epistemologically, than in Perény's analyses, Koster's dissertation is nevertheless representative of a trajectory which begins in philosophy, Koster's original discipline, and moves toward socio-anthropology.

This form of alliance between the philosophy of technology and sociology is representative of the work developed by the CETCOPRA laboratory (Gras, Bensaude Vincent, Guchet). The thesis follows a counter-trajectory to that of Perény: it begins not from techniques of the image, but from players; it attempts to characterize pleasures and experiences in order to assess them according to more general standards or structures, those of the "society of technicity" or "technical macro-systems."

These two concepts are explicitly borrowed from the writings of Alain Gras, the founder and director of the CETCOPRA laboratory. "The use of the notion of 'technical macro-systems' by Alain Gras allows for the characterization of cultural meanings attached to the organization of the society of technicity, 'the ways of being in the world' of technical macro-systems. (…) Each technical macro-system can thus be attached to anthropological values which determine ways of thinking, feeling and acting." (Koster 2013: 119) The task, based on descriptions of video game experiences provided by players in a series of highly enriching interviews (149 interviews), is thus to decrypt the normative frameworks and symbolic values which are the basis of such game practices.

These standards and values are masked in the game experience, which implies a form of "illusion" and "forgetting of technical aspects." (Guchet 2005) "The pleasure of the game (…) is a kind of illusion: it is necessary to forget the process of the object's conception in order to concentrate on the final result (…). The transparency of the object keeps its usage at a distance from the veritable political and social issues which govern its conception, its production, and its commercialization." (Koster 2013: 114) This epistemological program does not, however, lead to a critical sociology which would seek to stalk myth and illusion on the part of the player, but to a "sociology of uses" which attempts to understand "how we play the games," the symbolic values gaming relies upon, and especially the prevalence given to the sensations and to the spectacular, arising from the industrialization of leisure activities.

The theoretical effort is thus diametrically opposed to that provided in Perény's work: it issues from "social over-determination" rather than from the intrinsic logic of technical concretization. The reference to Simondon is, in fact, of a critical nature: "the 'appropriation' of the technical object observed in the sociology of uses goes beyond the simple framework of 'a margin of indetermination' of the technical object conceptualized by Simondon, to take into account the cultural significations which the user attaches to his practice." (Koster 2013: 118)

This symmetrical effect can also be observed in "blind spots," insofar as technicity is not referred to in the analysis. It is above all a question of "observing in a tangible way the subjective manifestations of leisure activities and of connecting them to cultural patterns and to wider frameworks." Through the interviews conducted by Koster, the analysis thus demonstrates several modalities of these leisure activities: mastery but also internalization and domestication of the constraints of the apparatus, immersion centered on "sensory pleasure, deployment of the imagination, and the advent of new forms of sociability."

7.7 Phenomenology of Artificial Perception

Elsa Boyer's work takes us to more speculative terrain, at a distance from the empirical dimension and disconnected from the sociology of uses. This return to philosophy is distinguished first of all by the object of the investigation, which focuses on the transcendental structures of conscience from a phenomenological perspective. Above all, it is concerned, along with "artificial perception," with the "auto-constitution of the flux of conscience" and with what it contains of exteriority and of technical prostheses. This discussion is pursued through a critical re-reading of the works of Husserl, Derrida and Stiegler.

In other words, video games in their empirical dimension do not constitute the first object of investigation. Similarly to the work of Pereny, the generalizing force of the study and the questioning of the artificialization of perception rely on a limited group of object-paradigms. In contrast to applications like the Aspen Movie Map found in Perény, Boyer proposes the work of the filmmaker Harun Farocki centering on the American army's use of "serious games" for training. These paradigms lead us away from video games as an object of mass consumption.

The video game appears in Boyer's analysis as one of the particular cases of "conflicts of perception" analyzed by Husserl: in addition to the ordinary perception of the player's immediate environment, the world of play superimposes itself, proposing a coherent universe with which to interact. However, the conflict between these two perceptions cannot be easily resolved according to the model of perceptive illusion, where two contradictory perceptions are briefly able to cohabit. Nevertheless, this conflict resembles other situations, such as daydreaming or free association of images within which, to quote Husserl, I let myself be drawn towards a "quasi-life." "Video game images provide visibility to an area which lies somewhere between technical apparatus, hallucination and fiction, which Husserl questions without confronting." (Boyer 2015: 18)

We are dealing here with a project in which the video game object encourages us not only to reevaluate certain philosophical descriptions, but also to rethink the place of technology in phenomenology at the level of structures of consciousness. The video game appears first as an unthinkable object for phenomenology, in which it plays the role of intruder *par excellence*, in the name of the joint exclusion of the non-originary in perception, technology as a mode of rationality bogged down in an intra-worldly perspective, and objects of popular culture.

But Derrida's and Stiegler's perspectives are also criticized for their difficulty in accounting for what is truly of a conflicting nature in the artificial perspective. Stiegler, in an alliance with Simondon, describes the technological dimensions of the temporal constitution of consciousness. But he prefers the critical perspective of "synchronization of stream of consciousness" provided by the mass media (Stiegler, Chap. 18 this volume). This notion of synchronization can indeed nurture our reflection on video games, but it does not account for the totality of perceptive situations, the discrepancies and de-synchronizations which can be observed in artificial perception.

The last chapters thus propose an unexpected return to Husserl based on the existence of these extreme cases, which are examined incidentally within the framework of phenomenology, but which invite us to rework the Husserlian theoretical framework from within. Boyer notes several situations and objects: the theater and aesthetic perceptions, wax museums, tinted glasses, stereoscopes, etc.

The examination of these extreme cases leads to a remarkable description of the sort of "constructed hallucination" which occurs in video games, when the perception of images is coupled with the feeling of being present in a real world. Video games retain the traits of the image freed of fantasy or of the fixity of the stereoscope. When Husserl evokes the imagining of a "Mars world," he explains that "it will be necessary to have power over sensitive fields, to be capable of changing the groups of phenomena" so that the situation of conflict cannot be easily resolved, according to the model of the return to originary perception. "Contrary to the examples of foreign perceptive fields used by Husserl, video games allow us to act directly on sensitive fields; they create a coherent sequence while the avatar can eventually serve as the intermediary for judgment and will." Thus for her, video game "offers, within the continuously varying limits provided by the program and the graphics used by the developers, a 'quasi-life' inside a world which plays out across several different media and supports." (Boyer 2015: 202)

Elsa Boyer's approach thus permits her to perceptively describe the superimpositions between different modes of presence in the video games situation. This superimposition culminates in a form of unresolved "conflict of perceptions," which reaches a crisis-point in "the untenable opposition between the originary and the non-originary" characteristic of the Husserlian philosophy.

One can nevertheless question the kind of knowledge project that this "local phenomenology" proposes. It has the obvious advantage of circumscribing a unique area for philosophy: that of the study of the forming of consciousness. Aside from the fact that the borderline with psychology as an empirical science is subject to debate, one can also wonder to what extent an articulation with empirical knowledge is possible. In other words, does not the examination of concrete game situations, in their very diversity, inevitably lead us away from the transcendental framework?

This is in fact one of the reproaches made by Boyer to Stiegler's work: that he moves from the strictly transcendental perspective of the heterogeneity of flow of consciousness to a reflection on the "exteriorization" of processes of consciousness, in the anthropological manner of Leroi-Gourhan (Lenay, Chap. 13 this volume). The meeting-point between phenomenology and philosophy of technology is situated on particularly unstable ground, which risks disintegration at the moment when the concreteness and variety of situations of play are reintroduced. Once again, one is faced with the issue of the type of knowledge that a philosophy of technology can possibly deliver, at the juncture between philosophy, psychology and anthropology.

7.8 Systems of Experience and Philosophy of Technical Environments

The publication of *Philosophie des jeux video* (Triclot 2011) coincided with the breakthrough, mentioned earlier, in French-language video game studies. The text gave rise to several research projects typical of the recent recognition of the video game as a legitimate object of academic study (Triclot 2013).

The text nevertheless falls within the tradition of "philosophy of technical milieus," to use the term entitling Jean-Claude Beaune's study (Beaune 1999). This tradition is certainly the least known abroad, in comparison with the rediscovery of Simondon's work. Yet this tradition falls within the history of French philosophy of the second half of the twentieth century, in the wake of the considerable influence generated by Georges Canguilhem.

The term "technical milieu" is borrowed from the sociology of work and anthropology. It first appeared in Leroi-Gourhan (1945) and Friedmann (1950) during the immediate post-war period. The philosophy of technical milieus adopted the term for its own purposes from the 1970s onward, when its use in the social sciences was declining (Triclot 2012b). But the definitions of "technical milieu" were not stabilized in Friedmann and Leroi-Gourhan (see Petit and Guillaume, Chap. 6 this volume). Friedmann uses "technical milieu" as an antonym for "natural milieu," whereas Leroi-Gourhan derives "technical milieu" from the notion of "internal milieu" in reference to the biological model of the cell articulated by Claude Bernard. In other words, Friedmann opposes technique and nature while Leroi-Gourhan conceives of the terms as a continuity, according to the specifically stated project of a "biology of technicity."

Nevertheless, in each case, the notion of "technical milieu" has the value of providing a framework for a "non-reductive technology," to use Beaune's expression: a techno-logy which "removes the technical object from its exteriority" and analyzes the interlacing of technology and subjectivity which is characteristic of human existence. This is the starting point for Beaune's epistemological program: "Why does a technical object, the first utensil in our life and our being, cause such strange 'sensations'? Or, to put it differently, are we capable of stating clearly what makes this object original when it is considered in and of itself and above all as being situated, in the *milieu* its presence animates, but which is often limited to the understanding we wish it to contain?" (Beaune 1999: 7)

In this perspective, video games comprise a particularly remarkable field of study, insofar as they create a form of "instrumented experience" due to the use of machines, where "strange sensations," created by the interactive coupling of the computer and the screen, can be cultivated for themselves. But the epistemological problem, as recognized by Beaune himself, is to know how to objectify the subjectivation processes at work throughout the technical milieu. Beaune, for his purposes, uses techniques which belong to literature, to the art of writing, giving life to the rhythms and time-spaces of the environment. But this issue concerns the work of Perény, Koster and Boyer as well. It is resolved alternatively by the notion of "coupling," (Perény) by accounts of experience thanks to qualitative interviews (Koster), or by the phenomenological analysis of perception (Boyer).

The philosophy of technical environments takes us down a historicist path. In *Philosophie des jeux video*, this path is embodied in the concept of "systems of experience," which aims to describe the different modes of involvement which have developed out of computer games. The notion itself is borrowed from a certain theory of cinema, which attempts to characterize the experience of the spectator in the movie theater rather than to decode the meaning of the films, in the tradition begun by Christian Metz (Metz 1978; Bellour 2009). Metz spoke about what he called "systems of desire" to designate the forms of contact between the subject in the movie theater and the cinematographic apparatus.

One of the central theses of the philosophy of video games is that the forms of experience which are given priority by the games depend on the milieus in which they are played. The shift from the university environment, in the 1960s and 70s, to the arcade, in the 70s and 80s, provides a paradigm of this change. From an experience centered on "the total control over a simulated universe" characteristic of university-developed games, we move to an experience of "loss of control," of "mechanical vertigo" in the arcades. Each of these experiences mirrors the properties of its milieu: Spacewar, the first video game, replays the issue of the control of air space, developed from the military apparatus of the SAGE anti-air defense program, with a requirement for complete mastery and visibility. On the contrary, arcade games borrow the vertiginous logic, the pleasure of near-accidents, the adrenaline rush of violently accelerating vehicles, from mechanical fun fair systems.

But how can these transfers between environment and systems of experience be described? Rather than defining a domain reserved for philosophy, which would be the transcendental domain – but where it is difficult to find the empirical data without betraying it –, the philosophy of milieus, whether that advocated by Beaune or by Leroi-Gourhan, promotes a style of necessarily incomplete analysis. "Analyses of a single object can thus be infinite: it means decomposing the simple act in which the internal milieu participates wholly." (Leroi-Gourhan 1964: 343)

This incompleteness argues in favor of a better relationship between the philosophy of technology, the humanities, and the social sciences. Philosophy is not condemned to a unique strategy of general explanations towards the decrypting of intentional attitudes. On the subject of game studies, this was, for example, Henriot's strategy, with the aim of studying play as a structure of consciousness in the manner of Sartre's famous analysis of self-deception (Triclot 2014). But philosophy can also be closely involved with the empirical, in order to invent modes of inquiry concerning what is at issue in the details of the situations.

From this point of view, video games provide an ideal object. Indeed, they permit the collecting of a considerable number of recordings and traces, which cannot be easily decoded. In order to do so I had the occasion to work on an analysis of filmed game sessions, which give rise to procedures of "phenomenographical" re-coding in the search for inflections in the modes of presence (Piette 2009) as well as to forms of rhythm analysis based on input capture on game controllers which reveal the manner in which the demands of the game apparatus are modulated by the players (Boutet et al. 2013).

The positioning which stems from the concept of "technical milieus" thus allows for an alliance with strategies of observation of a pragmatist-inspired anthropology, such as they are typically represented in the French-language context in the work of Manuel Boutet (Boutet 2006, 2012). The philosophy of "technical milieus" shares with this tradition not only the central reference to Leroi-Gourhan but also an attention to the production of lived experiences, everyday life, and the gestural and rhythmical dimension of activities, rather than mere symbolic and language-based investments. This alliance thus opens out new technical and conceptual possibilities with which to analyze the "strange sensations" of the environment and to pursue the infinite inquiry along new lines.

7.9 Conclusion

In examining the role played by philosophy of technology in the field of French-language studies devoted to video games, I have highlighted a diversity of approaches and theoretical positions in a context dominated by the question of play, of forms of experience, and of coupling with machines. The use of philosophical references thus appears as representative of the major tendencies in French-language "game studies."

Philosophies of technology operate within this area of inquiry, difficult to negotiate, at the junction between subjects and their devices, but also between philosophy and the humanities and social sciences. While their main role can be seen as the application of a "theoretical framework" from which an ontology or ready-to-use modes of description can be expected, the video game object nevertheless brings a particular intensity to the question of the relationship between the subject and the device. In this sense, the video game can be seen as a "technical milieu" in miniature where, to use Jean-Claude Beaune's beautifully-phrased description, "a technical milieu is first and foremost a place where we play out our lives and thought with our bodies, in successive degrees and osmoses, on the borderline between nature and artifice where we balance incessantly like dancers on a tightrope." (Beaune 1999: 11)

Acknowledgements A preliminary version of this text has been discussed by Sacha Loeve and Stéphane Vial during the authors' workshop held in Paris Sorbonne on 22–23 June 2015 for the preparation of the collective book.

References

Aarseth, E. (2001). Computer game studies, year one. *Game Studies*, 1–1.
Amato, E. A. (2008). *Le jeu vidéo comme dispositif d'instanciation: du phénomène ludique aux avatars en réseau.* PhD thesis, Université Paris 8.
Amato, E. A., & Pereny E. (2008). Comment le premier cybermédium a pu un temps échapper aux SIC ?. *16e Congrès de la Société Française des Sciences de l'Information.*
Amato, E. A., & Pereny, E. (2010). L'heuristique de l'avatar: polarités et fondamentaux des hypermédias et des cybermédias. *Revue des Interactions Humaines Médiatisées, 11*(1), 87–115.
Amato, E.A., & Pereny, E. (2011). De l'Image interactive au Jeu vidéo: rétro-prospective des transversalités d'un objet cybernétique. *Communication Game Studies ? A la française!.*
Amato, E. A., Gaon, T., & Boutet, M. (2011). *Games Studies? à la française! (Call for communications).* http://manuel.boutet.free.fr/GSALF2011.pdf. Accessed 18 Oct 2015.
Audureau, W. (2014). *Pong et la mondialisation.* Châtillon: Pix'n Love.
Auray, N., & Craipeau, S. (2003). Les jeux en ligne. *Les cahiers du numérique, 4*, 2.
Beaune, J. C. (1999). *Philosophie des milieux techniques.* Seyssel: Champ Vallon.
Bellour, R. (2009). *Le corps du cinéma: Hypnoses, Émotions, Animalités.* Paris: P.O.L.
Berry, V. (2012). *L'expérience virtuelle.* Rennes: PUR.
Berry, V., Coavoux, S., Rufat, S., & Ter Minassian, H. (2013). Qui sont les joueurs de jeu vidéo en France? In O. Lejade & M. Triclot (Eds.), *La fabrique des jeux vidéo* (pp. 172–177). Paris: La Martinière.

Blanchet, A. (2010). *Des pixels à Hollywood*. Châtillon: Pix'n Love.

Boellstorff, T. (2008). *Coming of age in second life: An anthropologist explores the virtually human*. Princeton: Princeton University Press.

Bogost, I. (2010). *Persuasive games: The expressive power of videogames*. Cambridge, MA: MIT Press.

Boutet, M. (2006). *De l'ordinateur personnel aux communautés en ligne. S'orienter dans les mondes informatiques*. PhD thesis, University of Nice – Sophia Antipolis.

Boutet, M. (2012). Jouer aux jeux vidéo avec style. Pour une ethnographie des sociabilités vidéoludiques. *Réseaux, 173–174*, 207–234.

Boutet, M., Carvajal, I., Ter Minassian, H., & Triclot, M. (2013). Au-delà du virtuel: interactions sociales et spatiales dans et autour d'un univers vidéoludique. *Revue MEI, 37*, 103–116.

Boyer, E. (2015). *Le conflit des perceptions*. Paris: MF.

Brougère, G. (2003). *Jouets et compagnie*. Paris: Stock.

Bruno, P. (1993). *Les Jeux vidéo*. Paris: Syros.

Caillois, R. (1958). *Les hommes et les jeux*. Paris: Gallimard.

Chateau, J. (1946). *Le jeu de l'enfant*. Paris: Vrin.

Coavoux, S., Rufat, S., & Ter Minassian, H. (2014). Jouer aux jeux vidéo en France. Géographie sociale d'une pratique culturelle. *Espace géographique, 43–4*, 308–323.

Donovan, T. (2010). *Replay: The history of video games*. Lewes: Yellow Ant.

During, E. (2006). Simondon au pied du mur. *Critique*, 706.

Friedman, G. (1950). *Où va le travail humain ?* Paris: Gallimard.

Genvo, S. (2006). *Le game design de jeux vidéo: approche communicationnelle et interculturelle*. Thèse de doctorat en SIC, Université de Metz.

Genvo, S. (2013). Penser les phénomènes de ludicisation à partir de Jacques Henriot. *Sciences du jeu, 1*.

Guchet, X. (2005). *Les Sens de l'évolution technique*. Paris: Léo Scheer.

Henriot, J. (1969). *Le jeu*. Paris: PUF.

Henriot, J. (1989). *Sous couleur de jouer*. Paris: Corti.

Jull, J. (2005). *Half-real: Video games between real rules and fictional worlds*. Cambridge, MA: MIT Press.

Kline, S., & Dyer-Witheford, N., & De Peuter, G. (2003). *Digital play. Digital play: The interaction of technology, culture, and marketing*. Quebec: McGill-Queen's University Press.

Koster, R. (2013). *Le jeu vidéo comme manière d'être au monde, Socio-anthropologie de l'expérience vidéoludique*. PhD thesis, University of Paris 1 Panthéon – Sorbonne.

Leroi-Gourhan, A. (1945). *Milieu et technique*. Paris: Albin Michel.

Leroi-Gourhan, A. (1964). *Le geste et la parole, Tome 1*. Paris: Albin Michel.

Mabillot, V. (2000). *Les mises en scène de l'interactivité: représentations des utilisateurs dans les dispositifs de médiations interactives*. PhD thesis, University Lyon 2 – Louis Lumière.

Markoff, J. (2006). *What the dormouse said: How the sixties counterculture shaped the personal computer industry*. London: Penguin Books.

Metz, C. (1978). *Le signifiant imaginaire*. Paris: Bourgois.

Montfort, N., & Bogost, I. (2009). *Racing the beam*. Cambridge: MIT Press.

Natkin, S. (2004). *Jeux vidéo et médias du XXIe siècle: quels modèles pour les nouveaux loisirs numériques ?* Paris: Vuibert.

Pereny, E. (2013). *Images interactives et jeu vidéo: de l'interface iconique à l'avatar numérique*. Paris: Questions Théoriques.

Piette, A. (2009). *L'acte d'exister: une phénoménographie de la présence*. Marchienne-au-Pont: Socrate Promarex.

Quezorla, J., & Verebely, F. (1983). Jeu, image et communication. *Réseaux, 1*, 3–17.

Rueff, J. (2008). Où en sont les Game Studies? *Réseaux, 151*(5), 139–166.

Rufat, S., & Ter Minassian, H. (2012). *Les jeux vidéo comme objet de recherche*. Paris: Questions Théoriques.

Servais, O. (2013). De l'ethnologie missionnaire à l'anthropologie prospective du virtuel. *Histoire, monde et cultures religieuses, 2*(26), 95–108.
Sidre, C. (2014). *L'objet vidéoludique et ses réseaux de distribution (1974–1988).* PhD thesis, École Nationale des Chartes, Paris.
Simondon, G. (1958). *Du mode d'existence des objets techniques*. Paris: Aubier.
Taylor, T. L. (2006). *Play between worlds: Exploring online game culture*. Cambridge, MA: MIT Press.
Tremel, L. (2001). *Jeux de rôles, jeux vidéo, multimédia. Les Faiseurs de mondes*. Paris: PUF.
Triclot, M. (2011). *Philosophie des jeux vidéo*. Paris: La découverte.
Triclot, M. (2012a). Jouer au laboratoire: le jeu vidéo à l'université (1962–1979). *Réseaux, 30*, 173–174, 177–205.
Triclot, M. (2012b). Milieu technique: généalogie d'un concept. In D. Parrochia & V. Tirloni (Eds.), *Formes, Systèmes et Milieux Techniques*. Jacques André: Lyon.
Triclot, M. (2013, December). Super Mario rentre au musée. *Le monde diplomatique*, 20–21.
Triclot, M. (2014). Game studies ou études du play? Une lecture croisée de Jacques Henriot et de Jesper Juul. *Sciences du jeu, 1*.
Turkle, S. (1984). *The second self: Computers and the human spirit*. Cambridge: MIT Press.
Weizenbaum, J. (1976). *Computer power and human reason*. San Fransisco: Freeman.
Wendling, T. (2010). Une joute intellectuelle au détriment du jeu ? Claude Lévi-Strauss vs Roger Caillois (1954–1974). *Ethnologies, 32*(1), 29–49.
Zabban, V. (2011). *"Ceci est un monde" Le partage des jeux en ligne: Conceptions, techniques et pratiques*. PhD thesis, Université Paris Est.
Zabban, V. (2012). Retour sur les Game Studies. *Réseaux, 30*(173–174), 141–176.

Part II
Coining and Reconfiguring Technoscience

Chapter 8
Technoscience: From the Origin of the Word to Its Current Uses

Gilbert Hottois

Abstract I have a long-standing relation with the noun "technoscience." In recent years, I have been concerned with its evolution and connotations, since the period when I first thought it up. This chapter presents a survey of the various uses, transfers and significations of the term. It makes a twofold claim (i) technoscientific research and development are conducted by a plural subject in need of a moral conscience; (ii) the study of technoscientific objects requires a methodological and operational materialism.

Augmented version for this volume of an essay first published in French (Hottois G. La technoscience: de l'origine du mot à ses usages actuels. In: Goffi J-Y (ed) *Regards sur les technosciences*, Vrin, Paris, pp. 23–38, 2006). Translation by John Stewart, University of Technology Compiègne.

Keywords Auto-transcendance · Ethics · Language · Materialism · Postmodernism · Technology · Technoscience · Transhumanism · Posthumanism

I have a long-standing relation with the word "technoscience," a relation that might be called "existential" and which is not without ambivalence.[1] In recent years, I have been concerned to examine what has become of this word, since the period when I first thought it up. This genealogical investigation was triggered by a symposium organized by Jean-Noël Missa in 1999 on "Technoscience."[2] "Technoscience" is not a peaceful reference: it arouses passions, technophobe and technophile, creationist and evolutionist, reactionary and futurist, globalizing and

[1] My science-fiction novel *Species Technica* (2002a) written in 1981 but published 20 years later, is evidence for this.

[2] On the occasion of an invitation from the University of Brussels of Dominique Lecourt who had just published a book devoted, one might say, to the imaginary dimension of technoscience (Lecourt 1996).

G. Hottois (✉)
Université Libre de Bruxelles, Bruxelles, Belgium
e-mail: ghottois@ulb.ac.be

S. Loeve et al. (eds.), *French Philosophy of Technology*, Philosophy of Engineering and Technology 29,
https://doi.org/10.1007/978-3-319-89518-5_8

alter-mondialist, modern and post-modern, caught up in the "science war" as well as in a war of cultures. In 2003, I chose "technoscience" as the main subject of a series of lectures at the Collège de France.[3] Since then, I have continued and extended my research on the genealogies of technoscience that I summarize in the epilogue to this essay.

8.1 Hypothetical Ancestors

When one acquires a history, one is promptly faced with the invention of numerous pre-histories, which go back until they are lost in the mists of time. Thus, quite recently, I read that technoscience has been traced back to the School of Alexandria (Ihde 1993: 6; 8). But at this rate, I asked myself, why not go back to the Ionians, to Thales himself who was a philosopher-engineer, a technical thinker as Gilbert Simondon reminds us… Most of the time, one is more modestly content to go back to Francis Bacon (1561–1626),[4] thus associating closely the birth of technoscience with that of modern science.

Is *The new scientific spirit* (1934) of Gaston Bachelard (1884–1962) the same as the spirit of technoscience? Many authors are convinced that it is, to the point of attributing to him, erroneously, the creation of the term.[5] Because the spirit of technoscience seems to be akin to that of quantum mechanics, which inspired Bachelard: operationally effective, but rebel, in spite of the efforts of Bernard d'Espagnat (1994) and many others, to any ontologically realistic interpretation. So it is not surprising that the famous 1955 conference of Werner Heisenberg on *Nature in contemporary physics* (1958) also evokes a number of characteristics typical of technoscience.[6] And of course there is also Heidegger… None of these authors actually uses the term, but the *idea* of technoscience – with all its implications for the conception of knowledge and human being – is something that they are groping after.

[3] Published in 2004 under the title *Philosophie des sciences, philosophies des techniques* (Hottois 2004a).

[4] With his *Novum Organum* (1620) and *New Atlantis* (1627).

[5] In a voluminous anthology *Philosophy of Technology* the editors Robert C. Sharff and Val Dusek attribute the paternity of the term "technoscience" to Bachelard, from whom Latour is supposed to have borrowed it (Sharff and Dusek 2003: 85).

[6] Heisenberg's "Nature in contemporary physics" includes a section on technology and constantly emphasizes the strong link between science and technology, as well as the transformation of a science that aims at representation into a science that is active and operational. The role of theory in this science-technology no longer concerns reality itself, but the interactions of the scientist with the real world. This text includes a radical re-evaluation of technology in its relation to science.

8.2 Inception and Hesitation

I myself began to use the word in the middle of the 1970s.[7] I employed it in the title of an article as early as 1978: "Ethique et Techno-Science," published in *La pensée et les hommes*, a Belgian journal on philosophy and "secular ethics" (Hottois 1978). The title is significant, because it suggests that if science is technoscience, it inevitably raises practical questions. However at that time I hardly addressed question of that order. By the term "technoscience" I wanted to designate what I considered to be the core of a problem that the dominant philosophies of the period seemed to positively wish to avoid. These philosophies positioned themselves almost exclusively at the level of language: linguistic analytical philosophy, late phenomenology and hermeneutics, structuralism, grammatology, archeology of the human sciences, philosophy of communicational interaction or dialogue, new rhetoric and philosophy of argumentation… I wanted to react against this inflation of language culminating in philosophical texts that were purely auto-referential and non-referential, cut off from reality. A reality which, as I emphasized, was no longer fundamentally natural, substantial and something to be represented but rather consisting of processes, something which could be operated on, something interactive and technological… My dissatisfaction also concerned those philosophies of science which only conceived the scientific enterprise as theoretical and discursive. The term "technoscience" aimed at emphasizing the operational dimensions – technological and mathematical – of contemporary science. My dissatisfaction extended further to those "philosophies of History" whose histories appeared to me to be blind and bereft of resources when faced with the cosmic an biological temporalities postulated by the contemporary technosciences. These philosophies of History comprised political and social philosophies – Marxisms and various leftist currents – which also dominated the intellectual scene of the 1960s and 1970s. To sum up, the term "technoscience" crystallized, for me, in the crucible of a deep internal dissatisfaction concerning what I knew of the philosophy of the period, and in opposition with the traditional philosophical conception of science.

In the course of the 1980s my use of the term "technoscience" became reserved, hesitant; I often preferred to speak of "*technique*" (Hottois 1984a, b), even if I continued to think of "technoscience." The use of this latter term almost automatically provoked reactions of incomprehension, irony and critical wrath on the part of philosophers and scientists alike, often on the basis of misunderstandings but also because of deep resistances. These reactions were particularly virulent in France where "technoscience" had become a sort of symbol of absolute evil, concentrating all the ills of the period: technicism and technocracy, multinational capitalism, neoliberal economics, pollution, exhaustion of natural resources, greenhouse effect, American imperialism, world injustice, disappearance of human values, etc. In the

[7] See the first texts reproduced in Hottois (1996). My use of the term « techno-science » is frequent in my doctoral thesis (Hottois 1976), published in abbreviated form under the title *L'inflation du langage dans la philosophie contemporaine* (Hottois 1979).

course of the 1990s the term nevertheless became more commonplace, in particular making its entry into the Larousse dictionary.[8]

8.3 Technoscience Travels to America

There have been two major French relays in the dissemination of the term: Jean-François Lyotard and Bruno Latour.

Lyotard came to use the term, probably after encountering it in *L'inflation du langage dans la philosophie contemporaine* (1979).[9] *Le postmoderne expliqué aux enfants* (1986) is a collection of papers which, to my knowledge, were the earliest where Lyotard spoke of "technoscience": the first dates from 1981 (in Italian; 1982 in French). The paper in question is "An Answer to the Question, What is the Post-modern?" In this paper Lyotard denounced the link between capitalism and technoscience,[10] as well as a subversion of modernity:

> Contemporary technoscience (...) fulfills the project of modernity: man makes himself master and possessor of nature. But at the same time contemporary technoscience profoundly destabilizes that project: For the term 'nature' must also include everything constituting the human subject: its nervous system, genetic code, cortical processor, visual and auditory receptors, communication systems (particularly linguistic), its organizations of group life, and so on. Its science and technoscience also end up being part of nature.[11] (Lyotard 1986: 20)

Lyotard had a significant role in the circulation of "technoscience" between phenomenology and STS (Science-Technology-Society).[12] He also disseminated the term on both sides of the Atlantic when he moved from University of Paris Vincennes[13] to University of California, Irvine, as Professor of "critical theory" and as visiting professor in other US universities. He continued to use the term "technoscience" on occasion, in particular in *L'inhumain: causeries sur le temps* (1988) and

[8] *Grand Larousse universel*, 1992, and the *Petit Larousse* in 1993. Titles including the term "technoscience" became more common in the francophone area (Breton et al. 1990; Prades 1992; Chirollet 1994)... I myself return to the term with *Entre symboles et technosciences* (Hottois 1996).

[9] *The Postmodern Condition* (1979) contains no occurrence of "technoscience."

[10] "But the victory of capitalist technoscience over the other candidates for the universal finality of human history is another means of destroying the modern project while giving the impression of completing it." (Lyotard 1986: 18).

[11] It is not exceptional that a philosophical reflection on technology brings one back to a philosophy of nature. But in the case it is a concept of nature which is profoundly transformed. From the viewpoint of technoscience, "rendering technological or operational" and "naturalizing" are complementary aspects of one and the same process.

[12] See the reference to the field of STS that Lyotard (1986: 21) associates with "the discovery of the subject's immanence in the object it studies and transforms."

[13] A university located in Saint-Denis, near Paris, and created in the aftermath of the political events of May 1968, where Gilles Deleuze, Félix Guattari and Alain Badiou, taught as well.

in *Moralités postmodernes* (1993). In this later book, the paper entitled "A Postmodern Fable," with its narrative of Humans leaving the Earth, is undoubtedly the one where Lyotard's version of technoscience finds its culminating expression. (Sebbah, Chap. 10 this volume) However, Lyotard does not find anything in this perspective to help him live and hope. Thus, it is in a postmodernism situated in the symbolic and esthetic registers, largely impermeable to the technosciences, that Lyotard sought refuge.

Rumor, propagated from book to book, attributes to Bruno Latour the introduction of the term "technoscience."[14] For example Dona Haraway, who makes abundant use of the term "technoscience," assumes that Bruno Latour was responsible for the adoption of the term "technoscience" in science studies and that he used it to attack the traditional divide between "science" and "society." (Haraway 1997: 279–280).

Latour has facilitated the diffusion of the term "technoscience" in the French-speaking world, and especially in the Anglo-American world. But he is not the author of the term, either in English or in French. It is in his 1987 book – *Science in action* – which appeared in French (*La science en action*) 20 years later, that he employs the term "technosciences." However, in a subsequent French edition (1995) he added a footnote reading:

> Since the term *technoscience* has unfortunately been taken by those who, following Heidegger, have forgotten that it is important to actually study scientific and technological productions before moaning about their 'absence of Being, of value, of beauty and truth', I only use the term in the plural and without any profound ontological connotation. (Latour 1995: 79)

The technosciences according to Latour refer to "science in action," the science that is actually done, paying due attention to the way it is done; and not to the mythically idealized or demonized philosopher's reconstruction of science. These technosciences can be briefly summed up by the following characteristics (Latour 1987):

(a) The technosciences are rhetorical in nature, purely eristic. The winner is the one who manages to impose his statements as "true" or "factual," and thus "real."
(b) They are complex enterprises which mobilize a network composed of various human agents but also non-human agents such as machines, systems of transport and communication, capital, laboratory animals, texts, etc. etc. The history of technosciences is, in large part, that of the extension of these networks.
(c) They are predominantly American and are performed as military enterprises, sometimes quite literally.

[14] For example, recently again, in *La revolucion tecnocientifica* by Javier Echeverria (2003).

8.4 Postmodern Technoscience

The term "technosciences" (plural) plays such an important role in *Science in Action* that it is surprising to find that Latour does not use it in his subsequent publications.[15] Twelve years will pass before he reuses the term in *Pandora's Hope. Essays on the Reality of Science Studies* (1999), where "technoscience" (in the singular) receives a limited meaning in the framework of the notion of "sociotechnical" (which designates any form of mixing of the social and the technological). Such mixtures are as old as mankind, but they have evolved, and Latour reconstructs this sociotechnical evolution in a mythical "pragmatogony."[16] The term "Technoscience" is used here to designate the penultimate level of sociotechnical, the one that precedes a fully-fledged political ecology which attributes a status of quasi-citizens to non-human entities on the grounds that they are an integral part of society. "Technoscience" designates a mixture of humans and nonhumans, but one which is deficient because it accords insufficient recognition to the "rights" of the latter.[17]

In the course of the 1990s, technoscience became the pretext for a considerable literature[18] in the sphere of socioconstructivism and postmodernism. The American Society for Social Studies of Science[19] entitled its newsletter: *Technoscience*.[20] What is meant by "technoscience" in the context of these "technoscience studies" to take up the expression of the philosophical sociologist John Law (2002)? I would say that "technoscience" became the keyword for a contemporary form of *entanglement*, the entanglement of processes. Dona Haraway provides one of the most colorful expressions of this: "the fiercely physical-semiotic world of technoscience (...) extravagantly exceeds the distinction between science and society, subjects and objects, and the natural and the artifactual that structured the imaginary time called modernity." (Haraway 1997: 1) The entanglement concerns equally the discourse or the disciplines in question, and the realities that are represented therein; it concerns

---

[15] In the two following works – *We Have Never Been Modern* (1991) and *Aramis or the Love of Technology* (1993) – it is question only of technology and science, and of research and development.

[16] On the model of "cosmogony," pragmatogony is a narrative recounting the genesis of *pragmata* (both things and public affairs, as well as matters of concern and of interest).

[17] "Through technoscience – defined for my purposes here, as a fusion of science, organization and industry – the forms of coordination learned through 'networks of power' (see level n°9) are extended to inarticulate entities. Nonhumans are endowed with speech, however primitive. (...) While in this model, the tenth meaning of sociotechnical, automata have no rights, they are much more than material entities; they are complex organizations." (Latour 1999: 203–204)

[18] Mary Tiles and Hans Oberdiek write that faced with the entanglement of science and technology, "it makes more sense to talk, as Bruno Latour does, of techno-science." (1995: 90) However, some doubts are occasionally expressed regarding the paternity of this term. Raphaël Sassouer attributes the invention of "technoscience" to Lyotard in 1982, but observes that its paternity remains an object of dispute (1995: 24).

[19] Founded in 1975 and often designated by the acronym 4S.

[20] Published three times a year, it goes back to the end 1980s. The latest news is that it ceased publication in 2004, being replaced by online information (http://www.4sonline.org/technoscience)

above all the *relations* between the discourses and their referents. Science studies, cultural studies, text studies, narrative studies, etc… slide into each other, and are collectively necessary to refer and to be able to act on the intermingled reality composed of sciences, technologies, economies, cultures, politics, arts, etc… In short, "there is nothing clearly and distinctly describable as science or culture or technology" Menser wrote (Aronowitz et al. 1996: 294).

The recognition of technoscience as an entanglement is not without relevance, insofar as it is an invitation to analyze complexity. The entanglement should not suggest that one does science in the same way that one does literature or politics. It refers only to the existence, in our civilization, of interactions which are strong, multiple and incessant between realms of what is symbolic and what is technoscientific. Whether it is a question of nuclear fusion, of genetically modified organisms or cloning, in a certain way – but a decisive one – the future of technosciences actually depends on their own image, an image which is constructed and manipulated, and which mixes up information and fantasy.

8.5 Postphenomenological Technoscience

The American philosopher Don Ihde has instigated for several years a seminar and a research group entitled "Technoscience" at the University SUNY at Stony Brook (Long Island). It is in *Instrumental realism* (1991), a book devoted to the question of the articulation between philosophy of science and philosophy of technology that he introduces, following Latour, the notion of "technoscience." In this book, Don Ihde argues for a philosophy of technology which would, so to say, have absorbed and digested the philosophy of science by revealing the technical body of science, and by showing that we have passed from "*science-driven technology*" to a "*technology-driven science*" which transforms and produces the world in which we live, while being concretely and institutionally established.

Don Ihde comes from the phenomenological tradition, and it is on this basis that he criticizes the Anglo-Saxon philosophy of science which sees Science only as language and theory, totally ignoring the embodied nature of every perception and every experience.[21] However, he rejects the phenomenological privilege accorded to any kind of experience which might be considered as fundamental because more original or more natural, a privilege associated with a tendency of phenomenology to be technophobic or indifferent to technology. This perspective of a phenomenology

[21] It is primarily in Husserl, Heidegger and Merleau-Ponty that Ihde finds the resources for a critique of Science as "theoretical," as well as the elements for an approach to science which emphasizes its dependence with respect to concretely situated perception and praxis: the *Lebenswelt* in Husserl, the phenomenology of perception and the body in Merleau-Ponty, and the technically equipped preoccupation of Heidegger from which theoretical objectivity is derived. One finds there sketches of the "reincorporation" or "re-embodiment" of science. See his *Technics and Praxis* (1979).

that explicitly considers the technically mediated character of perception and intentionality is what he later labelled "postphenomenology." (Idhe 1995)

More recently, Ihde has co-edited a collective volume entitled *Chasing Technoscience: Matrix for Materiality* (Idhe and Selinger 2003) comprising interviews and articles followed by discussions and comments, centred around four major representatives of technoscience: Ihde himself, Latour, Haraway and Andrew Pickering (although the latter barely uses the term "technoscience"). Their "backgrounds" are highly diverse: a philosopher of hermeneutical phenomenology; a philosopher of anthropological sociology; a biologist who is a critical historian of science; and a physicist who is a sociologist. Their "family resemblance" lies both in what they all reject, and more positively on certain references and characteristics that they share to a greater or lesser extent. They reject the divisions and the rigid conceptual hierarchies of modernity, which belong to the humanist tradition, as well as the conception of science that is attached to them. They share a vague philosophical reference to Whitehead, and less consistently to several other authors (Nietzsche, Deleuze, Foucault, and more rarely to the American pragmatists…). They have an approach which tends to be fully materialist, but without reductionism. The phenomena of meaning, signs, are also material; and objects, technologies and material artifacts are also meaningful. In short, their materialism is semiotic and their semiotics is material. They think in terms of relations, networks, functions, processes, interactions, construction, complexity, open pluralism, and they attribute great importance to the social dimension, to collective activity. Their ultimate frame of interpretation appears to be political; but the oscillation or hesitation between descriptive and normative approaches is fairly constant. This collection of characteristics is unevenly present; some of these traits are not explicitly present for all these authors. And Ihde himself is the most singular, perhaps because he is the one who remains the closest to traditional (phenomenological) philosophy.

8.6 Technoscience is Political

In his book *La technique* (1994), the French philosopher Jean-Pierre Séris (1941–1994) devotes a chapter[22] to the term "technoscience" where he exposes his reservations with respect to this neologism, which, according to him, expresses only "amalgam, agglutination, confusion, collusion": a "fusion" of concepts which is intellectually indefensible because it is "crude and interested. (Séris 1994: 240) "Technoscience" is sometimes the sign of a technophobia inspired by Ellul,[23]

[22] Seris (1994: 201–243), Chap. 5 "Technique et science."

[23] Séris assumed that Ellul inspired me, since he refered principally to *Le signe et la technique* (Hottois 1984a). But his view is not unambiguous: in one place (1994: 215) he credited me with of the authorship of the term ("The neologism 'technoscience' forged by G. Hottois"), But elsewhere (1994: 373) he seemed to attribute it to Ellul ("'Technoscience', the 'elegant' neologism, based on the corresponding adjective, invented by J. Ellul"). Ellul's Preface to *Le signe et la technique*

sometimes the expression of a devaluation of science as purely instrumental, utilitarian and technocratic. If we follow Séris, the insoluble problem is that it would be necessary to breathe science – with its idealism, its reference to the infinite – into technology, and not simply breathe technology – with its operational effectiveness – into science. "The term 'techno-science' does retain the idea of a task at the scale of humanity, but this task cannot be infinite to the extent that it does not bear on idealities." (1994: 215–216) Séris fears that with technoscience, all one will succeed in doing is to destroy the spirit of the scientific enterprise by rendering it subservient to finite subsidiary tasks. He considers that it is not possible that technoscience could absolutely require that the infinite scope of science be internalized in technological effectiveness. He thus puts his finger on an issue that is quite essential for our technoscientific civilization. I will come back to this point in my conclusions.

With *La revolucion tecnocientifica*, the Spanish philosopher Javier Echeverría (2003) offers us a conception of technoscience that is both more subtle and more systematic. He situates it in the continuation of the *macrociencia* – the *Big Science* – which finds its prototypical expression in the famous report "Science, the Endless Frontier"[24] produced at the request of Franklin D. Roosevelt (1944) by Vannevar Bush, President of the Office of Scientific Research and Development, and submitted to President Truman in 1945. This text was foundational for the American science policy during the first decades of the second half of the twentieth century; it has also inspired scientific policy outside the United States. It proposes a linear model of progress:

(1) Fundamental research is carried out in the universities; it is unpredictable, it should be free and financed by the State.
(2) It makes it possible to discover the laws of nature.
(3) These laws lead to the invention of new technologies and products.
(4) The latter promote the competitive development of enterprises.
(5) Enterprises ensure full employment and at the same time a better standard of living for all (health, comfort, physical and mental well-being).

The Bush report endeavored to transpose into peacetime the fruitful experiment of wartime Research and Development (R&D), in particular the Manhattan project. This *Big Science* prefigured the "technoscience" of the last decades of the twentieth century, but the overly simplistic nature of the linear model does not describe it adequately. The passage from *Big Science* to technoscience in the proper sense of the term involves a number of factors:

(1) The critique and the calling into question of the ideals and values of modernity, including the collusion between the State and *Big Science* (a movement which developed in the 1960s).
(2) The development of neo-liberal and ultra-liberal doctrines.

---

unfortunately contributed to distort the meaning and the scope of this book, perverting it in the direction of technophobia.

[24] United States Government Printing Office, Washington, 1945.

(3) The increasing privatization of R&D and its structural mode of financing, which depends on industries and thus on the market (in the course of the Reagan years).[25]
(4) Taking seriously into account all the consequences of the reality of Modern Science which is essentially operational, active and productive, which leads to the subordination and instrumentalization of its cognitive finality and the values which are associated with it (truth, universality, objectivity, disinterested, etc.).
(5) The development of Information and Communication Technology (ITC) as the "formalism of technoscience."[26]

The focalization on products and artifacts, on actions which transform nature and society, as well as the development of means of information and communication, all lead to the fact that science increasingly appears as something which concerns everybody: each citizen in a democracy, each consumer or user on the market. Whether the financing of research is public or private, there are *people* from the beginning to the end of the process. Technoscience, its actions and its products, result from the collaboration of a host of agents: research scientists from many disciplines, engineers and entrepreneurs, fund-raisers and share-holders, lawyers and economists, commercial and marketing agents, etc. An essential aspect is that the *subject* of technoscience – the actor, the motor and even the inventor – has become irreducibly *plural*: complex, interactive and inevitably *conflictual*. The real subject of technoscience is a far cry from the Cartesian or Kantian subject of modern science, who was supposedly rational, universal, and animated by purely cognitive intentions.

In a way, the sciences and scientists are victims of their own success: the things they make it possible to do interest everybody. And with globalization, this "everybody" tends to coincide indeed with the whole of humanity, whose immense diversity of cultures and their inequalities of living conditions render the classical formulation of the problem of "two cultures" by Charles Percy Snow (1961) a considerable over-simplification. Technoscientific culture is not a simple unit, and neither is traditional symbolic culture.

The plurality of the subject of contemporary technoscience – or TSRD (Techno-Scientific Research and Development), as I prefer to say – is more or less extensive, depending on the diversity of the interests one wishes to take into account. There is the core of those who are directly associated with R&D as such (research scientists and technicians, public and private fund-raisers with their experts in scientific policy, in economics and law); but there is also the whole set of all those who are potentially concerned, who will be affected by R&D, many of whom cannot share certain technoscientific choices because they orient society in a direction that they

[25] The fact that R&D enterprises enter the stock market, the creation of the NASDAQ (National Association of Securities Dealers Automated Quotations), the importance of patenting, all express this evolution.

[26] Echeverría insists strongly on this aspect that, together with private funding, he considers a major characteristic of technoscience as distinct from *Big Science*. In this evolution, the European Center for Nuclear Research (CERN) acted as a precursor (Echeverría 2003: 71; 105; 146).

do not wish to take, on the basis of their values and interests, their fears and hopes both rational and imagined. The subject of TSRD can no longer be simply identified with the classical "scientific community." And, as Echeverría says (2003: 225), according to the situation and the course of events what dominates in this plural subject of technoscience may be subjective, intersubjective or objective; and they are not always easy to distinguish. The subject of technoscience is neither unequivocal nor value-neutral; it is at the crossroads of values, which are irreducibly plural and quite often conflictual. When the financing of research is private, it imposes a capitalist value-system where profit is the dominant factor and the market is the norm (whether the consumer will buy or not); when the financing is public, the value-system in play is inspired by the public good, but is largely dictated by the perception of the public which is heavily influenced by lobbies and the strategies of the political parties.

However, within the framework of this plural subjectivity, scientific communities with their own values (rigor, objectivity, honesty, truth, transparency, etc…) do continue to have considerable importance. This is so, because if research scientists and experts were to allow themselves to be excessively influenced, in confusion, by beliefs and interests which were foreign or contrary to science,[27] the whole system would rapidly collapse.

In such a context, it is not difficult to understand the deep unease of a large portion of the scientific community. Since this community is part of a heterogeneous plural subject that it does not control and whose ends it does not determine, the cognitive and creative values which animate it and the work it produces are instrumentalized to the benefit of values and interests which are foreign, and which the scientist does not share. This situation corresponds precisely to the very definition of alienation. It is experienced in very different ways; a certain number of scientists have willingly and successfully converted themselves into heads of enterprises, managers and shareholders. But if they wish to obtain funding for their research projects, all of them – willy-nilly, like it or not – are obliged to take into account the values, norms, expectations, interests, fears and hopes of the other members of the plural subject of TSRD. The justification of their research projects in terms of purely cognitive values and aims are generally less convincing than the proclamation of goals and applications in economic and/or therapeutic terms, for example. It is only under the cover of this sort of rhetoric and a careful "marketing" – which may not be entirely unfounded but which have become indispensable – that a research project which also has a cognitive aim has any chance of finding a sponsor. The possibility of patenting the research results has thus become a virtual obligation for a large number of projects.

However it is important not to lose sight of the fact that the system as a whole is rich in interactions and feedback mechanisms, so that a seductive presentation of a project by scientists who are themselves principally concerned with the advancement of knowledge, can also instrumentalize in turn those who provide private and

[27] Profit, media narcissism, power, secret, various personal advantages…., and having recourse to means such as mercenary motives, dissimulation, cheating, faking, etc.

public funding. Many researchers quickly grasp how to fill out forms and satisfy the values, interests, desires and fantasies of the other members of the plural subject of technoscience. What haven't they promised, over the last few decades, in the fields of biotechnology, biomedicine and nanotechnology for example. But all's fair in love and war, at least as long as these strategies do not alter the intrinsic quality of the research and the results which are obtained, whether they were anticipated or not. These negotiated strategies can even improve the work and the results, as long as they bear traces of the complexity of the world where they are produced.

The problems with the choices posed by contemporary R&D do not come only from the plurality of the cultures and values of the subject of technoscience, which brings into interaction scientists, industrialist, financiers, politicians, users/consumers (including citizens with quite diverse cultural backgrounds). The problems also stem from the over-abundance of technoscientific possibilities: many interesting research paths, many creative and innovative technical developments can be envisaged; but the great majority of them would require financial and human investments that are far beyond the available means. We may think of space exploration and conquest; of robotics and Artificial Intelligence; of gigantic nuclear accelerators or nuclear fusion; but also of "soft" chemistry[28] or nano-techno-sciences; and of course the humanities and the necessity of discovering and preserving the immense cultural (and not only natural) riches we have inherited from the past… The explosion of what technoscience has made possible is another consequence of the success, of the immense fertility, of modern science and technology. Just as the plurality of the subject of TSRD is characterized by the absence of shared, common criteria for choosing, so the necessity to make choice amongst the over-abundance of what technoscience has made possible inevitably raises conflicts.[29] If the Manhattan project was the paradigmatic example of *Big Science*, the Human Genome Project of the 1990s plays a similar role for technoscience. We find here all its key features: public and private funding; cognitive stakes largely instrumentalized for economic, political, legal (patents) purposes, etc.; a pluralistic subject, complex and conflictual at the level of values and interests; digitalization; mobilization of the humanities and social sciences in Ethical, Legal, and Social Implications (ELSI) programs.[30]

[28] "soft chemistry' (*chimie douce*) is a phrase coined by French chemist Jacques Livage in the 1970s. It refers to the investigation of chemical reactions conducted at ambient temperature and low pressure. It includes sol-gel chemistry and bio-inspired chemistry.

[29] Echeverría emphasizes the structurally conflictual nature of the subject of technoscience. These conflicts cannot always be reduced to peaceful controversies and debates; there are also oppositions and incompatibilities in modes of life, in very concrete interests and social projects, which can become physically violent. (Echeverría 2003: 176)

[30] Echeverría recalls that at the start of the Human Genome Project its first director, James Watson, decided to allocate 5% of the budget to research on the ethical, legal and social implications of the Project (2003: 139).

8.7 Provisional Conclusions

The world described by *technoscience studies* is not especially attractive, unless, maybe, at a considerable postmodern esthetic distance. This world looks like a chaos of polymorphic material forces, more or less violently interacting, which presents a few unpredictable islands and phases of ephemeral organization and creation. The plural and conflictual subject of technoscience appears as both *material* and *unconscious*. This state of affairs leads me to two considerations by way of conclusion.

1. *The subject of technoscience is in need of a conscience.* Not just a mirror-conscience a merely descriptive one: a *moral* conscience, capable of deliberating and judging. A plural subject wants a plural conscience. I formulate the hypothesis that this deep need for a plural conscience of technoscience is emerging – among other ways, there is no monopoly here – through the multiplication of Ethical Committees (in particular, bio-ethics) in recent years. To support this hypothesis, I wish to emphasize two aspects of this development.

    (a) Ethical committees have been progressively set up at all levels of complexity and extension: local, national, more or less international (European Union, European Council), worldwide (Unesco).
    (b) These committees are, or at least they should be, genuinely pluridisciplinary and pluralistic, including representatives of all the associations of interest that make up society.

This multidisciplinarity includes the humanities and social sciences which contribute to inform the plural subject of technoscience as to its own nature; it also includes disciplines such as philosophy, theology or law, inviting the plural subject to put forward views and to discuss about values and norms, and to formulate an opinion. The highly pluralistic composition of the committee has the consequence that their own report will often be only partly consensual. But the irreducible divergences will at least have had the opportunity to express themselves by stating their presuppositions, their beliefs, concerns, values and the reasons that underlie them. Such a result constitutes an important step forward when compared with a blind conflict of forces and desires, without any conscience other than that of their desire to win and perceiving the others only as means or obstacles. The Ethical committee as a form of conscience is the instance where the plural subject of technoscience discusses instead of tearing itself apart. It is also the instance where it can acquire a technoscientific "trans-culture" and a "meta-culture" of multi-culturalism, sensitive to others and to diversity. There would be much to say as to the methodology appropriate to Ethical committees (Hottois and Missa 2001, Hottois 2004b). However, this is not the subject of this article. The complex institutional phenomenon of Ethics committees is not a panacea. I do however see in it an important space where the subject of technoscience, otherwise largely unconscious and structurally conflictual, can begin to develop a moral conscience.

2. *The objects of technoscience require a methodological materialism.* Technoscience and *technoscience studies* are exemplars of an approach, which is more and more completely *materialist and operational.*[31] It is not a doctrinal, metaphysical materialism, which would be based on an unequivocal definition of matter and which would undertake an ontological reduction of all reality to this material basis. Rather, it is a *methodological* materialism whose aim is not to represent, but to act and to operate, to produce and to transform. "Everything is material" in this sense means that everything is the result of an operation, and can be operated upon in turn; and that this open operability is if not totally without rules, at any rate without pre-defined meta-rules. It is empirical and not subject to the metaphysical or transcendental constraints forged a priori by theologians and idealist philosophers.

This methodological materialism is reflexive as it also concerns human beings (there is a whole tendency in *technoscience studies* which obliterates the difference between human and non-human). But it is not systematically anti-spiritualist: that would itself be a metaphysical position. It only takes seriously the empirical evidence of the absence of spirits that might exist independently of human brains in communicational interaction. Hence follows the working hypothesis that the extension or the intensification of mind and consciousness are also dependent on the operation of their empirical material conditions. This raises the ethical question, with political, religious and philosophical dimensions, of *anthropo-technics*, i.e. the operational and progressive auto-transcendence of certain fractions of the human species. This question is at the passionate crux of the unconscious and the conscience of the plural subject of technoscience, a subject which is violently divided on this issue. This is the question of mankind, posed for a future (which is already under way) in the course of which the exploration of this question risks being less exclusively symbolic, a question of discourse, hermeneutical interpretation and representation, and increasingly techno-physical, experimental and operational. The horizon is no longer a creationist or ontological order, but an evolution – possibly multiple – that is not yet decided and cannot be anticipated by narration or speculation. Its import is no longer measured by the world (the horizon of the Earth), nor by History (end-of-the world scenarios, be they secular or religious). It is a matter of cosmic spaces and times. This is another reason why humanity – the plural subject of technoscience – should cultivate a consciousness that it has the whole of time before it (unless there is a cosmic accident), and the greatest prudence is in order (Hottois 2002b, 2004a, 2005).

These mind-boggling perspectives are rarely evoked as such in *technoscience studies*, where the horizon is generally socio-political and at close range. They do however respond to the legitimate anxiety expressed by Jean-Pierre Séris when he fears that, with technoscience, we are exclusively encouraging a short-sighted utilitarian materialism for a society having lost its sense of the infinite which is still

[31] As *Matrix for Materiality*, the subtitle of the aforementioned collective work edited by Idhe and Selinger (2003) suggests.

present in the idea of Science (Séris 1994). But Séris would probably hesitate before an invitation to introduce the infinite desire of science, religion and philosophy into the progressive technical operation of reality and the human species.

My two concluding points – moral conscience and material effectiveness – are complementary: indeed, the operational auto-transcendence of the plural subject of technoscience should be accompanied by a highly careful conscience. Without reintegrating the meta-viewpoints of over-arching wisdom so dear to theological, metaphysical and transcendental consciences, the plural subject of technoscience would often gain by taking a step back before entering, in conscience, into ethical discussion. To encourage it to do so remains a function – humbly modest – of philosophy.

8.8 Epilogue

The first version of this essay was written in 2006[32]; it has been reproduced here without any substantial modification. Since then, I have continued my research on the origin and the evolution of the significance and uses of "technoscience," and these subsequent findings have not altered these earlier claims.

A first, fundamental finding is that the term "technoscience" was introduced independently in American and in French during the 1970s. Although the perspectives were different, this synchronicity is interesting in itself. Concerning the American version, the most developed expression is due to W. Henry Lambright, in his book *Governing Science and Technology* (1976). His was the approach of a political analyst, illuminating in his own domain, but lacking any philosophical dimension. He was mainly concerned with the best possible management of R&D, essential for the military capacity, the economic power, national prestige and the well-being of civil society, whether it be a question of full employment or urbanism and transports… This approach of a political analyst does not question the classical philosophical and epistemological frameworks which separate "science" and "technology." The expression "technoscience" as used here is a factual amalgam, which reflects the situations, the circumstances, the contexts in which scientific research and technologies are associated, to varying degrees. As the case may be, one can go back without any problem to the separation: on one side science, on the other the technological applications. Lambright hardly ever talked about technoscience or technosciences as such: he only considered "technoscience agencies" (such as the NASA, the NSF, the NIH, etc.): the term "technoscience" and its meaning were attached to research institutions or agencies. Technoscience had no autonomy, conceptual or other. It could neither be praised nor criticized. The existence of "« technoscience agencies »," left untouched the traditional distinction between science

[32] It was initially published in French in Jean-Yves Goffi, Ed. (2006).

and technology; both were ultimately in the service of society and the betterment of the human condition, that they do not put into question.[33]

As for my own introduction of the *French* term (it so happens that the French and English words are the same, which tends to obscure the independence of the neologisms), I have given a very detailed account of the circumstances, insisting on the context of the philosophy of the 1970s, and making explicit the imaginary matrix (science-fictional) of the notion of technoscience, in *Généalogies philosophique, politique et imaginaire de la technoscience* (2013). The Science-fiction literature of the period expressed a deep understanding and image of "contemporary science" that was, in some ways, deeper than the ones provided by philosophers of science. Here is the very first sentence in which the term "technoscience" appeared in my works: "The hypothesis (which my thesis defends) is this: what is barred from contemporary philosophy is techno-science, the cosmic confrontation deprived of authentic light which occurs there, the cosmos with trans-human possibilities." (Hottois 1976) At the time, I also used the terms "post-human, ab-human." Although these terms do not appear in the above 2006 essay, the implicit links between the concept of "technoscience" and the trans/post-humanist movements (Hottois 2014, Hottois et al. 2015, Hottois 2017) remain in what is said about "*anthropo-technology,* i.e. the operational and progressive auto-transcendence of certain fractions of the human species." These links were originally constitutive of the concept of "technoscience" as I first imagined it, now 40 years ago.

[33] Lambright was Professor of "Public Administration and International Affairs, and Political Science" at the Maxwell School of Syracuse University (NY). In the course of the 1970s previous to the substantial use of the term by Lambright (1976), I have managed to find several publications (Caldwell 1970; Erber 1970; Rosenthal 1973; Micklin 1973) containing occurrences of the noun "technoscience" (a dozen) in works on social, political or environmental sciences that are more or less politically committed. "Technoscience" appears as an autonomous noun designating a Western reality which is the object of anxiety and criticism. The term either evokes a bunch of environmental concerns (Caldwell, Rosenthal, Micklin), or a bureaucratic concern for the appropriate management of science and technology, or urban planning (Erber). It is not impossible that one or other of these authors used the term at the very end of the 1960s. But to recap, these previous uses of "technoscience" and "technoscientific" were aimed at "coloring" a discourse with a number of suggestive connotations, rather than setting forth a new – and yet to be thought – *concept.* Let us also point out the first occurrence in Danish of a term – "Teknovitenskap" – that was later translated into English as "techno-science," in Edgar N. Schieldrop (1956), "*På skilleveien i dette angstens og håpets århundre*" ("A Century of Fear and Hope at the Crossroads"): a speech pronounced at the Danske Ingeniørforenings on the occasion of the 70th anniversary of Niels Bohr, and translated in *Mechanical Engineering* in 1959. Here is the translated context of this occurrence: "At this critical stage we are bound to ask if the human race, with the vast power techno-science has placed in its hands, really understands how watchful it must be if the world is not plunged into a disaster surpassing all our nightmares." It is an interesting hapax that remained at the time without influence as far as we know.

References

Aronowitz, S., Martinsons, B., & Menser, M. (Eds.). (1996). *Technoscience and cyberculture*. New York/London: Routledge.

Bachelard, G. (1934). *Le nouvel esprit scientifique*. Paris: Alcan. English edition; Bachelard, G. (1985). *The new scientific spirit* (A. Goldhammer, Trans.). Boston, MA: Beacon Press.

Bacon, F. (1620). *De Novum Organum Scientiarum*.

Bacon, F. (1627). *New Atlantis. A work unfinished.*

Breton, P., Rieu, A.-M., & Tinland, F. (1990). *La techno-science en question*. Seyssel: Champ Vallon.

Caldwell, L. K. (1970). *Environment: A challenge for modern society*. Garden City: Natural History Press.

Chirollet, J.-C. (1994). *Esthétique et technoscience*. Liège: Mardaga.

d'Espagnat, B. (1994). *Le réel voilé - Analyse des concepts quantiques*, Paris: Fayard. English edition; d'Espagnat, B. (2003). *Veiled reality: An analysis of quantum mechanical concepts*. Boulder: Westview Press.

Echeverría, J. (2003). *La revolución tecnocientífica*. Madrid: Fondo de Cultura Económica de España.

Erber, E. (1970). *Urban planning in transition*. New York: Grossman.

Goffi, J.-Y. (2006). *Regards sur les technosciences*. Paris: Vrin

Haraway, D. (1997). *Modest_Witness@Second_Millennium. FemaleMan_Meets_ OncoMouse™: Feminism and technoscience*. New York: Routledge.

Heisenberg, W. (1958). The representation of nature in contemporary physics (1955). *Daedalus, 87*(3), 95–108.

Hottois, G. (1976). Essai sur les causes, *les formes et les limites de l'inflation du langage dans la philosophie contemporaine*. PhD thesis, Université Libre de Bruxelles, Bruxelles.

Hottois, G. (1978). Ethique et techno-science. *La pensée et les hommes, 22*, 111–116.

Hottois, G. (1979). *L'inflation du langage dans la philosophie contemporaine*. Bruxelles: Presses de l'Université de Bruxelles.

Hottois, G. (1984a). *Le signe et la technique. La philosophie à l'épreuve de la technique*. Paris: Aubier.

Hottois, G. (1984b). *Pour une éthique dans un univers technicien*. Bruxelles: Presses de l'Université de Bruxelles.

Hottois, G. (1996). *Entre symboles et technosciences. Un itinéraire philosophique*. Seyssel: Champ Vallon.

Hottois, G. (2002a). *Species technica*. Paris: Vrin.

Hottois, G. (2002b). *Technoscience et sagesse?* Nantes: Pleins Feux.

Hottois, G. (2004a). *Philosophie des sciences, philosophie des techniques*. Paris: Odile Jacob.

Hottois, G. (2004b). *Qu'est-ce que la bioéthique?* Paris: Vrin.

Hottois, G. (2005). *La science, entre valeurs modernes et postmodernité*. Paris: Vrin.

Hottois, G. (2006). La technoscience : de l'origine du mot à ses usages actuels. In J.-Y. Goffi (Ed.), *Regards sur les technosciences* (pp. 23–38). Paris: Vrin.

Hottois, G. (2013). *Généalogies philosophique, politique et imaginaire de la technoscience*. Paris: Vrin.

Hottois, G. (2014). *Le transhumanisme est-il un humanisme ?* Bruxelles: Editions de l'Académie Royale de Belgique.

Hottois, G. (2017). *Philosophie et idéologies trans/posthumanistes*. Paris: Vrin.

Hottois, G., & Missa, J.-N. (Eds.). (2001). *Nouvelle encyclopédie de bioéthique: médecine, environnement, biotechnologie*. Bruxelles: Éditions De Boeck Université.

Hottois, G., Missa, J.-N., & Perbal, L. (Eds.). (2015). *Encyclopédie du transhumanisme et du posthumanisme*. Paris: Vrin.

Ihde, D. (1979). *Technics and praxis*. Boston/Dordrecht/London: Reidel.

Ihde, D. (1991). *Instrumental realism: The interface between philosophy of science and philosophy of technology*. Bloomington: Indiana University Press.

Ihde, D. (1993). *Philosophy of technology*. New York: Paragon House.

Ihde, D. (1995). *Postphenomenology: Essays in the postmodern context*. Northwestern: Northwestern University Press.

Ihde, D., & Selinger, E. (Eds.). (2003). *Chasing technoscience. Matrix for materiality*. Bloomington/Indianapolis: Indiana University Press.

Lambright, W. H. (1976). *Governing science and technology*. Oxford: Oxford University Press.

Latour, B. (1987). *Science in action. How to follow scientists and Engineers through society*. Harvard: Harvard University Press. French edition; Latour, B. (1995). *La science en action. Introduction à la sociologie des sciences (1989)*. Paris: La Découverte.

Latour, B. (1991). *Nous n'avons jamais été modernes*. Paris: La Découverte. English edition; Latour, B. (1993). *We have never been mod*ern (C. Porter, Trans.). Cambridge, MA, Harvard University Press.

Latour, B. (1993). *Aramis ou l'amour des techniques*. Paris: La Découverte. English edition; Latour, B. (1996). *Aramis or the love of technology* (C. Porter, Trans.). Cambridge, MA, Harvard University Press.

Latour, B. (1999). *Pandora's hope. Essays on the reality of science studies*. Cambridge, MA: Harvard University Press.

Law, J. (2002). *Aircraft stories. Decentering the object of technoscience*. Durham: Duke University Press.

Lecourt, D. (1996). *Prométhée, Faust, Frankenstein: Les Fondements imaginaires de l'éthique*. Paris: Les Empêcheurs de penser en rond.

Lyotard, J.-F. (1979). *La condition postmoderne: rapport sur le savoir*. Paris: Éditions de Minuit. English edition; Lyotard, J.-F. (1984). *The Postmodern Condition: A Report on Knowledge* (G. Bennington & B. Massumi, Trans.). Minneapolis: University of Minnesota Press.

Lyotard, J.-F. (1986). *Le Postmoderne expliqué aux enfants: Correspondance, 1982–1985*. Paris: Galilée. English edition: Lyotard, J.-F. (1993). *The Postmodern Explained: Correspondence, 1982–1985* (D. Barry, Trans.). Minneapolis, MI: University of Minnesota Press.

Lyotard, J.-F. (1988). *L'Inhumain: Causeries sur le temps*. Paris: Galilée. English edition: Lyotard, J.-F. (1991). *The Inhuman: Reflections on Time* (G. Bennington & R. Bowlby, Trans.). Stanford: Stanford University Press.

Lyotard, J.-F. (1993). *Moralités postmodernes*. Paris: Galilée. English edition: Lyotard, J.-F. (1997). *Postmodern Fables* (G. Van Den Abbeele, Trans.). Minneapolis: University of Minnesota Press.

Micklin, M. (1973). *Population, environment, and social organization: Current issues in human ecology*. Hinsdale: Dryden Press.

Prades, J. (Ed.). (1992). *La technoscience*. Paris: L'Harmattan.

Rosenthal, A. H. (1973). *Public science policy and administration*. Albuquerque: University of New Mexico Press.

Sassauer, R. (1995). *Cultural collisions. Postmodern technoscience*. New York/London: Routledge.

Scharff, R. C., & Dusek, V. (Eds.). (2003). *Philosophy of technology: The technological condition – An anthology*. Malden/Oxford/Melbourne: Wiley-Blackwell.

Schieldrop, E. N., (1956). På skilleveien i dette angstens og håpets århundre. *Teknisk ukeblad, 42*, 17–18. (E. N. Schieldrop, Trans.). (1959). A century of fear and hope at the crossroads, *Mechanical Engineering, 81*(3), 44–45.

Séris, J. P. (1994). *La technique*. Paris: Presses Universitaires de France.

Snow, C. P. (1961). *The two cultures and the scientific revolution: The Rede Lecture, 1959*. New York: Cambridge University Press.

Tiles, M., & Oberdiek, H. (1995). *Living in a technological culture: Human tools and human values*. New York: Routledge.

Chapter 9
Cybernetics Is an Antihumanism. Technoscience and the Rebellion Against the Human Condition

Jean-Pierre Dupuy

Abstract There is no science that does not rest on a metaphysics, though typically it remains concealed. It is the responsibility of the philosopher to uncover this metaphysics, and then to subject it to criticism. What I have tried to show is that cybernetics, far from being the apotheosis of Cartesian humanism, as Heidegger supposed, actually represented a crucial moment in its demystification, and indeed in its deconstruction.

Keywords Antihumanism · Converging technologies · Cognitive science · Cybernetics · Descartes · Heidegger · Nanotechnology · Sartre · Synthetic biology · Technoscience

I chose the topic of my contribution after I discovered, first with amazement, then with wonder, N. Katherine Hayles's beautiful book (1999), *How we became posthuman. Virtual Bodies in Cybernetics, Literature, and Informatics*. Amazement because she and I worked on the same fairly confidential corpus, in particular the proceedings of the Macy conferences, which were the birthplace of cybernetics and, I have claimed, of cognitive science, we celebrate the same heroes, in particular Warren McCulloch, Heinz von Foerster and Francisco Varela, and, in spite of these

---

This text has been first delivered as a paper for the "Transhumanism and the meanings of progress" workshop, Arizona State University, Tempe, April 24–25, 2008, and originally published in *The Global Spiral*, June 5, 2008

J.-P. Dupuy (✉)
Ecole Polytechnique, Stanford University, Paris, France
e-mail: jpdupuy@stanford.edu

S. Loeve et al. (eds.), *French Philosophy of Technology*, Philosophy of Engineering and Technology 29,
https://doi.org/10.1007/978-3-319-89518-5_9

shared interests and passions, we apparently never heard of each other. She and I live and work worlds and languages apart. The world is still far from being a close-knit village. Wonder at realizing how from the same corpus we could arrive at interpretations that, although compatible or even complementary, are so richly diverse or even divergent.

My book on the Macy conferences and the origins of cybernetics and cognitive science, *Sur l'origine des sciences cognitives*, was first published in French (Dupuy 1985); a second and completely revised edition followed in 1994 (Dupuy 1994); the first English-language edition, an extensively revised and amplified version of the latter, came out in 2000 (Dupuy 2000). It is with shame that I acknowledge that during all this time, I never came across Ms. Hayles' work, published in book form in 1999. It is with great sadness that I realize that there is no longer any way that I could ask my two great friends, Heinz von Foerster and Francisco Varela, two men of communication, why they never put us in touch. The Chilean neurophilosopher Francisco Varela was the cofounder of the theory of autopoietic systems (Varela and Maturana 1980); he chose to come to France and work in my research institution after he was expelled from his country. Heinz von Foerster, a Viennese Jewish immigrant to the United States, after serving as secretary to the Macy Conferences, went on to found what was to be called second-order cybernetics. Francisco and Heinz play important roles in the story that I tell in my book. The former passed away in 2000; the latter in 2002. I miss them both terribly.

My book seeks to disabuse readers of a number of ideas that I consider mistaken. Cybernetics calls to mind a series of familiar images that turn out on closer inspection to be highly doubtful. As the etymology of the word suggests, cybernetics is meant to signify control, mastery, governance. In short, the philosophical project associated with Descartes, who assigned mankind the mission of exercising dominion over the world, and over mankind itself. Within the cybernetics movement, this view was championed by Norbert Wiener—unsurprisingly, perhaps, since it was Wiener who gave it its name. But this gives only a very partial, if not superficial idea of what cybernetics was about, notwithstanding that even a philosopher of such penetrating insight as Heidegger was taken in by it.

In my work I have relied on the notion, introduced by French philosopher of science Émile Meyerson (1921) and adopted by Karl Popper (1976), of a *metaphysical research program*, which is to say a set of presuppositions about the structure of the world that are neither testable nor empirically falsifiable, but without which no science would be possible. For there is no science that does not rest on a metaphysics, though typically it remains concealed. It is the responsibility of the philosopher to uncover this metaphysics, and then to subject it to criticism. What I have tried to show is that cybernetics, far from being the apotheosis of Cartesian humanism, as Heidegger supposed, actually represented a crucial moment in its demystification, and indeed in its deconstruction. To borrow a term that has been applied to the structuralist movement in the human sciences, cybernetics constituted a decisive step in the rise of *antihumanism*. Consider, for example, the way in which cybernetics conceived the relationship between man and machine. The philosophers of consciousness were not alone in being caught up in the trap set by a question such as "Will it be possible one day to design a machine that thinks?' The cybernetician's answer, rather in the spirit of Molière, was: "Madame, you pride yourself so on thinking. And yet, you are only a machine!" The aim of cognitive science always was—and still is today—the mechanization of the mind, not the humanization of the machine.

"Continental" political philosophy has yet to acknowledge the notion of posthumanism. On the other hand, the notion of antihumanism has been debated for at least four decades. My contribution will bear on the latter only. My hope is that it will enable us to explore the possible connections between the two notions and, beyond, perhaps, bridge the gap between two cultural worlds so far apart.

9.1 Heidegger's Error

I will start with a classic question: can the idea that we have of the human person, which is to say of ourselves, survive the forward march of scientific discovery? It is a commonplace that from Copernicus to molecular biology, and from Marx to Freud along the way, we have had steadily to abandon our proud view of ourselves as occupying a special place in the universe, and to admit that we are at the mercy of determinisms that leave little room for what we have been accustomed to consider our freedom and our reason. Is not cognitive science now in the process of completing this process of disillusionment and demystification by showing us that just where we believe we sense the workings of a mind, there is only the firing of neural networks, no different in principle than an ordinary electric circuit? The task in which I have joined with many others, faced with reductive interpretations of scientific advance of this sort, has been to defend the values proper to the human person, or, to put it more bluntly, to defend humanism against the excesses of science and technology.

Heidegger completely inverted this way of posing the problem. For him it was no longer a question of defending humanism but rather of indicting it. As for science and technology, or rather "technoscience" (an expression meant to signify that science is subordinated to the practical ambition of achieving mastery over the world through technology), far from threatening human values, they are on Heidegger's view the most striking manifestation of them. This dual reversal is so remarkable that it deserves to be considered in some detail, even—or above all—in a reflection on the place of cybernetics in the history of ideas, for it is precisely cybernetics that found itself to be the principal object of Heidegger's attack.

In those places where Heideggerian thought has been influential, it became impossible to defend human values against the claims of science. This was particularly true in France, where structuralism—and then poststructuralism—reigned supreme over the intellectual landscape for several decades before taking refuge in the literature departments of American universities. Anchored in the thought of the three great Germanic "masters of suspicion"—Marx, Nietzsche, and Freud—against a common background of Heideggerianism, the human sciences *à la française* made antihumanism their watchword (Ferry and Renaut 1990), loudly celebrating exactly what humanists dread: the death of man. This unfortunate creature, or rather a certain image that man created of himself, was reproached for being "metaphysical." With Heidegger, "metaphysics" acquired a new and quite special sense, opposite to its usual meaning. For positivists ever since Comte, the progress of science had been seen as forcing the retreat of metaphysics; for Heidegger, by contrast, technoscience represented the culmination of metaphysics. And the height of metaphysics was nothing other than cybernetics.

Let us try to unravel this tangled skein. For Heidegger, metaphysics is the search for an ultimate foundation for all reality, for a "primary being" in relation to which all other beings find their place and purpose. Where traditional metaphysics ("onto-theology") had placed God, modern metaphysics substituted man. This is why modern metaphysics is fundamentally humanist, and humanism fundamentally metaphysical. Man is a subject endowed with consciousness and will: his features were described at the dawn of modernity in the philosophy of Descartes and Leibniz. As a conscious being, he is present and transparent to himself; as a willing being, he causes things to happen as he intends. Subjectivity, both as theoretical presence to oneself and as practical mastery over the world, occupies center stage in this scheme—whence the Cartesian promise to make man "master and possessor of nature." In the metaphysical conception of the world, Heidegger holds, everything that exists is a slave to the purposes of man; everything becomes an object of his will, fashionable as a function of his ends and desires. The value of things depends solely on their capacity to help man realize his essence, which is to achieve mastery over being. It thus becomes clear why technoscience, and cybernetics in particular, may be said to represent the completion of metaphysics. To contemplative thought—thought that poses the question of meaning and of Being, understood as the sudden appearance of things, which escapes all attempts at grasping it—Heidegger opposes "calculating" thought. This latter type is characteristic of all forms of planning that seek to attain ends by taking circumstances into account. Technoscience, insofar as it constructs mathematical models to better establish its mastery over the causal organization of the world, knows only calculating thought. Cybernetics is precisely that which calculates—computes—in order to govern, in the nautical sense (Wiener coined the term from the Greek *xvbepvntns*, meaning "steersman"): it is indeed the height of metaphysics.

Heidegger anticipated the objection that would be brought against him: "Because we are speaking against humanism people fear a defense of the inhuman and a glorification of barbaric brutality. For what is more *logical* than that for somebody who negates humanism nothing remains but the affirmation of inhumanity?" (Heidegger 1947: 225) Heidegger defended himself by attacking. Barbarism is not to be found where one usually looks for it. The true barbarians are the ones who are supposed to be humanists, who, in the name of the dignity that man accords himself, leave behind them a world devastated by technology, a desert in which no one can truly be said to dwell.

Let us for the sake of argument grant the justice of Heidegger's position. At once an additional enigma presents itself. If for him cybernetics really represented the apotheosis of metaphysical humanism, how are we to explain the fact that the human sciences in France, whose postwar development I have just said can be understood only against the background of Heidegger's philosophy, availed themselves of the conceptual toolkit of cybernetics in order to deconstruct the metaphysics of subjectivity? How is it that these sciences, in their utter determination to put man as subject to death, each seeking to outdo the other's radicalism, should have found in cybernetics the weapons for their assaults?

From the beginning of the 1950s—which is to say, from the end of the first cybernetics—through the 1960s and 1970s, when the second cybernetics was investigating theories of self-organization and cognitivism was on the rise, the enterprise of mechanizing the human world underwent a parallel development on each side of the Atlantic. This common destiny was rarely noticed, perhaps because the thought of any similarity seemed almost absurd: whereas cognitive science claimed to be the avant-garde of modern science, structuralism—followed by poststructuralism—covered itself in a pretentious and often incomprehensible philosophical jargon. What is more, it was too tempting to accuse French deconstructionists of a fascination with mathematical concepts and models that they hardly understood. But even if this way of looking at the matter is not entirely unjustified, it only scratches the surface. There were very good reasons, in fact, why the deconstruction of metaphysical humanism found in cybernetics an ally of the first order.

At the beginning of the 1940s, a philosopher of consciousness such as Sartre could write: "The inhuman is merely … the mechanical." (Sartre 1943) Structuralists hastened to adopt this definition as their own, while reversing the value assigned to its terms. Doing Heidegger one better, they made a great show of championing the inhuman—which is to say the mechanical.[1] Cybernetics, as it happened, was ready to hand, having come along at just the right moment to demystify the voluntary and conscious subject. The will? All its manifestations could apparently be simulated, and therefore duplicated, by a simple negative feedback mechanism. Consciousness? The "Cybernetics Group" (Heim 1991) had examined the Freudian unconscious, whose existence was defended by one of its members, Lawrence Kubie, and found it chimerical. If Kubie often found himself the butt of his colleagues' jokes, it was not because he was thought to be an enemy of human dignity. It was rather because the postulation of a hidden entity, located in the substructure of a purportedly conscious subject, manifesting itself only through symptoms while yet being endowed with the essential attributes of the subject (intentionality, desires, beliefs, presence to oneself, and so on), seemed to the cyberneticians nothing more than a poor conjuring trick aimed at keeping the structure of subjectivity intact.

It is remarkable that a few years later the French psychoanalyst Jacques Lacan, along with the anthropologist Claude Lévi-Strauss and the Marxist philosopher Louis Althusser one of the founders of structuralism, should have adopted the same critical attitude toward Freud as cybernetics. The father of psychoanalysis had been led to postulate an improbable "death wish"—"beyond the pleasure principle," as he put it—as if the subject actually desired the very thing that made him suffer, by

[1] "To render philosophy inhuman" – thus the task Jean-François Lyotard set himself in 1984 (Lyotard 1984).

voluntarily and repeatedly placing himself in situations from which he could only emerge battered and hurt. This compulsion (*Zwang*) to repeat failure Freud called *Wiederholungszwang*, an expression translated by Lacan (1956) as "automatisme de répétition," which is to say the *automatism* of repetition. In so doing he replaced the supposed unconscious death wish with the senseless functioning of a machine, the unconscious henceforth being identified with a cybernetic automaton. Interestingly enough, it is also rendered as "*death drive*" in current English. The alliance of psychoanalysis and cybernetics was neither anecdotal nor fortuitous: it corresponded to a radicalization of the critique of metaphysical humanism.[2]

There was a deeper reason for the encounter between the French *sciences de l'homme* and cybernetics, however. What structuralism sought to conceive—in the anthropology of Lévi-Strauss, for example, and particularly in his study of systems of exchange in traditional societies—was a *subjectless cognition*, indeed cognition without mental content. Whence the project of making "symbolic thought" a mechanism peculiar not to individual brains but to "unconscious" linguistic structures that automatically operate behind the back, as it were, of unfortunate human "subjects," who are no more than a sort of afterthought. "It thinks" was destined to take the place once and for all of the Cartesian cogito. Now cognition without a subject was exactly the unlikely configuration that cybernetics seemed to have succeeded in conceiving. Here again, the encounter between cybernetics and structuralism was in no way accidental. It grew out of a new intellectual necessity whose sudden emergence appears in retrospect as an exceptional moment in the history of ideas.

9.2 The Self-Mechanized Mind

It is time to come back to our enigma, which now may be formulated as a paradox. Was cybernetics the height of metaphysical humanism, as Heidegger maintained, or was it the height of its deconstruction, as certain of Heidegger's followers believe? To this question I believe it is necessary to reply that cybernetics was both things at once, and that this is what made it not only the root of cognitive science, which finds itself faced with the same paradox, but also a turning point in the history of human conceptions of humanity. The title I have given to this section—the self-mechanized mind—appears to have the form of a self-referential statement, not unlike those strange loops the cyberneticians were so crazy about, especially the cyberneticians of the second phase. But this is only an appearance: the mind that carries out the mechanization and the one that is the object of it are two distinct (albeit closely related) entities, like the two ends of a seesaw, the one rising ever higher in the heavens of metaphysical humanism as the other descends further into the depths of its deconstruction. In mechanizing the mind, in treating it as an artifact, the mind presumes to exercise power over this artifact to a degree that no psychology claiming to be scientific has ever dreamed of attaining. The mind can now hope not only

[2] See also Lyotard's cybernetic reading of Freud (Sebbah, Chap. 10 this volume).

to manipulate this mechanized version of itself at will, but even to reproduce and manufacture it in accordance with its own wishes and intentions. Accordingly, the technologies of the mind, present and future, open up a vast continent upon which man now has to impose norms if he wishes to give them meaning and purpose. The human subject will therefore need to have recourse to a supplementary endowment of will and conscience in order to determine, not what he can do, but what he ought to do—or, rather, what he ought not to do. These new technologies will require a whole ethics to be elaborated, an ethics not less demanding than the one that is slowly being devised today in order to control the rapid development and unforeseen consequences of new biotechnologies. But to speak of ethics, conscience, the will—is this not to speak of the triumph of the subject?

The connection between the mechanization of life and the mechanization of the mind is plain. Even if the Cybernetics Group snubbed biology, to the great displeasure of John von Neumann, it was of course a cybernetic metaphor that enabled molecular biology to formulate its central dogma: the genome operates like a computer program. This metaphor is surely not less false than the analogous metaphor that structures the cognitivist paradigm. The theory of biological self-organization, first opposed to the cybernetic paradigm during the Macy Conferences before later being adopted by the second cybernetics as its principal model, furnished then—and still furnishes today—decisive arguments against the legitimacy of identifying DNA with a "genetic program." Nonetheless—and this is the crucial point—even though this identification is profoundly illegitimate from both a scientific and a philosophical point of view, its technological consequences have been considerable. Today, as a result, man may be inclined to believe that he is the master of his own genome. Never, one is tempted to say, has he been so near to realizing the Cartesian promise: he has become—or is close to becoming—the master and possessor of all of nature, up to and including himself.

Must we then salute this as yet another masterpiece of metaphysical humanism? It seems at first altogether astonishing, though after a moment's reflection perfectly comprehensible, that a German philosopher following in the tradition of Nietzsche and Heidegger, Peter Sloterdijk, should have recently come forward, determined to take issue with the liberal humanism of his country's philosophical establishment, and boldly affirmed that the new biotechnologies sound the death knell for the era of humanism. Unleashing a debate the like of which is hardly imaginable in any other country, this philosopher ventured to assert: "The domestication of man by man is the great unimagined prospect in the face of which humanism has looked the other way from antiquity until the present day." And to prophesy:

> It suffices to clearly understand that the next long periods of history will be periods of choice as far as the [human] species is concerned. Then it will be seen if humanity, or at least its cultural elites, will succeed in establishing effective procedures for self-domestication. It will be necessary, in the future, to forthrightly address the issue and formulate a code governing anthropological technologies. Such a code would modify, a

> posteriori, the meaning of classical humanism, for it would show that *humanitas* consists not only in the friendship of man with man, but that it also implies … , in increasingly obvious ways, that man represents the supreme power for man.[3]

But why should this "superhuman" power of man over himself be seen, in Nietzschean fashion, as representing the death of humanism rather than its apotheosis? For man to be able, as subject, to exercise a power of this sort over himself, it is first necessary that he be reduced to the rank of an object, able to be reshaped to suit any purpose. No raising up can occur without a concomitant lowering, and vice versa.

Let us come back to cybernetics and, beyond that, to cognitive science. We need to consider more closely the paradox that an enterprise that sets itself the task of naturalizing the mind should have as its spearhead a discipline that calls itself artificial intelligence. To be sure, the desired naturalization proceeds via mechanization. Nothing about this is inconsistent with a conception of the world that treats nature as an immense computational machine. Within this world man is just another machine—no surprise there. But in the name of what, or of whom, will man, thus artificialized, exercise his increased power over himself? In the name of this very blind mechanism with which mankind is identified? In the name of a meaning that humans claim is mere appearance or phenomenon? Man's will and capacity for choice are now left dangling over the abyss. The attempt to restore mind to the natural world that gave birth to it ends up exiling the mind from the world and from nature. This paradox is typical of what the French sociologist Louis Dumont, in his magisterial study of the genesis of modern individualism, called

> the model of modern artificialism in general, the systematic application of an extrinsic, imposed value to the things of the world. Not a value drawn from our belonging to the world, from its harmony and our harmony with it, but a value rooted in our heterogeneity in relation to it: the identification of our will with the will of God (Descartes: man makes himself master and possessor of nature). The will thus applied to the world, the end sought, the motive and the profound impulse of the will are [all] foreign. In other words, they are extra-worldly. Extra-worldliness is now concentrated in the individual will. (Dumont 1986: 56)

The paradox of the naturalization of the mind attempted by cybernetics, and today by cognitive science, then, is that the mind has been raised up as a demigod in relation to itself.

Many of the criticisms brought against the materialism of cognitive science from the point of view either of a philosophy of consciousness or a defense of humanism miss this paradox. Concentrating their (often justified) attacks on the weaknesses and naiveté of such a mechanist materialism, they fail to see that it invalidates itself by placing the human subject outside of the very world to which he is said to belong. The recent interest shown by cognitive science in what it regards as the "mystery" of consciousness seems bound to accentuate this blindness.

[3] Peter Sloterdijk, "On the Rules of the Human Fleet," a paper delivered at a conference on Heidegger at Elmau Castle, Upper Bavaria, on July 17, 1999, and presented as a reply to Heidegger's "Letter on Humanism." (Sloterdijk 2009).

9.3 The Nanotechnological Dream

I want now to broach not so much the intellectual evolution of cognitive science itself as its embodiment by new technologies, or, as one should rather say, its instantiation by ideas for new technologies. For the moment at least these technologies exist only as projects, indeed in some cases only as dreams. But no matter that many such dreams will acquire physical reality sooner or later, the simple fact that they already exist in people's minds affects how we see the world and how we see ourselves.

Since my book was first published, I have thought a great deal about the philosophical foundations of what is called the NBIC Convergence—the convergence of nanotechnology, biotechnology, information technology, and cognitive science—and about the ethical implications of this development (Dupuy and Grinbaum 2004, Dupuy 2007a, b, 2008). Here I have found many of the same tensions, contradictions, paradoxes, and confusions that I discerned first within cybernetics, and then within cognitive science. But now the potential consequences are far more serious, because we are not dealing with a theoretical matter, a certain view of the world, but with an entire program for acting upon nature and mankind.

In searching for the underlying metaphysics of this program, I did not have far to look. One of the first reports of the National Science Foundation devoted to the subject, entitled "Converging Technologies for Improving Human Performance," (Roco and Bainbridge Eds. 2002: 13) summarizes the credo of the movement in a sort of haiku:

> If the Cognitive Scientists can think it,
> The Nano people can build it,
> The Bio people can implement it, and
> The IT people can monitor and control it.

Note that cognitive science plays the leading role in this division of labor, that of thinker—not an insignificant detail, for it shows that the metaphysics of NBIC Convergence is embedded in the work of cognitive scientists. It comes as no surprise, then, that the contradictions inherent in cognitive science should be found at the heart of the metaphysics itself.

One of the main themes of my book is the confrontation between Norbert Wiener and John von Neumann, Wiener embodying the ideas of control, mastery, and design, von Neumann the ideas of complexity and self-organization. Cybernetics never succeeded in resolving the tension, indeed the contradiction, between these two perspectives; more specifically, it never managed to give a satisfactory answer to the problems involved in realizing its ambition of *designing* an autonomous, self-organizing machine. Nanotechnology—whose wildest dream is to reconstruct the natural world that has been given to us, atom by atom—is caught up in the same contradiction.

The most obvious element of the nanotechnological dream is to substitute for what French biologist François Jacob called *bricolage*, or the tinkering of biological evolution, a paradigm of *design*. Damien Broderick, the Australian cultural theorist and popular science writer, barely manages to conceal his contempt for the world that human beings have inherited when he talks about the likelihood that "nanosystems, designed by human minds, will bypass all this Darwinian wandering, and leap straight to *design success.*" (Broderick 2001: 118) One can hardly fail to note the irony that science, which in America has had to engage in an epic struggle to root out every trace of creationism (including its most recent avatar, "intelligent design") from public education, should now revert to a logic of design in the form of the nanotechnology program—the only difference being that now it is mankind that assumes the role of the demiurge.

Philosophers, faced with the ambition of emerging technologies to supersede nature and life as the engineers of evolution, the designers of biological and natural processes, may suppose that they are dealing with an old idea: Descartes' vision of science as the means by which man may become the master and possessor of nature. Again, however, this is only part of a larger and more complicated picture. As another influential visionary, the American applied physicist Kevin Kelly, revealingly remarked, "It took us a long time to realize that the power of a technology is proportional to its inherent *out-of-controlness*, its inherent ability to surprise and be generative. In fact, unless we can worry about a technology, it is not revolutionary enough."[4] With NanoBioConvergence, a novel conception of engineering has indeed been introduced. The engineer, far from seeking mastery over nature, is now meant to feel that his enterprise will be crowned by success only to the extent that the system component he has created is capable of surprising him. For whoever wishes ultimately to create a self-organizing system—another word for life—is bound to attempt to reproduce its essential property, namely, the ability to make something that is radically new.

In her masterful study of the perils facing mankind, *The Human Condition*, Hannah Arendt (1958) brought out the fundamental paradox of our age: whereas the power of mankind to alter its environment goes on increasing under the stimulus of technological progress, less and less do we find ourselves in a position to control the consequences of our actions. I take the liberty of giving a long quotation here whose pertinence to the subject at hand cannot be exaggerated—keeping in mind, too, that these lines were written 50 years ago:

> To what extent we have begun to *act into nature*, in the literal sense of the word, is perhaps best illustrated by a recent casual remark of a scientist [Wernher von Braun, December 1957] who quite seriously suggested that "*basic research is when I am doing what I don't know what I am doing.*"
>
> This started harmlessly enough with the experiment in which men were no longer content to observe, to register, and contemplate whatever nature was willing to yield in her own appearance, but began to prescribe conditions and to provoke natural processes. What then developed into an ever-increasing skill in *unchaining elemental processes,* which, without the interference of men, would have lain dormant and perhaps never have come to pass, has finally ended in a veritable art of *"making" nature,* that is, of creating "natural" processes

[4] Kevin Kelly, "Will Spiritual Robots Replace Humanity by 2100?", Conference at Stanford University April 2000.

> which without men would never exist and which earthly nature by herself seems incapable of accomplishing ....
>
> [N]atural sciences have become exclusively sciences of process and, in their last stage, *sciences of potentially irreversible, irremediable "processes of no return"*.... (Arendt 1958: 231)

The sorcerer's apprentice myth must therefore be updated: it is neither by error nor terror that mankind will be dispossessed of its own creations, but by *design*—which henceforth is understood to signify not mastery, but non-mastery and out-of-controlness.

9.4 The Rebellion Against the Human Condition

Arendt began the same, decidedly prescient book with the following words:

> The human artifice of the world separates human existence from all mere animal environment, but life itself is outside this artificial world, and through life man remains related to all other living organisms. For some time now, a great many scientific endeavors have been directed toward making life also "artificial," toward cutting the last tie through which even man belongs among the children of nature....
>
> This future man, whom the scientists tell us they will produce in no more than a hundred years, seems to be possessed by *a rebellion against human existence as it has been given,* a free gift from nowhere (secularly speaking), which he wishes to exchange, as it were, for something he has made himself. (Arendt 1958: 2–3)

The nanotechnological dream that began to take shape only a few decades after the utterance of Arendt's prophesy amounts to exactly this revolt against the finiteness, the mortality of the human condition. Human life has an end, for it is promised to death. But not only do the champions of NBIC Convergence oppose themselves to fate, by promising immortality; they quarrel with the very fact that we are born. Their revolt against the given is therefore something subtler and less visible, something still more fundamental, than the revolt against human mortality, for it rejects the notion that we should be brought into the world for no reason.

"Human beings are ashamed to have been born instead of made." Thus the German philosopher Günther Anders (Arendt's first husband and himself a student of Heidegger) characterized the essence of the revolt against the given in his great book, *Die Antiquiertheit des Menschen*—The Antiquatedness (or Obsolescence) of the Human Being. (Anders 1956) One cannot help recalling here another philosophical emotion: the nausea described by Jean-Paul Sartre, that sense of forlornness that takes hold of human beings when they realize that they are not the foundation of their own being. The human condition is ultimately one of freedom; but freedom, being absolute, runs up against the obstacle of its own contingency, for we are free to choose anything except the condition of being *un*free. Discovering that we have been *thrown* into the world without any reason, we feel abandoned. Sartre acknowledged his debt to Günther Anders in expressing this idea by means of a phrase that was to become famous: man is "to freedom condemned." (Sartre 1946)

Freedom, Sartre held, never ceases trying to "nihilate" that which resists it. Mankind will therefore do everything it can to become its own maker; to owe its freedom to no one but itself. But only things are what they are; only things coincide with themselves. Freedom, on the other hand, is a mode of being that never coincides with itself since it ceaselessly projects itself into the future, desiring to be what it is not. Self-coincidence is what freedom aspires to and cannot attain, just as a moth is irresistibly attracted to the flame that will consume it. A *metaphysical self-made man*, were such a being possible, would paradoxically have lost his freedom, and indeed would no longer be a man at all, since freedom necessarily entails the impossibility of transforming itself into a thing. Thus Anders' notion of "Promethean shame" leads inexorably to the obsolescence of man.

Had they lived to see the dawn of the twenty-first century, Sartre and Anders would have found this argument resoundingly confirmed in the shape of the NBIC Convergence—a Promethean project if ever there was one. For the aim of this distinctively metaphysical program is to place mankind in the position of being the divine maker of the world, the demiurge, while at the same time condemning human beings to see themselves as out of date.

At the heart of the nanotechnological dream we therefore encounter a paradox that has been with us since the cybernetic chapter in the philosophical history of cognitive science—an extraordinary paradox arising from the convergence of opposites, whereby the overweening ambition and pride of a certain scientific humanism leads directly to the obsolescence of mankind. It is in the light, or perhaps I should say the shadow, of this paradox that all "ethical" questions touching on the engineering of mankind by mankind must be considered.

9.5 "Playing God" Versus the Blurring of Fundamental Distinctions

In 1964, Norbert Wiener published an odd book with the curious title *God and Golem, Inc.: A Comment on Certain Points where Cybernetics Impinges on Religion.* In it one finds this:

> God is supposed to have made man in His own image, and the propagation of the race may also be interpreted as a function in which one living being makes another in its own image. In our desire to glorify God with respect to man and Man with respect to matter, it is thus natural to assume that machines cannot make other machines in their own image; that this is something associated with a sharp dichotomy of systems into living and non-living; and that it is moreover associated with the other dichotomy between creator and creature. Is this, however, so? (Wiener 1964: 12)

The rest of the book is devoted to mobilizing the resources of cybernetics to show that these are false dichotomies and that, in truth, "machines are very well able to make other machines in their own image." (Wiener 1964: 13)

In recent years, the enterprise of "making life from scratch" has been organized as a formal scientific discipline under the seemingly innocuous name of synthetic biology. In June 2007, the occasion of the first Kavli Futures Symposium at the University of Greenland in Ilulissat, leading researchers from around the world gathered to announce the convergence of work in synthetic biology and nanotechnology and to take stock of the most recent advances in the manufacture of artificial cells. Their call for a global effort to promote "the construction or redesign of biological systems components that do not naturally exist"[5] evoked memories of the statement that was issued in Asilomar, California, in 1975, by the pioneers of biotechnology. Like their predecessors, the founders of synthetic biology insisted not only on the splendid things they were poised to achieve, but also on the dangers that might flow from them. Accordingly, they invited society to prepare itself for the consequences, while laying down rules of ethical conduct for themselves. We know what became of the charter drawn up at Asilomar. A few years later, this attempt by scientists to regulate their own research had fallen to pieces. The dynamics of technological advance and the greed of the marketplace refused to suffer any limitation.

Only a week before the symposium in Ilulissat, a spokesman for ETC Group, an environmental lobby based in Ottawa that has expanded its campaign against genetically modified foods to include emerging nanotechnologies, greeted the announcement of a feat of genetic engineering by the J. Craig Venter Institute (JCVI) in Rockville, Maryland with the memorable words, "For the first time, God has competition." In the event, ETC had misinterpreted the nature of the achievement.[6] But if the Ilulissat Statement is to be believed, the actual synthesis of an organism equipped with an artificial genome ("a free-living organism that can grow and replicate") will become a reality in the next few years. Whatever the actual timetable may turn out to be, the process of fabricating DNA is now better understood with every passing day, and the moment when it will be possible to create an artificial cell using artificial DNA is surely not far off.[7]

The question arises, however, whether such an achievement will really amount to *creating life*. In order to assert this much, one must suppose that between life and non-life there is an absolute distinction, a critical threshold, so that whoever crosses it will have shattered a taboo, like the prophet Jeremiah and like Rabbi Löw of Prague in the Jewish tradition, who dared to create an artificial man, a *golem*. In the view of its promoters and some of its admirers, notably the English physicist and science writer Philip Ball (2007), synthetic biology has succeeded in demonstrating that no threshold of this type exists: between the dust of the earth and the creature that God formed from it, there is no break in continuity that permits us to say

[5]The Ilulissat Statement, Kavli Futures Symposium, "The merging of bio and nano: towards cyborg cells," 11–15 June 2007, Ilulissat, Greenland.

[6]Carole Lartigue's JCVI team had succeeded in "simply" transferring the genome of one bacterium, *Mycoplasma mycoides*, to another, *Mycoplasma capricolum*, and showing that the cells of the recipient organism could function with the new genome. In effect, one species had been converted into another. See Lartigue et al. (2007).

[7]Indeed, this feat has been accomplished by the JCVI team no later than 2010 (Gibson et al. 2010) [note of the editors].

(quoting *Genesis* 2:7) that He breathed into man's nostrils the breath of life. And even in the event that synthetic biology should turn out to be incapable of fabricating an artificial cell, these researchers contend, it would still have had the virtue of depriving the prescientific notion of life of all consistency.

It is here, in the very particular logic that is characteristic of dreams, that nanotechnology plays an important symbolic role. It is typically defined by the scale of the phenomena over which it promises to exert control—a scale that is described in very vague terms, since it extends from a tenth of a nanometer to a tenth of a micron.[8] Nevertheless, over this entire gamut, the essential distinction between life and non-life loses all meaning. It is meaningless to say, for example, that a DNA molecule is a living thing. At the symbolic level, a lack of precision in defining nanotechnology does not matter; what matters is the deliberate and surreptitious attempt to blur a fundamental distinction that until now has enabled human beings to steer a course through the world that was given to them. In the darkness of dreams, there is no difference between a living cat and a dead cat.

Once again, we find that science oscillates between two opposed attitudes: on the one hand, vainglory, an excessive and often indecent pride; and on the other, when it becomes necessary to silence critics, a false humility that consists in denying that one has done anything out of the ordinary, anything that departs from the usual business of normal science. As a philosopher, I am more troubled by the false humility, for in truth it is this, and not the vainglory, that constitutes the height of pride. I am less disturbed by a science that claims to be the equal of God than by a science that drains one of the most essential distinctions known to humanity since the moment it first came into existence of all meaning: the distinction between that which lives and that which does not; or, to speak more bluntly, between life and death.

Let me propose an analogy that is more profound, I believe, than one may at first be inclined to suspect. With the rise of terrorism in recent years, specifically in the form of suicide attacks, violence on a global scale has taken a radically new turn. The first edition of this book belongs to a bygone era, which ended on 11 September 2001. In that world, even the most brutal persecutor expressed his attachment to life, because he killed in order to affirm and assert the primacy of his own way of living. But when the persecutor assumes the role of victim, killing himself in order to maximize the number of people killed around him, all distinctions are blurred, all possibility of reasoned dissuasion is lost, all control of violence is doomed to impotence. If science is allowed, in its turn, to continue along this same path in denying the crucial difference that life introduces in the world, it will, I predict, prove itself to be capable of a violence that is no less horrifying.

Among the most extreme promises of nanotechnology, as we have seen, is immortality (or "indefinite life extension," as it is called). But if there is thought to be no essential difference between the living and the non-living, then there is nothing at all extraordinary about this promise. Yet again, Hannah Arendt very profoundly intuited what such a pact with the devil would involve:

---

[8] A nanometer is one-billionth of a meter.

> The greatest and most appalling danger for human thought is that what we once believed could be wiped out by the discovery of some fact that had hitherto remained unknown; for example, it could be that one day we succeed in making men immortal, and everything we had ever thought concerning death and its profundity would then become simply laughable. Some may think that this is too high a price to pay for the suppression of death. (Arendt 2005: 1)

The ETC Group's premonitory observation—"For the first time, God has competition"—can only strengthen the advocates of the NBIC Convergence in their belief that those who criticize them do so for religious reasons. The same phrases are always used to sum up what is imagined to be the heart of this objection: human beings do not have the right to usurp powers reserved to God alone; *playing God* is forbidden. Often it is added that this taboo is specifically "Judeo-Christian."

Let us put to one side the fact that this allegation wholly misconstrues the teaching of the Talmud as well as that of Christian theology. In conflating them with the ancient Greek conception of the sacred--the gods, jealous of men who have committed the sin of pride, *hubris,* send after them the goddess of vengeance, Nemesis—it forgets that the Bible depicts man as co-creator of the world with God. As the French biophysicist and Talmudic scholar Henri Atlan notes with regard to the literature about the Golem:

> One does not find [in it], at least to begin with, the kind of negative judgment one finds in the Faust legend concerning the knowledge and creative activity of men "in God's image." Quite to the contrary, it is in creative activity that man attains his full humanity, in a perspective of *imitatio Dei* that allows him to be associated with God, in a process of ongoing and perfectible creation. (Atlan 1999: 45)

Within the Christian tradition, authors such as G. K. Chesterton, René Girard, and Ivan Illich see Christianity as the womb of Western modernity, while arguing that modernity has betrayed and corrupted its message. This analysis links up with the idea, due to Max Weber, of the desacralization of the world—its famous "disenchantment"—in regarding Christianity, or at least what modernity made of it, as the main factor in the progressive elimination of all taboos, sacred prohibitions, and other forms of religious limitation.

It fell to science itself to extend and deepen this desacralization, inaugurated by the religions of the Bible, by stripping nature of any prescriptive or normative value. It is utterly futile, then, to accuse science of being at odds with the Judeo-Christian tradition on this point. Kantianism, for its part, conferred philosophical legitimacy on the devaluation of nature by regarding it as devoid of intentions and reasons, inhabited only by causes, and by severing the world of nature from the world of freedom, where the reasons for human action fall under the jurisdiction of moral law.

Where, then, is the ethical problem located, if in fact there is one here? It clearly does not lie in the transgression of this or that taboo sanctioned by nature or the sacred, since the joint evolution of religion and science has done away with any such foundation for the very concept of a moral limitation, and hence of a transgression. But that is precisely the problem. For there is no free and autonomous human society that does not rest on some principle of self-limitation. We will not find the limits we desperately need in the religions of the Book, as though such limits are imposed on us by some transcendental authority, for these religions do nothing more than confront us with our own freedom and responsibility.

The ethical problem weighs more heavily than any specific question dealing, for instance, with the enhancement of a particular cognitive ability by one or another novel technology. But what makes it all the more intractable is that, whereas our capacity to act into the world is increasing without limit, with the consequence that we now find ourselves faced with new and unprecedented responsibilities, the ethical resources at our disposal are diminishing at the same pace. Why should this be? Because the same technological ambition that gives mankind such power to act upon the world also reduces mankind to the status of an object that can be fashioned and shaped at will; the conception of the mind as a machine—the very conception that allows us to imagine the possibility of (re)fabricating ourselves—prevents us from fulfilling these new responsibilities. Hence my profound pessimism.

9.6 Alcmena's Paradox

To pay Heinz von Foerster a final homage, I would like to conclude by recounting a very lovely and moving story he told me, one that has a direct bearing on the arguments developed here.

The story takes place in Vienna toward the end of 1945, and it concerns another Viennese Jew, the psychiatrist Viktor Frankl, whose celebrated book *Man's Search for Meaning* was to be published the following year. Frankl had just returned to Vienna, having miraculously survived the Auschwitz-Birkenau camp; in the meantime he had learned that his wife, his parents, his brother, and other members of his family had all been exterminated. He decided to resume his practice. Here, then, is the story as my friend Heinz told it:

> Concentration camps were the setting for many horrific stories. Imagine then the incredulous delight of a couple who returned to Vienna from two different camps to find each other alive. They were together for about six months, and then the wife died of an illness she had contracted in the camp. At this her husband lost heart completely, and fell into the deepest despair, from which none of his friends could rouse him, not even with the appeal "Imagine if she had died earlier and you had not been reunited!" Finally he was convinced to seek the help of Viktor Frankl, known for his ability to help the victims of the catastrophe.
>
> They met several times, conversed for many hours, and eventually one day Frankl said: "Let us assume God granted me the power to create a woman just like your wife: she would remember all your conversations, she would remember the jokes, she would remember every detail: you could not distinguish this woman from the wife you lost. Would you like me to do it?" The man kept silent for a while, then stood up and said, "No thank you, doctor!" They shook hands; the man left and started a new life.
>
> When I asked him about this astonishing and simple change, Frankl explained, "You see, Heinz, we see ourselves through the eyes of the other. When she died, he became blind. But when he *saw* that he was blind, he could see!"[9]

[9] Translated from the German ("Wir sehen uns mit den Augen des anderen.... Als er aber erkannte, daß er blind war, da konnte er sehen!"). See Heinz von Foerster (1993).

This, at least, is the lesson that von Foerster drew from this story--in typical cybernetic fashion. But I think that another lesson can be drawn from it, one that extends the first. What was it that this man suddenly saw, which he did not see before? The thought experiment that Frankl invited his patient to perform echoes one of the most famous Greek myths, that of Amphitryon. In order to seduce Amphitryon's wife, Alcmena, and to pass a night of love with her, Zeus assumes the form of Amphytryon.

> All through the night, Alcmena loves a man whose qualities are in every particular identical to those of her husband. The self-same description would apply equally to both. All the reasons that Alcmena has for loving Amphitryon are equally reasons for loving Zeus, who has the appearance of Amphitryon, for Zeus and Amphitryon can only be distinguished numerically: they are two rather than one. Yet it is Amphitryon whom Alcmena loves and not the god who has taken on his form. If one wishes to account for the emotion of love by appeal to arguments meant to justify it or to the qualities that lovers attribute to the objects of their love, what rational explanation can be given for that "something" which Amphitryon possesses, but that Zeus does not, and which explains why Alcmena loves only Amphitryon, and not Zeus?. (Canto-Sperber 2004: 41)

When we love somebody, we do not love a list of characteristics, even one that is sufficiently exhaustive to distinguish the person in question from anyone else. The most perfect *simulation* still fails to capture something, and it is this something that is the essence of love—this poor word that says everything and explains nothing. I very much fear that the spontaneous ontology of those who wish to set themselves up as the makers or re-creators of the world know nothing of the beings who inhabit it, only lists of characteristics. If the nanobiotechnological dream were ever to come true, what still today we call love would become incomprehensible.

References

Anders, G. (1956). *Die Antiquiertheit des Menschen: über die Seele im Zeitalter der zweiten industriellen Revolution* (Vol. 1). Munich: Beck.

Arendt, H. (1958). *The human condition*. Chicago: University of Chicago Press.

Arendt, H. (2005). *Journal de pensée (1950–1973), 2 vols*. (S. Courtine-Denamy, Trans.). Paris: Seuil.

Atlan, H. (1999). *Les étincelles du hasard, vol. 1: Connaissance spermatique*. Paris: Seuil.

Ball, P. (2007, June 28). Meanings of 'life': Synthetic biology provides a welcome antidote to chronic vitalism. *Nature, 447*, 1031–1032.

Broderick, D. (2001). *The spike: How our lives are being transformed by rapidly advancing technologies*. New York: Forge.

Canto-Sperber, M. (2004). Amour. In M. Canto-Sperber (Ed.), *Dictionnaire d'éthique et de philosophie morale* (4th ed.). Paris: Presses Universitaires de France.

Dumont, L. (1986). *Essays on individualism: Modern ideology in anthropological perspective*. Chicago: University of Chicago Press.

Dupuy, J.-P. (1985). L'essor de la première cybernétique (1943–1953). *Cahiers du CREA, 7*, 9–139.

Dupuy, J.-P. (1994). *Aux origines des sciences cognitives*. Paris: La Découverte.

Dupuy, J.-P. (2000). *The mechanization of the mind* (M. B. DeBevoise, Trans.). Princeton: Princeton University Press, 2000. Revised edition, Dupuy, J.-P. (2008). *On the origins of cognitive science. The Mechanization of the mind* (M. B. DeBevoise, Trans.). Cambridge, MA: MIT Press.

Dupuy, J.-P. (2007a). Some pitfalls in the philosophical foundations of nanoethics. *Journal of Medicine and Philosophy, 32*(3), 237–261.

Dupuy, J.-P. (2007b). Complexity and uncertainty: A prudential approach to nanotechnology. In J. Weckert et al. (Eds.), *Nanoethics: the ethical and social implications of nanotechnology* (pp. 119–131). Hoboken: Wiley.

Dupuy, J.-P. (2008). The double language of science, and why it is so difficult to have a proper public debate about the nanotechnology program. In F. Allhoff & P. Lin (Eds.), *Nanotechnology & society: Current and emerging ethical issues (foreword)*. Dordrecht: Springer.

Dupuy, J.-P., & Grinbaum, A. (2004). Living with uncertainty: Toward a normative assessment of nanotechnology. *Techné: Research in Philosophy and technology/Hyle: International Journal for Philosophy of Chemistry, 8*(2), 4–25.

Ferry, L. & Renaut, A. (1990). *French Philosophy of the Sixties: An Essay on Antihumanism* (M. H. S. Cattani, Trans.). Amherst, MA: University of Massachusetts Press.

Gibson, D. G., Glass, J. I., Venter, J. C., et al. (2010). Creation of a bacterial cell controlled by a chemically synthesized genome. *Science, 329*(5987), 52–56.

Hayles, K. N. (1999). *How we became Posthuman. Virtual bodies in cybernetics, literature, and informatics*. Chicago: Chicago University Press.

Heidegger, M. (1947). *Brief über den Humanismus*. Bern: A. Francke Verlag. English edition: Heidegger, M. (1977). Letter on Humanism (trans: Capuzzi, F. A). In *Basic writings* (pp. 190–242). New York: Harper and Row.

Heims, S. (1991). *The cybernetics group*. Cambridge, MA: MIT Press.

Lacan, J. (1956). Le séminaire sur la lettre volée. *La Psychanalyse*, 2, 1–44. Reprinted in Lacan, J. (1966). *Écrits* (p. 11–60). Paris: Seuil. English Edition: Lacan, J. (1972). Seminar on the "purloined letter" (J. Mehlam, Trans.). *Yales French Studies, 48*, 39–72.

Lartigue, C., Glass, J. I., Venter, J. C., et al. (2007). Genome transplantation in bacteria: Changing one species to another. *Science, 317*(5838), 632–638.

Lyotard, J.-F. (1984). *Le tombeau de l'intellectuel et autres papiers*. Paris: Galilée.

Meyerson, E. A. (1921). *De l'explication dans les sciences*. Paris: Payot.

Popper, K. R. (1976). *Unended quest: An intellectual autobiography*. London/Fontana: Collins.

Roco, M. C., & Bainbridge, W. S. (Eds.). (2002). *Converging technologies for improving human performance: Nanotechnology, biotechnology, information technology, and cognitive science*. Washington, DC: National Science Foundation.

Sartre, J.-P. (1943). Explications de *l'Etranger. Critiques littéraires*. Reprinted in Sartre, J.-P. (1947). *Situations I*, Paris: Gallimard. English Edition: Sartre, J.-P. (1955). *Literary and philosophical essays* (A. Michelson, Trans.). New York: Criterion Books.

Sartre, J.-P. (1946). *L'Existentialisme est un humanisme*. Paris: Nagel.

Sloterdijk, P. (2009). Rules for the human zoo: A response to the letter on humanism. *Environment and Planning D: Society and Space, 27*(1), 12–28.

Varela, F., & Maturana, H. (1980). *Autopoiesis and cognition: The realization of the living*. Boston: Reidel.

von Foerster, H. (1993). Mit den Augen des anderen. In S. J. Schmidt (Ed.), *Wissen und Gewissen. Versuch einer Brücke* (pp. 350–363). Frankfurt: Suhrkamp Verlag.

Wiener, N. (1964). *God and Golem, Inc.: A comment on certain points where cybernetics impinges on religion*. Cambridge, MA: The MIT Press.

Chapter 10
Lyotard on the (In)Humanity of Technoscience

François-David Sebbah

Abstract Known for being a major representative of the French postphenomenology, Jean-François Lyotard's also provides an important contribution to the development of the notion of "technoscience." This chapter endeavors to establish the originality of Lyotard's understanding of technoscience, one that brings into play notions such as performativity, force, as well as precarity, sensibility, and vulnerability, and further still an original understanding of the "inhuman." It is also a matter of showing that if the notion only appears relatively late in Lyotard's work, one can bring out developmental strata in all of his work that, from a certain point of view, prepare the way for it.

Keywords Efficiency · Inhuman · Material/immaterial · Negentropy · Originary supplement · Performance · Postmodern · Postphenomenology · Survival · Technoscience

This chapter attempts to work out Lyotard's *specific* contribution to the development of the notion of technoscience. It lies, however, within a much larger preoccupation with the developments of the notion of technoscience in the context of the so-called contemporary French "post-phenomenology" running approximately from 1970 and 2000. I am thinking in particular (though not exclusively) of the work of Jacques Derrida and Michel Henry, besides that of Jean-François Lyotard (Sebbah 2010).

I am using the notion of "postphenomenology" here in a different sense than the one popularized by American philosopher Don Ihde. This latter sense designates a phenomenology that has ceased being "foundational" and that describes intentionality as always already contextualized, and in particular as always already

---

This chapter has been written in French for this volume and translated by Daniel J. Palumbo†, The Pennsylvania State University. These pages are dedicated to his memory.

F.-D. Sebbah (✉)
Département de philosophie, University of Paris Nanterre, Nanterre Cedex, France

S. Loeve et al. (eds.), *French Philosophy of Technology*, Philosophy of Engineering and Technology 29,
https://doi.org/10.1007/978-3-319-89518-5_10

technologically mediated. Following this, Peter-Paul Verbeek insists for his part on the mediating role of technology as much for subjectivity as for objectivity, and in particular on the process of co-formation between subjectivity and reality, a process in which technologies are not neutral but *shape* the very relation to reality (which does not preexist this hermeneutic relation). In a way, it is phenomenology after the "empirical turn," phenomenology turned into an empirical enquiry through the systematic taking into account of technical mediation.

In France, "Post-phenomenology" rather designates a theoretical context that presupposes both a productive inheritance of the phenomenological tradition and method (that of Husserl and Heidegger), and a critical twist—even a rupture, in the way that rupture presupposes that with which it essentially breaks—whether in the sense of Derrida's deconstruction, or of Henry's "radical phenomenology," or of a relativization of phenomenology as a "language game" by Lyotard, among others. Comparing "postphenomenology" in the sense I am using here and "postphenomenology" in Don Ihde's sense—especially as this latter "postphenomenology" thematized the notion of "technoscience" in an exemplary way—constitutes an important task. However, it would exceed the scope of this chapter.

Lyotard borrows the term "technoscience" from Gilbert Hottois (1984, 2004; Chap. 8 this volume), but he shifts its meaning. In many ways, "technoscience" for Lyotard gathers together and concatenates certain strata of significations that he elaborated throughout his oeuvre well before he takes up the term; such that the "phenomenon" of technoscience seems in the end to crystallize and to organize many of these traits with a singular intensity. This chapter examines these different strata of significations of "technoscience" in Lyotard's work.

10.1 First Stratum

At the beginning of the 1970s, Lyotard was attentive to the themes of pulsion and desire.[1] Taking off from Freud, he contributed to enlighten the transformation of pulsion into desire (as passage through limitation, establishment of a relation regulated by *the absence* of the object of desire, passage through language, etc.). However, it would not betray Lyotard to say that a large part of his effort consists in emphasizing "the pulsional" as such, and in this way to mark, for example, that which remains an irreducible drive in desire itself. The pulsional, according to Lyotard reading Freud, is of force or of energy. And if energy does not dissipate into pure expenditure, it is because it is regulated by that which is, as a result, an

[1] [The French terms "pulsion" and "pulsionnel," which translate Freud's "*Trieb*," do not have clear English equivalents. Iain Hamilton Grant, in his translation of Lyotard's *Libidinal Economy* (1974), prefers to transliterate the term as "pulsion" and "pulsional," except when, "for reasons of euphony," he chooses to use "drive" (xi). I've more or less followed this convention here, though have tried to use "drive" and its variants when possible. – Tr.]

apparatus [*un appareil*]: energy makes the apparatus function, the apparatus regulates the energy (and is nothing but that).

As we know, Freud thinks the psyche according to the model of an apparatus. Lyotard stresses this point. What counts first of all in Freud's discovery is this: from the individual to the collective, from dream to culture, there is a flow of energy regulated by an apparatus. The model is even more precisely cybernetic and not merely mechanical: this regulation connects an inside to an outside or an environment, and the "dispositif" [*dispositif*] is sensitive to feed-back. That said, Lyotard notes a certain number of decisive differences between a machine (even a cybernetic one) and the psychic apparatus. In the latter case, energy and excitation do not come from outside; the drive pushes from "within." He also notes that the relation between drive and dispositif is not described solely occurring between force and that which it animates and which in turn regulates this force, but that the apparatus belongs to "the figural" [*le figural*] as well: the figural by which the pulsional—always older than every representation, every sign and even every good perceptual form—finds out how to express itself and to mark itself out, to "figure" itself. The pulsional only escapes from pure expenditure, from the death-drive (an explosive destruction that already extends to the inanimate) insofar as the dispositif regulates it while "figuring" it: such is the "originary" process that makes us human from the inhuman. The originality of the Lyotardian "dispositif" compared with the Heideggerian *Gestell* or with the dispositif according to Foucault (and Foucault commented on by Deleuze), is that it owes itself to the "stream," to the "flux" of the drive or to the force that the dispositif "figures." And vice versa: to regulate and to figure the force that draws from its own source—that would be expended too much and immediately, that would not be expressed, without it—is what characterizes the "dispositif," which, in turn, is nothing but regulation and figuration of forces. What one must name with Lyotard the "pulsional dispositif" [*dispositif pulsionnel*]—that by which the human is made human against the background of the inhuman, and by which "the stream brings life to the apparatus, and carries it away to death" (Lyotard 1994)—most likely constitutes the core of every description of the living human being.

For this reason, the phenomena that one claims fall within technoscience are only one region (among other justifiable ones) of a description in terms of "pulsional dispostifs." And to claim that every pulsional dispositif can be said in a technoscientific sense would by extension lose the precise meaning of technoscience on the one hand, and would bring nothing to the understanding of what Lyotard calls "pulsional dispositif" on the other. Nevertheless, reading the descriptions of "technoscience" that Lyotard produced in his later work (while remembering that, with the "dispositif," it is a matter of the drive's flow animating and pushing toward lifelessness [*le mort*] in one and the same movement), allows us to see that one can legitimately make the following hypothesis: that technoscientific phenomena are *exemplarily* and *intensely* "pulsional dispositifs." What, then, is technoscience, if not this pure power (this life that affirms itself) and this pure death (the lifelessness of self-regulating systems; the death against which the life drive struggles in the very movement that is the pure transport toward death)? Life and death, contradictories that cannot be abstracted from one another, which would collapse apart from one another, and the relations of which do not resolve in a dialectically surmountable opposition. Does not technoscience testify exemplarily to this reciprocal imbrication?

10.2 Second Stratum

In *The Postmodern Condition* (Lyotard 1979), Lyotard, without yet using the term, describes science becoming technoscience. Science is essentially characterized as pure performativity, entirely in the efficient production of effects. It is incommensurable with the constative and the descriptive (the order of the "true"), and is cut off from every "narrative" [*récit*] likely to take hold of it again and to legitimate it. Or, more exactly, Lyotard describes the process by which performativity is made into a criterion of legitimization received in the postmodern condition (even when it is, when laid bare, the very crisis of legitimization): efficiency begins to measure the true. I will note, without dwelling on it here, that this description of technoscience tends to subordinate the true, completely reconceived in terms of forces of production, to a moment in the circulation of capital: from this point of view, technoscience finds its essence in capitalism. What's more, it occurs to Lyotard to evoke "techno-capitalism," which could seem to make technoscience secondary and regional by relating it to economic coordinates that allow it to describe capitalism….

But Lyotard—we saw this right away while reconstructing and commenting on the notion of "pulsional dispositif" (and it will be more and more evident as his oeuvre develops)—*also* always gives the means to describe *directly* something of technoscience as closely as possible to the great enigma that incessantly starts thought back up again. Thus here it is a matter of the self-manifestation of force [*la puissance*], as much through capitalism as through technoscience.

10.3 Third Stratum

A third implicit elaboration of technoscience can be found in *The Differend* (Lyotard 1984). Lyotard's attention shifts from the force of desire toward performativity, inspired by the "pragmatic" theories of language: a taking into consideration of the pure production of effects, the effect being situated nowhere else but in its effectuation. As for the notion of event, one notes a shift in emphasis there as well: it is less a matter of the drive that brings about the event than of a reunderstanding of the event on the basis of Heidegger's notion of *Ereignis*. All givenness is inseparable from a withdrawal of Being; the event is inseparable from a withdrawal of Being. Consequently, Lyotard writes, the event falls within the unpresentable.

To this echoes the "inexpressible" character of the "wrong" [*tort*]. "Damages" can be expressed; they can be expressed in the language [*langue*] of the court so that justice can be served. The "differend" is impossibility of a common language [*langage*] or metalanguage in which what is consequently no longer a "damage" but a "wrong" could be expressed. The differend, then, calls for an absolutely new language that breaks with the previous one. A "new phrase," in Lyotard's words, must link on to the preceding one in order not to leave the victim stuck in the suffering of an inexpressible wrong. But this linkage must in no case be produced in the dominant meta-language: such a linkage would accomplish the wrong as such by, so to speak, "robbing" the victim of it. This necessity for a linkage that does not link up in the dominant, already available language makes the "Differend" properly an appeal in the direction of what cannot be programmed in the already-there; it is an appeal in the direction of that which announces itself as absolutely undetermined and contingent: the "is it happening?" None of the genres that imply mastery (theory, narrative, etc.) are thus suitable to the event that plays out in the Differend's appeal in the direction of the "is it happening?"

"Technoscience," which is not named as such,[2] emerges then in the "site" of a major Differend, of *the* major Differend; a Differend so intensely a Differend that it ceases to be one; that what is at stake infinitely exceeds the "category" of "Differend." It is absolutely impossible to link onto it. Technoscience appears at the point of articulation, which is also and first of all the point of crisis, between "Differend" and "Event": Auschwitz. The event as Differend maximized to the point of bursting the Differend's form; the event annulled as major Differend, and already beyond every Differend: the event without witness. As one knows, the extermination of the Jews of Europe was eminently technoscientific: mobilization of efficient rationality as administrative management, organization of transportation flows, technological innovations of destruction… In many ways, the Shoah can seem to fall under Heidegger's denunciation of modern technology as *Gestell*, as "dispositif" or "enframing" [*arraisonnement*][3] of everything, as unconcealment of every being as "making available for …": one could think that the Shoah makes up its darkest part, at least up until now. But even if such an assessment is not illegitimate at a certain level of description, it is precisely important to Lyotard to show that this in no way accounts for "Auschwitz," that there is even something scandalous in Heidegger's famous claim in one of the Bremen lectures, the one entitled "Positionality" [*Dispositif*], the *Gestell*: "Agriculture is now a mechanized food industry, in essence the

[2]According to Hottois (2004), it is in "Response to the question, 'What is the post-modern?'" (1982), text collected in the English version (1984) of *The Postmodern Condition* (Lyotard 1979), that the notion appears explicitly for the first time. The sentence is: "There is no denying the dominant existence today of techno-science, that is, the massive subordination of cognitive statements to the finality of the best possible performance, which is; the technological criterion." (Lyotard 1982: 76–77)

[3][*"Arraisonnement"* is A. Préau's translation of *Gestell* in his French translation of Heidegger's *Question concerning Technology* ("La question de la technique"), collected in French in *Essais et conférences*. Like *Gestell* in German, *"arraisonner"* is a common French word meaning to inspect (as in to inspect a ship), but clearly contains the word *"raisonner,"* "to reason." Françoise Dastur, along with S. Jollivet, prefers to translate *Gestell* as *"Dispositif."* But since we've seen how Lyotard (and Sebbah following him) use *"dispositif"* in a specific sense, and since Sebbah retains the word *Gestell* in German throughout the essay, I've decided to translate *"arraisonner"* by the standard English translation of *Gestell*, "enframing." – Tr.]

same as the production of corpses in the gas chambers and extermination camps." (Heidegger 2012) It is important for Lyotard to not allow Auschwitz to be only an occurrence (even if an exemplary one) of the *Gestell*. It is important for him to hold that the absolute singularity of the Shoah is in no way reducible to an expression of technoscience. Upon this major clarification another implicit one is made, one which completes the symmetry: technoscience is not Auschwitz (even if it is a part of it). Auschwitz is neither the destiny nor the truth of technoscience (Lyotard 1988a; Lacoue-Labarthe 1987; de Fontenay 2006). Technoscience is at once unpresentable event and enframing calculation [*calcul arraisonnant*]: the first on the occasion of the second. And its ambivalence is absolutely irreducible: it *can* be (1) that technoscience is shown exclusively as enframing *Gestell*, and (2) that through the latter, "Auschwitz" unfolds its terrifying performance of destruction. But none of these two linked possibilities is an ineluctable necessity. Lyotard warns of the flattening out of the singularity of "Auschwitz" into the generality of the *Gestell*, and, inversely, warns of the dissolution of technoscience as such in "the event" that annuls the event, in Auschwitz.

Nevertheless, the register of technoscience, register of pure performance and pure efficiency, is by definition untranslatable into another register of phrases – that of the true (unless this had itself been dissolved into that of performance) and especially that of the just. No more inclusive "narrative" can make these registers commensurable. It can be that the event comes "from" technoscience, that it implies a rupture that makes the linkage in a common language impossible; and it can be that this event is "fortunate" just as it can be that it is "unfortunate." But it can also be that "to link" onto a technoscientific production is impossible – in Lyotard's words, that the "is it happening?" is abolished, and thus the event collapses into the very impossibility of the event. But "it can be" … it is also *possible* – although the tone of Lyotard's texts is more often pessimistic than optimistic – that in its intrinsic ambivalence technoscience brings about a fortunate event.

10.4 Fourth Stratum

Lyotard's description of technoscience is thus exemplarily sensitive to its *ambivalence*: his texts about it are strewn with doubled *Gestalts* ("forms").

Technoscience is most likely transformation – or rather unconcealment – of science as pure performative force because it is calculating, intimately tied to economic force, such as capitalism. It is "technocapitalism": this is the first Lyotardian lesson that we have encountered. And the meditation on the "*mancipium*" of the "*mainmise*" in the texts of the final years of the 1980s no doubt continues this line of thought.[4]

[4]For the reference to Lyotard's texts and for their commentary, cf. the subchapter "the technoscientific mainmise" in Alberto Gualandi's (1999: 146).

But if the coherence of this profile of technoscience is unchanging for Lyotard, it is a kind of "diagnosis": it neither denounces nor condemns without remainder, and for several reasons.

In particular, Lyotard is attentive to the artistic experiences made possible by new technologies. If technoscience tends to produce reality without remainder (exemplarily so in the digital realm), then from this point of view (and from only this point of view) technoscience aims to abolish finitude. Artistic work no longer finds givens that are already there in the finitude implied by the a priori forms of sensibility. The material is made immaterial, if "matter" can mean that which I encounter directly without mediation and without "filter," without having chosen it, and which resists my activity. But Lyotard is at the same time extremely sensitive to the fact that, with artistic productions tied to new technologies, it is a matter of new dimensions of experience, that is, of new ways of being affected[5]: the constant ambivalence of technoscience.[6]

Here I'll take up another characteristic of technoscience for the later Lyotard, one which never reduces technoscience to the performative force that it nonetheless is as well. Technoscience marks humanity's final narcissistic wound. While Lyotard can show us in certain texts that it is the erasure of finitude (as we've just seen), in another sense it states humanity's contingency and finitude in a final way – in a final way because it states that finitude itself is finite. If for a certain Heidegger, for example, "being-toward-death" opens up the very dimension of an access to Being, and so steals the human from her finite naturality in the very gesture that radicalizes finitude, then Lyotard, for his part, notes that all life (including human life), *is only* negentropic: the tireless task of pushing back one's ultimate disappearance. In a sense, secretly true to his reflection on the intertwining of the death-drive and the life-drive developed in *Des dispositifs pulsionnels*, he proposes a decisive shift in emphasis: the very life of our sun is finite; to live is always to survive, here in the sense of "continuing on by other means;" and this from the beginning, for from the beginning the question will always only have been one of survival. From birth (at the ontogenetic as much as the phylogenetic level), living always and irreducibly amounts first of all – whatever life may be besides – *to postponing* death. Technoscience, then, will be the necessary complexification that will allow the privileged few among human beings to leave earth when the Sun dies in order to try to survive elsewhere, though without any guarantee.[7] (There is no other plane, whether

[5] Cf. for example (and exemplarily) the exhibition "The immaterials" [*Les Immateriaux*], Centre Georges Pompidou, 1985, of which Lyotard was the head curator. This exhibition proposed to "make sensible" its subject, that which is immaterial. Cf. "Argument 2: Immateriality" in the text from the exhibition's introduction. See also Jean-Louis Déotte (2012).

[6] Profound ambivalence of the *conatus*: is perseverance first of all affirmation and growth, or always basically and first of all preservation?

[7] Humans leaving the dying earth and its dead sun in a spaceship in order to ensure the survival of humanity elsewhere in the Universe—this is the drama, or the "fable" on the basis of which a decisive dimension of the later Lyotard's thinking of technoscience (among other things) unfolds. This fable appears in numerous texts, for example "Ticket for A New Stage" in *The Post-Modern Explained* (1986), or in different pages of *The Inhuman* (1988b), and it is developed in "A Postmodern Fable" (1993).

transcendental, or essential, or purely spiritual, by which the human will always already be saved from the radical material contingency whose law is entropy, and where the struggle with no other hope than indefinite deferral is negentropy.)

To understand what is offered explicitly as a "postmodern fable," one can make several remarks. First, that in a sense the technical is "primary" and "constitutive," since every living thing – even the most primitive – is, within the framework of this radical materialism of finite contingency, fundamentally *nothing other* than the means for its survival. It in no way precedes, in no matter what kind of purity or autonomy one wants, its prosthesis; it completely coincides with its prosthesis. From this point of view, there is no place at all to mark a rupture between the living and its prosthesis, the human and the tool: the reign of what one calls techniques is only the ineluctable continuation, by complexification, of a prosthetic process that will *always already* have begun, that will have begun since "the beginning" (according to which, therefore, the living being is in fact *nothing* as a pure origin).

Let us insist on this: this thought of the "always already" of the continuity of living beings with the technical – far from thinking the tool as a "projection of the organ," (Kapp 1877) as the continuation of the living body that would invest it with its abilities – thinks the living being as the always already of the (its) prosthesis. So the human brain is nothing other than, in the strict sense, an organ, a tool, one particularly complex and subtle so as to assure survival, to delay death. An organ: an originary prosthesis that, consequently, acquires other always more subtle prostheses in this indefinite course beyond hope and despair (one will never find a way out from this once and for all, but the indefiniteness of this extension is still possible). Spaceships that will allow one to leave the dying earth, and the technologies allowing one to inhabit other worlds – a certain idea of technoscience, then – are the ineluctable prosthetic complexification to come.

I will note two things. First, this merciless entropic materialism does not deny the dimension of the spirit nor reduce it to matter, but states it exactly as it is: the *effect* of the living being's complexification always already continued in its prostheses (that is to say, in this case, it is in a sense "preceded" by them). Consequently, one glimpses what is *also* a face of technoscience: the more complex it becomes, the more human thought and action, as such sensitive to the incalculable, extend to new dimensions. Through which Lyotard explicitly encounters Bergson: the Bergsonian notion of "supplement of soul" [*supplement d'âme*][8] will in fact [*en effet*] never have meant that all technics or technology, in its frailty, required that one grant it a spiritual dimension or meaning that would be heterogeneous or superior to it. Rather, it meant very precisely that the more there is technical complexification, the more there will "effectively" [*en effet*] be "spiritual" complexification and novelty. And this even if only *analogy* can be set up between one and the other order once

[8] "The body, now larger, awaits a supplement of soul." (Bergson 1932: last chapter). No doubt the style of Lyotard's reflection on the relations between the body and the technical to which we refer here exceed, radicalize, complicate, and displace—even "invert"—Bergson's on the same subject. But Lyotard's reflection never ceases to be indebted to it. Taken up again in a disillusioned postmodern tone, unveiled as negentropy that is always already prosthetic, Bergson's vitalism endures in Lyotard's writings in which desire, even when recognized as struggle against the ineluctable, remains desire…

one compares and describes them, inasmuch as analogy in no way diminishes the abyss that it reveals even while bridging it.[9] Moreover, Lyotard makes us see that the genitive in the expression "supplement of soul" must be heard as a subjective genitive as much as an objective genitive: not so much a surplus of soul for that which would be previously deprived of it (a supplement *of what*? of soul) as the soul *as* supplement. Which means that in order to give "soul," there must be an originary lack: in this sense, the technical calls for, incites, the soul. The soul would not emerge without the technical lack that calls for the supplement; the soul is always already supplementary. And to say that the soul is always already supplementary, brought about by the lack (as technical), is to say at the same time that the soul is always already supplemented, that the technical is equally well always already supplement (originary supplement, if you like).[10]

That said, technoscience is no less a terrible wound if it shows to the human its final illusion as such, if it reveals the human to itself as surviving on borrowed time and so as on its way out: it is necessary to leave the Earth; and, on the scale of cosmic time, the preparations have already begun. More radically, the "always already" of survival's negentropic becoming is brought to light, the negentropic becoming that is only assured as complexification: this negentropic complexification of matter itself (recognized as energy beneath every individual stability) will have always already begun, and the human will only ever have been a moment of it. The human as *effect* of this complexification; the human as traversed by it, the human as "transformer" [*transformateur*][11] of material complexification into "spirit," that is, into survival. The human, deposed from its position as origin, is not even a result. As Lyotard tells the story, the question is not so much one of knowing how the human and its brain will abandon the dying earth, but rather of knowing how the brain, with its human, will carry out such a task. Technoscience; or how negentropic complexification continues through the human as transformer. Consequently, the human depends on so many "inhumans." There is the positively connoted inhuman of the incalculable (which, according to Lyotard, belongs par excellence to childhood), and the merciless inhuman of calculation as complete determinism, as suppression of the incalculable. Technoscience does not present itself solely under this second, somber face, as the terrifying inhuman. Not that it ceases to be so when seen from a certain angle, or taken from a particular end, but because it appears that, *as such*, it will have always been the *only resource* for the inhuman as incalculable and as childhood. Let us say this another way by calling up a third figure of the inhuman: the inhuman of the negentropic complexification always already at work, in a sense always already technical, is transformed through the human into technoscience, as

[9] Cf., "As a material ensemble, the human body hinders the separability of this intelligence, hinders its exile and therefore survival. But at the same time the body, our phenomenological, mortal, perceiving body is the only available *analogon* for thinking a certain complexity of thought," in "Can Thought go on without a Body" collected in Lyotard (1988b: 22).

[10] Let us note that we are using a conceptual matrix here – that of the originary supplement, of the prosthesis, of the originary lack, of constitutive technicity – that is hardly Lyotard's and much more Derrida's or Bernard Stiegler's. This translation does not seem a betrayal.

[11] On Lyotard's notion of "transformer," cf. C. Enaudeau, J.F. Nordmann, J.-M. Salanskis, F. Worms, ed. (2008).

it were. In this way science's means (formalization and experimentation) show themselves for what they are: mastery and performativity, certainly, but basically nothing other than the difficulty of survival, never safe from death.[12]

One can of course find in the late Lyotard several gloomy denunciations of technoscience as "mainmise," as annulment of the incalculable. One can of course also detect here and there something like the remainder of a somewhat "vitalistic" admiration and trust in negentropic complexification. But the basic tone – "postmodern" – will be the one that, beyond hope and despair, claims to state a human that is nothing other than the precarious transformer of negentropic complexity: a complexification that, in the earth's final hour of agony, tries to "save its neck" as technoscience. One can never say enough that technoscience is nothing other than the most advanced stage of a precarious and vulnerable survival, one that is always pending. In a way, it is not an exaggeration to say that technoscience is the most advanced, familiar figure of vulnerable survival. And in addition, although the denunciation of relentless and triumphant technoscience is perceptible here and there; although the admiring faith in the *élan vital* that continues (though as negentropic, as the *élan* of *sur*vival) is also perceptible; these two affects – fundamentally that of modernity's beginning (the trust in human life continued by the sciences and technology) and that of modernity's coming to an end (disillusionment and denunciation of technoscience) – are set aside in favor of the perspicacity, which is no doubt disillusioned but without bitterness, that characterizes the post-modern. It is never just the curses against technoscience, nor primarily the hope of a tomorrow that sings (no matter the score), but a gaze of a clearsightedness without concession cast on this negentropic complexification that passes through the human, and the performative arrogance of which always in fact betrays the radical precariousness – the vulnerability – that requires attention and care.

10.5 Conclusion

In this chapter I have passed through four strata of the Lyotardian notion of techno-science as it takes shape in his oeuvre (even before he *explicitly* uses the term). (1) It pertains with the analysis of "pulsional dispositif" in the early works. (2) It is

[12] I will not insist on it here, but it's necessary to point out that Lyotard diagnoses this vulnerability of "technoscience" in another site of knowledge: he interprets the so-called "foundational crisis" in mathematics that marked the twentieth century as the impossibility of formulating a meta-language that is able to justify itself, and which is able to include all languages as its principle of absolute justification. The diversity of scientific languages makes them incommensurable with one another, makes the reduction of differends impossible: the delegitimizing precariousness of the absence of foundation. If it is a consistent feature of the various thinkings of technoscience to reveal the most theoretical and formalized scientific practices as themselves also coming – in the same way as "materialized" technologies – from performativity, or even from a making-available, then this operation has, for Lyotard, the at-first-glance unexpected effect of being revealed (for example in mathematics) as a precarity (should one say a "contingency"?), a survival that requires care...

approached (but not exhausted) as pure performativity under the name of "techno-capitalism." (3) It is an appeal to the "is it happening?" maintaining the tension between the "Event," and the "Differend." And (4), it designates the "originary prosthesis" that bears witness of life as being always already survival.

Thus for Lyotard the figure of technoscience is always profoundly ambivalent: the human depends on so much that is inhuman, an inhuman that runs through the human. In particular, even though Lyotard does not stop referring to its performative force and its *mainmise*, he has given us the means: (1) to understand technoscience as the exemplary manifestation of the life-drive as much as the death-drive, and especially of their irreducible entanglement; (2) to understand that its performative force was also the invention of new "experiences," of new ways of being affected; (3) finally, as we've insisted, that triumphant performativity was also precariousness, even vulnerability itself: the vulnerability of the human, that is to say, of the inhuman survival to which the human owes itself, and which carries on by means of the human – among others.

References

Bergson, H. (1932). *Les deux sources de la morale et de la religion*. Paris: Presses Universitaires de France. English Edition: Bergson, H. (1935). *The two sources of morality and religion* (R. A. Andrà, C. Brereton & W. H. Carter, Trans.). New York: Henry Holt and Co., Inc.

de Fontenay, E. (2006). *Une toute autre histoire – Questions à Jean-François Lyotard*. Paris: Fayard.

Déotte J.-L. (2012). *Les immatériaux* de Lyotard (1985): un programme figural. *Revue Appareil* [online], 10/2012, URL : http://appareil.revues.org/797; https://doi.org/10.4000/appareil.797.

Enaudeau, C., Nordmann, J.-F., Salanskis, J.-M., & Worms, F. (Eds.). (2008). *Les transformateurs Lyotard*. Paris: Sens et Tonka.

Gualandi, A. (1999). *Lyotard*. Paris: Les Belles Lettres.

Heidegger, M. (2012). *Bremen and Freiburg lectures* (A. J. Mitchell, Trans.). Bloomington: Indiana University Press.

Hottois, G. (1984). *Le signe et la technique*. Paris: Aubier.

Hottois, G. (2004). *Philosophies des sciences, philosophies des techniques*. Paris: Odile Jacob.

Kapp, E. (1877). *Grundlinien einer Philosophie der Technik. Zur Entstehungsgesschichte des Kultur as neuen Gesichtspunkten*. Brunswick: Westermann.

Lacoue-Labarthe, P. (1987). *La fiction du politique*. Paris: Christian Bourgois éditeur. English Edition : Lacoue-Labarthe P. (1990). *Heidegger, art, and politics: The fiction of the political* (C. Turner, Trans.). Oxford: Blackwell.

Lyotard, J.-F. (1974). *Economie libidinale*. Paris: Minuit. English Edition: Lyotard, J.-F. (1993). *Libidinal economy* (I. H. Grant, Trans.). Bloomington: Indiana University Press.

Lyotard, J.-F. (1979). *La condition postmoderne*. Paris: Minuit. English Edition: Lyotard, J.-F. (1984). *The postmodern condition. A report on knowledge* (G. Bennington & B. Massumi, Trans.). Minneapolis: University of Minnesota.

Lyotard, J.-F. (1982). Réponse à la question: qu'est-ce que le postmoderne ?. *Critique* (419), 357–367. Reprint in Lyotard, J.-F. (1984). *The Postmodern Condition. A report on knowledge* (R. Durand, Trans., pp. 71–82). Minneapolis: University of Minnesota.

Lyotard, J.-F. (1984). *Le Différend*. Paris: Minuit. English Edition: Lyotard, J.-F. (1988) *The differend* (G. Van Den Abbeele, Trans.). Minneapolis: University of Minnesota.

Lyotard, J.-F. (1986). *Le post-moderne expliqué aux enfants*. Paris: Galilée. English Edition: Lyotard, J.-F. (1992). *The post-modern explained* (B. Don, B. Maher, J. Pefanis, V. Spate & M. Thomas, Trans.). Minneapolis: University of Minnesota.

Lyotard, J.-F. (1988a). *Heidegger et « les juifs »*. Paris: Galilée. English Edition: Lyotard, J.-F. (1990). *Heidegger and "the jews"* (A. Michael & M. S. Roberts, Trans.). Minneapolis: University of Minnesota.

Lyotard, J.-F. (1988b). *L'inhumain, causeries sur le temps*. Paris: Galilée. English Edition: Lyotard, J.-F (1991). *The inhuman: Reflections on time* (G. Bennington & R. Bowlby, Trans.). Cambridge: Polity Press.

Lyotard, J.-F. (1993). *Moralités postmodernes*. Paris: Galilée. English Edition: Lyotard, J.-F. (1997). *Postmodern fables* (G. Van Den Abbeele, Trans.). Minneapolis: University of Minnesota Press.

Lyotard, J.-F. (1994). *Des dispositifs pulsionnels*. Paris: Galilée.

Sebbah, F.-D. (2010). *Qu'est-ce que la technoscience ?* Paris: Encre Marine/Les Belles Lettres.

Chapter 11
Toward a Philosophy of Technosciences

Bernadette Bensaude Vincent and Sacha Loeve

Abstract The term "technoscience" gained philosophical significance in the 1970s but it aroused ambivalent views. On the one hand, several scholars have used it to shed light on specific features of recent scientific research, especially with regard to emerging technologies that blur boundaries (such as natural/artificial, machine/living being, knowing/making and so on); on the other hand, as a matter of fact "technoscience" did not prompt great interest among philosophers. In the French area, a depreciative meaning prevails: "technoscience" means the contamination of science by management and capitalism. Some even argue that "technoscience" is not a concept at all, just a buzzword. In this chapter, on the contrary, we make the case for the constitution of a philosophical *concept* of technoscience based on the characterization of its objects in order to scrutinize their epistemological, ontological, political and ethical dimensions.

Keywords Epistemology · Ethics · Design · History and philosophy of technoscience · Objects · Ontology · Philosophy of technology · Technoscience · Science and technology studies · Values in science and technology

Technosciences have a bad reputation. Despite the inflation of the use of this term over the past decades – indicated by Google n-Gram viewer – it is neither a taxonomic category referring to a class of disciplines nor a well-defined concept. The compound term has been used as a qualifier ("technoscientific societies") in a course of public policy science as early as the 1960s (Caldwell and Deville 1968).

B. Bensaude Vincent (✉)
Centre d'Étude des Techniques, des Connaissances et des Pratiques (CETCOPRA), Université Paris 1 Panthéon-Sorbonne, Paris Cedex 05, France
e-mail: bernadette.bensaude-vincent@univ-paris1.fr

S. Loeve
Faculté de Philosophie/Institut de Recherches Philosophiques de Lyon (IRPhiL - EA 4187), Université Lyon 3 Jean Moulin, Lyon, France
e-mail: sacha.loeve@univ-lyon3.fr

S. Loeve et al. (eds.), *French Philosophy of Technology*, Philosophy of Engineering and Technology 29,
https://doi.org/10.1007/978-3-319-89518-5_11

However, as François-David Sebbah (2010) rightly noticed, it has never been appropriated by the actors of technoscientific research, who chose to label the research fields they initiated "materials science and engineering," "biotechnology" or "genetic engineering," rather than "molecular technoscience" or "biotechnoscience." Scientist, engineers, designers, science-policy makers, research managers are reluctant to use this term. Technoscience appears to be a practice without (explicit) practitioners.

While the term did not catch up among scientific communities, it did not raise greater enthusiasm among philosophers. Some of them are very critical: technoscience is just a distortion of pure science resulting from its contamination by ideology (Seris 1994; Bunge 2012) or a post-modernist buzzword (Raynaud 2016). Accordingly it would be better to forget about it and return to more serious topics! Even scholars in Science and Technology Studies (STS) and philosophers of science who are more engaged with actual scientific practices do not care for a distinction between science and technoscience, since technical intervention is a necessary condition of all knowledge production in modern science. The close interaction between science and technology, between representing and intervening in Ian Hacking's terms (1983) is a major feature of all scientific practice. So there is no need for a special philosophical investigation of technosciences.

This chapter takes the concept of technoscience seriously and makes the case for the relevance and legitimacy of a philosophy of technoscience. It uses it as a philosophical tool for better understanding current trends in scientific research and shows how it can acquire a descriptive and analytic value despite and along with its polemic charge. Following a brief historical survey of the evolution of the notion of technoscience and its uses since its coinage, this chapter will outline three *profiles* of technoscience, (i) epistemological, (ii) ontological, (iii) politico-ethical. These three ways of "portraying" technoscience highlight some of its major, albeit nonexclusive and non-exhaustive, philosophical characteristics:

(i) a way of knowing through making, which can be seen as a further development of Ian Hacking's characterization of laboratory style in a "design" mode;
(ii) a specific mode of existence of objects which commands a shift from epistemic pluralism to ontological pluralism;
(iii) an intrinsic value-ladenness which requires shifting our views of objects from ethical neutrality to ethical ambiguity.

11.1 Genealogy of a Concept

The term "technoscience" has been coined by Belgian philosopher Gilbert Hottois to initiate a philosophical program in the 1980s. Undoubtedly, the term had been in use long before. As French philosopher Dominique Raynaud carefully highlighted (Raynaud 2016), "technoscience" came out in the immediate post-war period and appears occasionally for over two decades in literature related to science policy

(Laswell 1946, 1948, 1957; Roherty 1960), or to environmental issues (Rudd 1964; Clement 1965), especially in the North American era. Hottois has thus no paternity *stricto sensu* on the *word* "technoscience." However, he certainly was the first to provide the *concept* of "technoscience" with high philosophical significance and impact.

He first used the compound term "techno-science" as a provocative phrase for waking up philosophers from their "linguistic slumber" (Hottois 1979). Hottois deplored that both analytic and hermeneutic philosophers had relinquished the handling of reality to the techno-sciences. By seeking refuge in a "metalinguistic," they condemned philosophy to a condition of "secondarity." Consequently, Hottois initiated a philosophy of "technoscience" (without hyphen), focused on the contemporary practices of science characterized by a new regime of research in which technology becomes the milieu, the driver and the finality of research (Hottois 1984). Technoscience thus referred to the reciprocal process of internalization of science into *technè* and of *technè* into science. This concept is close to Don Idhe's views on instrumental materiality developed in *Instrumental Realism: The Interface between Philosophy of Science and Philosophy of Technology* (Idhe 1991).

However, as Hottois's Chap. 8 in this volume clearly demonstrates, this genealogy of technoscience is only part of the story because of the close association of this notion with postmodernism, initiated by Jean-François Lyotard who popularized the term "postmodern." In *La condition postmoderne. Rapport sur le savoir* (1979), Lyotard reflected on the status of knowledge in postindustrial and computerized societies. Two major features struck him: technoscience is ruled by the norms of performativity (see Sebbah, Chap. 10 in this volume) and it demises the grand narratives (*grands récits*). He argued that the traditional relationship between science and technology was reversed. Technology was taking the lead in scientific research. As knowledge and power became two sides of the same coin, postmodern culture could be characterized by the collapse of the modern ideal of emancipatory science. Technoscience thus mainly referred to a reversal of the values between science and technology, which marks an epochal break (Forman 2007).

In the context of early STS, Bruno Latour gave technoscience a quite different meaning. He used "the word technoscience to describe all the elements tied to the scientific contents no matter how dirty, unexpected or foreign they seem, and the expression 'science and technology', in quotation marks, to designate *what is kept of technoscience* once all the trials of responsibility have been settled." (Latour 1987: 174)[1] In analyzing the construction of both scientific facts and technological objects along the networks of a multitude of heterogeneous actors, Latour claimed to uncover the complex alliances between human and non-human actors, nature and society, that are usually obfuscated by the work of "purification" of scientific facts. Contrarily to Hottois or Lyotard, Latour did not refer technoscience to a new epoch of research. It is nothing but "science in the making": The true expression of the real, impure and mixed practices of the sciences as they are made. Accordingly,

[1] One may notice that this use of "technoscience" makes of the label STS ("Science and Technology Studies") a misnomer!

science has always been technoscience, and today's explicitly impure technosciences – such as nanotechnology or synthetic biology – are nothing more than the "speaking-truth" of science, the sign that we are ceasing to believe that we have once been "modern."

Donna Haraway (1997) embodied technoscience in her famous figure of the cyborg, which combines biological processes with social and cultural patterns, thus making an actual mixture of heterogeneous components that challenges all attempts at purification. Just like the phrase "cyborg" results from the agglutination of the two incompatible notions of "cybernetic machine" and "biological organism," in her prose, "technoscience" expresses one of the "wobbly couplings" of the contemporary condition (nondualistic, nonmonistic), and it is the same for naturecultures, oncomouse, and FemaleMan. Haraway added to the notion of technoscience a heavy load of irony and blasphemy as she incorporated it in the corpus of feminist and post-modernist literature. When she moved from the figure of the cyborg to domesticated dogs, she added to the notion of impurity that of mutual invention (Haraway 2003). Just as dogs and humans invent each other, every one shapes her identity through interactions with otherness.

The tremendous success of feminist and postmodernist studies jeopardized further philosophical investigations of technoscience.[2] Hottois himself switched to bioethics when he realized that the phrase "technoscience" had become a fashionable term widely used to refer to a vague entanglement of science, utilitarianism and capitalism (Hottois 1996), where he couldn't recognize his own notion. Don Idhe, by contrast, organized a seminar at Stony Brook University on technoscientific research in an attempt to engage a dialogue between the various perspectives on technoscience. Interesting comparisons came out of this attempt at "*chasing technoscience*." (Don Idhe 2003) A single common feature emerged from the panorama sketched in this volume: all contributors shared a concern with the materiality of scientific or human practices in general. On this basis, however, it seems difficult to design a common research agenda in order to prevent the dissolution of technoscience in a vague postmodernist rhetoric.

The emergence of nanotechnology and converging technologies (NBIC for nano-bio-info-cogno technologies) in the 2000s prompted a renewed interest in technosciences among philosophers (Bensaude Vincent 2009; Sebbah 2010; Nordmann 2010; Guchet 2011; Dupuy, Chap. 9 in this volume).[3] These ambitious programs were associated with national funding initiatives all over the world and gave rise to an entire economy of promises and fears. They attracted attention to a regime of production of science in which research is conducted in a context of application

[2] According to a bibliometric study based on a Google N-gram enquiry (Raynaud 2015) 41.69% of the citations using the phrase "technoscience" refer to Donna Haraway's, *Modest_Witness@Second_Millenium. FemaleMan©_Meets_OncoMouse™: Feminism and Technoscience* (Haraway 1997).

[3] Recently a comprehensive and epistemologically informed history of technoscience has been published by David F. Channel (2017). He argues that the roots of technoscience can be traced to the nineteenth and early twentieth centuries in chemical industry, electrical lighting, and telephone and radio research.

(Carrier and Nordmann 2010), where the setting of research priorities mimics the dynamics of markets while the production of knowledge mimics the industrial production of commodities (Pestre 2003). Yet this current regime of research deeply affects the status of knowledge. While it is now dominant, this regime of research is, however, not new. As a category referring to a style of research characterized by the sociotechnical shaping and production of scientific objects, technoscience can well be traced back at least to eighteenth-century chemistry (Klein 2005). The notion of technoscience may help disentangle the epistemological implications of such research practices provided it is viewed as an idealtype rather than as a new paradigm or an epochal break (Nordmann et al. 2011).

Going beyond the "nothing new" statement (science has always been technoscience) and the too historically simplistic claim of an epochal break was one of the objectives of the French-German project GOTO, "The Genesis and Ontology of Technoscientific Objects" (2010–2014).[4] Far from claiming that all science could be exposed as technoscience, this research project was based on the assumption that one can clarify the distinction between science and technoscience by shifting the attention from the subjects to the *objects* of knowledge and clarifying their ontology. Scientific ontologies are typically made of facts, laws, and causal dispositions; they orient the cognitive practice towards the acquisition of a kind of knowledge that takes form of propositions, theories, hypotheses, models, explanations, representations, or predictions that are *about* the world.[5] By contrast, technoscientific research seeks to establish demonstrable capacities of construction and control by functionalizing objects, implementing new capacities and enhancing their value. Far from denying any difference between science and technology, the investigators of the project argued that while science and technology are two distinct albeit interacting spheres *in the idealtype of "science,"* they are indistinguishable *in the idealtype of "technoscience."* Far for declaring the work of scientific "purification" of facts futile and meaningless, they argued that a technoscientific object is encountered when such purification proves impossible or unnecessary (Bensaude Vincent et al. 2011). The project invited philosophers, STS and historians to engage with these questions not only in order to appreciate the difference between science and technoscience, but also to draw their attention toward the modes of existence of research objects (Bensaude et al. 2017).

[4]The GOTO program, funded jointly by ANR (France) and DFG (Germany), gathered Bernadette Bensaude Vincent and Sacha Loeve in France together with Alfred Nordmann and Astrid Schwarz in Germany.

[5]Of course, from the epistemological perspective of instrumentalism, scientific representation does not reach an unobservable mind-independent reality, but it makes it observable *if one carries out certain actions*.

11.2 Epistemological Profile: Research in a Design Mode

If intervening rather than representing the world, captures the epistemological credo of Nancy Cartwright and Ian Hacking, then technoscientific research could bring grist to the mill of the so-called "Stanford School of philosophy." Hacking convincingly argued that laboratory experimentations were not just a way to control postulation through observation and measurement and to represent nature. Cartwright (1999) emphasized the virtues and limitations of the experimental settings specifically constructed by physicists or economists to fit in their theoretical models and providing some understanding and control of phenomena. "Nomological machines" – as she named these arrangements used to capture regularities and formulate laws – are suited to serve cognitive and predictive functions but as they provide an idealized picture, they are not robust enough to encounter the real world.

Technoscientific research is full of machines – electronic devices, sensors, actuators, microscopes, molecular machines – but they are not used as typical scientific instruments. They are not used to test hypotheses, to control postulation through observation and measurement and to represent nature. They rather belong to a specific style of laboratory experiments aimed at manipulating objects (Hacking 1983). They do not operate as Cartwright's "nomological machines" revealing regularities and laws. They are *enabling machines*, or tools to *make* something. They violate the scientific imperative of distance to secure objectivity. For instance, the scanning tunneling microscope (STM), the icon of nanotechnology research, challenges the epistemic gospel of modern science. First, the STM and other related techniques of near-field microscopy, which approach the object as closely as possible in order to pick-up the information at the surface of the sample, induce a "collapse of distance" (Nordmann 2006). Moreover, the STM is as much an instrument of observation as an apparatus of manipulation. It is not just because, as Hacking argued, there is no visualization without intervention. The STM does not visualize but probes the atomic surface with its tip in a kind of machinic "touch." It even connects itself to a molecular adsorbate. It does not just scan the structure under the tip. It actually constructs new structures through its intervention. The STM is both an instrument and a tool for designing new materials and machines (Nordmann 2010; Loeve 2011a).

Technoscientific research is mainly oriented toward design.[6] "Redesigning life" or "shaping the world atom by atom," the slogans of synthetic biology and nanotechnology initiatives suggest that everything, from molecules to organisms can be designed (see Loeve, Chap. 22 this volume). "Materials by design," i.e. materials intentionally built up for specific purposes and for performing specific tasks are the core-model of materials research. With the mass diffusion of composite materials in automotive industry, aeronautics and sport articles, materials ceased to be the precondition for technological projects. Materials Science and Engineering sub-

[6] "Design" is such a fashionable term that is also prevails in STS. Significantly, design was the thematic topic of the 2012 joint meeting of the 4S and the European Association for the Study of Science and Technology (EASST) with 1600 papers.

verted the linear model of innovation – from basic science to applied science to industry and market – and developed a systems approach with close collaborations between a variety of scientists and engineers (Bensaude Vincent 2001).

The view that materials were no longer a constraint has been reinforced in the 2000s by the notion of "bottom-up design" spread in nanotechnology initiatives. Designing functional objects and organisms is the major achievement in technoscience. As the focus of research shifts from the correlation between structures and properties to performances and process, the object of design is no longer a sample representing general phenomena or a theoretical model embodied in matter. It is a thing with an intrinsic value, an end in itself rather than a means toward an end. In this context, atoms and molecules, genes and genomes, which were once considered as the basic constituents of matter and living beings, are re-conceptualized as devices to make nanomotors, nanocars, nano-wheelbarrows, etc. Similarly, yeasts, bacteria, viruses are being reprogrammed, re-engineered, or redesigned to perform a number of tasks such as synthesizing therapeutic molecules, biofuels, or decontaminating toxic sites.

Does it mean that science would be sacrificed on the altar of technological innovations and utilitarianism? Actually, technoscientific researchers are often content to publish proofs-of-principles. By constructing a biological device or a molecular machine in the well-controlled conditions of the laboratory, they aim to show that such technology is possible. Such a proof manifests a capacity and opens up a possible future, but the effective realization of this possible is not a matter of concern in technoscientific research (Nordmann 2006), which in this regard cannot be confounded with applicative research. From the perspective of applied science or "pure engineering," a proof-of-principle is only a temporary and limited result that calls for further research and development efforts in order to be scaled-up. From the perspective of technoscience, it is genuine and valuable knowledge-production, knowledge about the possible rather than about the actual. Thus, that technoscience is not "pure science" does not mean that it is "pure engineering," nor that it is simply an "impure" hybrid of science and engineering.

Behind the rhetoric of promises used by technoscientists in their search for funding sources, their research practices are actually driven by cognitive goals. "What I cannot create, I do not understand," this remark by theoretical physicist Richard Feynman has been used again and again by synthetic biologists to describe their endeavor. Through making synthetic chromosomes or metabolic circuits, they seek knowledge about the fundamental workings of life, or *possible* life – extent life being often considered too "provincial" (i.e. too particular) by synthetic biologists to support fundamental biological knowledge (Attwater and Holliger 2014). For instance, the construction of minimal cells is explicitly aimed at two intermingled objectives: tackling the fundamental question of the origin of life and providing a standard "chassis" on which various functionalities can be implemented for predictably delivering specific performances on demand. Similarly, micro-machines are designed on the model of cell motility for the dual purpose of better understanding the complex behavior of living cells and guiding tiny robots within the body for diagnostic or therapeutic actions. To design such micro-robots researchers do not hesitate to practice a reverse engineering of natural cells in order to extract information about their behavior and design their machines on the basis of this information (Arroyo et al. 2012).

Within this epistemological framework where knowing and making are intermingled, nature itself comes to be viewed as a designer, whether it be an insuperable engineer (Jones 2004) or an awkward tinkerer whose work needs to be superseded (Endy 2005; Marliere 2009). This view is underlying the boom of biomimetic strategies in chemical industries and robotics where research programs on soft machines and soft robots are conducted. A remarkable example is plant robotics, aimed at growing plant-like robots compliant to environment. In a program inspired by plant roots, roots are viewed as "soft sensors & actuators," with "distributed intelligence." Plant roots are a fascinating model because they are capable of odor detection without nose, breathing without lungs, movement without muscles, light perception without eyes, as well as of decision without brain, and communication without mouth (Mazzolai 2014). Through design practices, this technoscientific program results in disclosing nature's capacities rather than increasing our technological control over natural phenomena. As research in a design mode, technoscience is not necessarily meant at enhancing human performances or increasing our domination over nature. It is more adequately characterized as an exploration of nature's capacities, whereby nature is reconfigured as a field of possibilities.

To be sure, digital computation has fostered the ambitions of "rational design." Computational chemistry, for instance, is using the basic rules of physics and chemistry to model the behavior of potential compounds. Similarly, system biology aims to provide guidelines for synthetic biology by modeling genetic and metabolic circuits. The alleged complementarity between systems and synthetic biology relies on the dichotomy between theoretical hypotheses and experimental testing. At first glance, it thus seems that research in a design mode is enhancing the control of knowing over making, of basic science over technology. Computation based on the most fundamental information about atoms and genes dispenses with the cost of synthesizing thousands of molecules or genomes for selecting the one with desirable properties. "Now you can find out how well a new compound works before he does," claimed the advertisement of a corporation of molecular design (*Chemical and Engineering News* 1983: 19). The production of artifacts seems to proceed from the interaction between algorithms and the basic laws of physics and would just be the materialization of these products of the mind.

This view of design is favored by "star" technoscientists. For instance, biologist Craig Venter proudly advertised the success of his research group in transplanting a synthetic chromosome (a replica of the natural genome of *Mycoplasma mycoides* less 25% "useless" genes) into another bacterial cell having its chromosome removed (*Mycoplasma capricolum*). The synthetic chromosome "takes control" over and "reprograms" the recipient cell, thus giving rise to a new species christened *Mycoplasma laboratorium* (Gibson et al. 2010). "This is the first self-replicating cell we have had on the planet whose parent is a computer." (*USA Today* 2010) In thus emphasizing the role of the information embedded in genetic sequences, Venter obscured the huge efforts, technical skills and years of trials and errors that this prowess required from an army of human and nonhuman collaborators for inserting an entire genome in a cell and getting the cell to express it. First, the "artificial" synthesis of the chromosome required the help of other bacteria: the chemically synthesized DNA cassettes had to be assembled and cloned in *Escherichia coli*, and reworked in yeast. Second, processing DNA requires preexisting molecular machinery such as DNA and RNA polymerases for replication and transcription, ribosomes and other expression factors. These helper molecules are not synthesized *de novo*, they are extracted from preexisting living cells. Finally, the expression of the transplanted genome was possible only because the two species chosen as donor and recipient were close cousins belonging to the same genus *Mycoplasma*. Because DNA requires proteins to make proteins, two too distant species could not make it, as they would present incompatible binding sites and binding factors. The painstaking technical work and the know-how displayed by laboratory workers (including the helper bacteria) were systematically kept in the backstage in order to overemphasize the conceptual and abstract part of the process. The design is reduced to the model computed by bioinformatics while the actual production of the artifact is supposed to be no more than the execution of a program or the material projection of a conceptual pattern. The priority conferred to the abstract pattern over the process of concretization is clearly in keeping with the old hylemorphic model of art as the imposition of forms created by the mind (or a computer) upon a material substrate (Simondon 2016).

A closer glimpse on the actual practices of design in synthetic biology and nanotechnology laboratories conveys a quite different view. It provides a window on a range of experimental practices aimed at exploring the world as a field of potentials. On the basis of a number of famous examples of design in synthetic biology laboratories, Maureen O'Malley (2009, 2011) convincingly argued that the laboratory practices of synthetic biologists are not the materialization of computer models. She describes the experimental practices of synthetic biologists as an open-ended exploration of complex phenomena through the construction of objects. Far from being a straightforward and smooth process, such experimental investigations are made of "epistemic iterations" (Chang 2004), of gradual corrections of the wrong assumptions embedded in the design of the device. And at each step, they involve a lot of trials-and-errors, patching, hacking, debugging and kludging.[7] In other terms, the image of rational design that the champions of synthetic biology have constructed is in stark contrast with the real skills that they have to mobilize to make do. In her conclusion O'Malley made an interesting suggestion:

> The rhetoric of pure engineering appears to function as a strategy of discipline formation, which needs to be contrasted against the technical achievements (quite remarkable) and failings (less advertised) of synthetic biology so far. This question of whether kludging can be overcome or whether it lies inseparably at the heart of both life and biological practice is perhaps the general research question that synthetic biology is addressing (even if the "field" does not see it that way). (O'Malley 2009, 386)

[7] The phrase "kludging" coined in information technology refers to an inelegant but successful solution to a problem in computer hardware or software. It is said to be an acronym made of three terms: klumsy, ugly and dumb.

The self-image of synthetic biologists as responsible engineers relying on sound and rational principles is at odds with their actual practice of astute and heterogeneous tinkering. In technoscientific research, failure is expected; failure is welcome because it is not perceived as the refutation of a conjecture or a model. It is rather seen as an invitation to a rapprochement between the ways of nature and the ways of human technology. For instance, Michael Elowitz who pioneered the design of genetic circuits "from scratch" on the basis of two engineering principles – decoupling and abstraction – built a genetic circuit, an oscillator that was meant to operate as independently as possible from the underlying cellular system (Elowitz and Leibler 2000). However, the device did not work because noise and interaction with the host cell contributed to the process (Nandagopal and Elowitz 2011). Although the failure threatens the basic assumptions of biopart engineering, it has not been considered as a refutation that could threaten the promises of this approach to synthetic biology. The negative result has been turned into a new opportunity to explore the role of noise and stochasticity in living cells.

Here points the fascinating perspective of a process of mutual learning between the object and the subject of investigation. When looking at the actual practices of research in a design mode, it is clear that it is nothing like the projection of abstract engineering principles on a passive matter.

11.3 Ontological Profile: In the Midst of Things

Let us turn now to the ontological assumptions underlying such research practices. Indeed ontology is not the major concern of technoscientific researchers. As mentioned above, they are not interested in representing the structure of matter, finding the ultimate particles, or even discovering the laws of nature. They are remarkably indifferent to the ontological structure of the world. As Peter Galison (2017) notices, the unconcern about ontological questions is a striking feature of current research, even in physics. Yet being indifferent to ontology does not mean being ontology-free.

One way of disentangling underlying ontological assumptions is by looking at the metaphors used by active scientists. In their discourses synthetic biologists use two favorite metaphors to describe what they are doing: assembling Lego® bricks into modules and reading and rewriting the code of life (Bensaude Vincent 2016). Both metaphors convey the view of intervention on passive material entities. Despite the popularity of the playful metaphor of the Lego® construction in the discourse of synthetic biologists, their practice is more like playing chess game with cells than assembling bricks to make a module. Listen for instance to Elise Cachat, a young scientist who is working on the design of mammalian cells in order to engineer tissues for kidney repair at Edinburgh University. She presents her work as "engineering self-organization in mammalian cells." (Cachat 2016) The title itself includes a paradox: if the cells are self-organized, the targeted arrangement proceeds from their own dispositions rather than from human intervention. Engineering in this case is not analogous to a design, with a designer informing matter and controlling its behavior. In a private interview Cachat acknowledges that she is uncomfortable with the term "chassis" borrowed from automotive industry, which suggests independent parts to be assembled along an assembly line. She says "my chassis often rebels. It is faster than me and responds before I can understand what's going on." In other terms, she plays and negotiates with the powers embedded in her object of design. Instead of looking at the object under scrutiny from a distance in order to objectify a phenomenon and control it, she operates in the middle of things, *in medias res* and strives to remain close to them.

While in the idealtype of "science" one always assumes a distance between knowing and being, technoscientific objects merge the epistemic and the ontological. Since scientific representations take the form of propositions, they assume that the world is composed of *facts* rather than of things or objects. Scientific propositions typically claim "that something is the case," (including dispositional properties), or that "this has been observed or measured," etc. By contrast, in the technoscientific model, the capacities of construction and control that objects demonstrate are not considered as confirmations or corroborations of propositions about them, but as knowledge in itself – "thing knowledge." (Baird 2004)[8] It could well be argued from a technoscientific perspective that the kind of knowledge synthetic biology displays shifts the focus away from epistemic agents to *objects as knowers*, as Axel Gelfert (2013) put it. He provokingly but convincingly argued that the micro-organisms themselves could be considered as the loci of knowledge, of a "living thing knowledge," already stabilized in the form of sequences, proteins, organelles, and metabolic pathways that function well together, while synthetic

[8] More materialistic than Latour, Baird criticizes both the semantic model of scientific knowledge as "justified true belief" and the semiologic model of the actor-network theory, with its text-producing black-boxes. Baird argues that scientific instruments do embed objective knowledge not so much because they are theory-laden (often they first function without a theory), but rather because of the analogy they draw between their technical functionning and the functional properties of truth. By studying the technicalities of instruments, Baird insists on "what truth does for us," assuming that the technical creation and stabilization of a new phenomenon is objective knowledge, even without theory or propositional knowledge. However, Baird's account of instrumental knowledge concerns science and matters of truth and falsehood, and not technoscience, on which he takes a critical sociological stance (i.e. technoscience means the contamination of the gift economy characteristic of scientific exchanges by the values of market economy). Accordingly, Baird does not go as far as considering a distinctively *technoscientific* "thing knowledge." Baird's thing knowledge is always about objective knowledge with a pretension to universality, not about local model/objects fittings in which a lot of technoscientific knowledge consists. Similarly Hans-Jörg Rheinberger (1997) defines "experimental systems" as the smallest integral working unit of research where the division between "epistemic thing" and "technical conditions" is relevant. "Epistemic things" are the material entities manipulated in experiments and they embody what researchers do not now or hope to know. When epistemic things become known, they are turned into standard techniques, tools for mundane mapping or commercial applications. They become "technical objects" embodying what has been known during the dialogue between the technical conditions and the epistemic thing. Talking about "technoscientific knowledge" would bypass the distinction between epistemic things and technological conditions that Rheinberger regards as the driver of experimental science. For Rheinberger, it would be talking about industrial development, not about research.

biologists are clumsily attempting to access the working knowledge encapsulated in living organisms by reverse engineering.

Most technoscientific research presupposes the assumption of powers and agencies in molecules and living entities. In this respect, it is closer to Leibniz's monadology than to Descartes's mechanism. While the latter banned all powers and qualities from nature and compared it to a clock, the former understood the clock as a restless and responsive mechanism (Riskin 2015). Technoscience naturalizes agency rather than transferring all the powers and agencies to a designer. It is a way of exploring the capacities of a wide range of objects – molecules, nanoparticles, materials, genes, proteins, neurons, circuits, networks, etc – and taking advantage of their inner powers and spontaneous movements. It is seizing opportunities and trying to cooperate with what molecules can afford in certain circumstances or under specific constraints.

This focus on capacities rather than on the regularities of general laws suggests that a philosophy of technoscience could encourage the trend of Neo-Aristotelianism in philosophy supporting a realist perspective on causal powers (Greco and Groff 2013). While modern science equates nature and artifact and merges them in the universal mechanism, technosciences are rather sensitive to the local potentialities of matter. Far from being homogeneous and passive, materials have "implicit forms" (Simondon 2005) that offer a range of opportunities to scientists and engineers. As Simondon emphasized, technological design succeeds provided that it fits in with these local forms. Although technosciences operate according to scientific laws (quantum physics for instance remains the general framework in nanophysics), they are not interested in nomological work. They above all consist in taking advantage of local dispositions and powers that fix the spectrum of what can be done, and that require adapted design strategies. Technosciences deal less with an homogeneous and universal nature, than with a broad range of *phuseis* that are of local relevance.

Although they invite us to overcome modern concepts and to update perhaps more antique ones, the fact remains that technosciences do not accurately fit in Aristotle metaphysics. Metaphysical concepts such as dispositions are adequate, but not for their explanatory power. They matter as agencies operating in the world rather than as causal powers. There is no attempt at identifying basic powers from which everything else could be derived. In stark contrast to laws-centered science, dispositions are introduced for practical reasons, for what they afford. So it is the pragmatist orientation of technoscience, which commands a new ontology.

Technoscientific objects do not fit in Aristotle's metaphysical framework for three additional reasons.

First, many of them (e.g. molecular devices and nanoparticles) are not adequately characterized by a stable structure and constitution as they continuously interact with the instrumental environment. They are better defined as relational entities. In this respect Rom Harré's redefinition of James J. Gibson's concept of affordance is adequate (Harré 2003). The interplay between the experimental setting and the causal powers of the world is the main attribute of affordances. Harré's emphasis on the ontological disparity between the instruments and nature also matters. Together they form a "World/apparatus complex" which conveys the view of the technoscientific world as a domesticated version of the wild world, like a farm, a space of culture.

Second, technoscientific objects such as nanoparticles could hardly be considered as metaphysical substances. In so far as they come into being through the intrinsic dynamic of material entities and endure in existence through interactions they are always in the making, waiting for realization. They challenge the distinction between *substantia* (what it is) and *potentia* (what it can do or become). Accordingly, they would rather require the kind of process ontology outlined by John Dupré (2012).

Finally, technoscientific objects challenge the ancient divide between *phusis* and *technè* as well as between *phusis* and *nomos*. They belong to no specific category and they exhibit multiple temporalities (cosmological, biological, social, technical, economical, …). From an ontological perspective, they do not appear as coherent entities. A narrative genre like an ontography seems more appropriate (Loeve and Bensaude Vincent 2017). Unlike ontology, ontography is an attempt to identify the modes of existence of particular entities and focuses on the multiplicity of modes. Not only it deflates the quest for the fundamental level underlying material entities (Lynch 2013) but it does not assume a causal chain between levels of being. In this perspective technoscientific objects appear as both real and historical. Real because as agencies they interact with the world and its causal powers; historical because they exist thanks to transitory associations of natural powers, technological instruments and people, and like any of us, they have a life trajectory.

11.4 Political-Ethical Profile: Value-Ladenness

Technoscience challenges the classical dichotomy between subjects and objects of knowledge. On the one hand, the transcendental ego gave way to a plurality of heterogeneous producers of knowledge including situated scientists and engineers, instruments, hackers, science policy makers. … on the other, the objects of knowledge are no longer defined by their relation to the representations of knowing subjects. Rather, they are defined by what they do, by their presence and performance in the world – while epistemic agents are redefined by their participation in the agency of objects. Technoscience thus challenges the classical dichotomy between subjects and objects not only from the viewpoint of knowledge but also from that of ethics.

Technoscientific objects shift from the modernist self-image of science as value neutral because they are *explicitly* value-laden, endowed with values that are epistemic as well as economic (e.g. competitiveness), sociopolitical or ethical (e.g. sustainable development). Whereas something comes to existence as a scientific object when it is considered as a matter of fact and kept away from the matters of interests it elicit, technoscientific objects have no clear-cut boundaries that would circumscribe their existence to the realm of pure research *versus* applications, or of facts *versus* values, object *versus* subject, properties *versus* uses, being *versus* becoming, etc. Contrary to the alleged neutrality or amorality of scientific research – its *applications* alone being considered value-sensitive – technoscientific objects are designed to acquire new capacities and functions in an ongoing process of valuation. Therefore, they have "unrestricted materiality": what counts as technoscientific object cannot be defined once and for all. As exemplified by carbon allotropes they may change their mode of existence by connecting with different entities (Loeve and Bensaude Vincent 2017). When traveling in different environments like nanoparticles technoscientific objects shift identity in vivo (Faadel et al. 2013, Albanese et al. 2014). Their materiality may spread far away from what designers initially planned – an issue that is of ethical concern since technoscientific objects may connect to other entities in unpredictable ways.

Technoscientific objects are value-laden not only because they are designed to suit human purposes for useful applications but rather because they are *mundane*. While they often behave in unfamiliar ways, elicit surprises or display uncanny properties, they are made familiar for lay publics through their merging with well-known objects such as the "molecular wheelbarrow" and so on (Loeve 2011b). They are invested with a variety of values and interests that make them worldly and talkative, meaningful for everyone. The ethos of disinterestedness, or the epistemological break that Gaston Bachelard described as constitutive of scientific activity, are abolished: Technoscientific objects such as the oncomouse or stem cells are close to the laypublic, they are both attractive and repulsive, they generate hopes and fears. Undoubtedly, the strong involvement of artists in bio- and nanotechnology highly contributed to this trend.

The heavy load of values carried by technoscientific objects is indeed due to the porous boundaries between science and society in what science policy analysts have labeled "Mode 2" research (Gibbons et al. 1994), or "post-normal science" (Funtowicz and Ravetz 1993). However, the load of values far exceeds the expectation of industrial applications or economic benefits. Technoscience is much more than application-oriented research. It is better defined by its axiological charge resulting from a process of investment by multiple actors. The intrinsic value-laden character of technoscientific objects can lead to dignify technoscience (it is not only about money, power and practical efficiency) as much as to critically engage with technoscientific objects (by showing how alternative values could or should be invested in their design). The two stances are not exclusive. All sorts of values – epistemic, technological, societal, economic, military, and environmental – are invested in technoscientific research so that, as Javier Echeverria (2003) argues, it is the conflict of values, which characterizes what he describes as the "technoscientific revolution." Technoscientific research consequently demands assessments and regulations that are both matters of concern in the community of philosophers of science.

11.5 Conclusion

In this chapter, we have made the case for the constitution of a philosophical concept of technoscience despite its polemic charge. As philosophers we are neither *pro nor contra* technoscience as we have no normative purpose. We do not even claim that technoscience took over science because, in our view, science and technoscience are not stable categories with fixed attributes and boundaries. Rather we consider technoscience as an idealtype of research practice focused on the design of objects which may still co-exist with the scientific idealtype. In labeling objects as "technoscientific" and scrutinizing their epistemological, ontological, and ethical status we hope to open up new philosophical perspectives.

The concept of technoscience is an incentive for philosophers of science to pay more attention to emerging technologies as well as for philosophers of technology and STS scholars to address knowledge issues (Houkes 2009). In this respect, it may act as a boundary concept between current STS and recent philosophical trends concerned with objects, things and modes of existence. While contributing to the "empirical turn" taken by the philosophy of technology (Kroes and Meijers 2001; Brey 2010), a philosophy of technoscience has the potential to open up the field of philosophy to the "political turn" taken by STS scholars in the last decade (Pestre 2004, 2008, 2010), as well as to the ontological twist given to the STS movement by its pioneers Woolgar and Latour (Woolgar and Lezaun 2013; Latour 2013). Finally, a philosophy of technoscience is vital to develop critical views about the cloud of buzzwords, which surround a number of big research projects, and more importantly to better understand what's going on in research laboratories and the kind of objects that come into existence.

References

Albanese, A., Walkey, C. D., Olsen, J. B., Guo, H., Emili, A., & Chan, W. C. (2014). Secreted biomolecules alter the biological identity and cellular interactions of nanoparticles. *ACS Nano, 8*(6), 5515–5526.

Arroyo, M., Heltai, L., Millán, D., & DeSimone, A. (2012). Reverse engineering the euglenoid movement. *Proceedings of the National Academy of Sciences, 109*(44), 17874–17879.

Attwater, J., & Holliger, P. (2014). A synthetic approach to abiogenesis. *Nature Methods, 11*(5), 495–498.

Baird, D. (2004). *Thing knowledge: A philosophy of scientific instruments*. Berkeley: University of California Press.

Bensaude Vincent, B. (2001). The construction of a discipline: Materials science in the U.S.A. *Historical Studies in the Physical and Biological Sciences, 31*(part 2), 223–248.

Bensaude Vincent, B. (2009). *Les vertiges de la technoscience. Façonner le monde atome par atome*. Paris: La Découverte.

Bensaude Vincent, B. (2016). The moral economy of synthetic biology. In J. Boldt (Ed.), *Synthetic biology. Metaphors, worldviews, ethics and law* (pp. 87–100). Springer VS: Wiesbaden.

Bensaude Vincent, B., Loeve, S., Nordmann, A., & Scharz, A. (2011). Matters of interest: The objects of research in science and technoscience. *Journal for General Philosophy of Science, 42*, 365–383.

Bensaude Vincent, B., Loeve, S., Nordmann, A., & Schwarz, A. (Eds.). (2017). *Research objects in their technological setting*. Abingdon: Routledge.

Brey, P. (2010). Philosophy of technology after the empirical turn. *Techné: Research in Philosophy and Technology, 14*(1), 36–48.

Bunge, M. (2012). *Evaluating philosophies*. New York: Springer.

Cachat, E. (2016). Engineering self-organization in mammalian cells, bioarchitecture symposium, In: *Living machines, 5th international conference in biomimetic and biohybrid systems*, Edinburg, July 19–22, 2016.

Caldwell, L. K., & Deville, W. B. (1968). *A syllabus for the study of science, technology and public policy*. Bloomington: Indiana University Press.

Carrier, M., & Nordmann, A. (Eds.). (2010). *Science in the context of application. Boston studies in the philosophy of science* (Vol. 274). Dordrecht: Springer.

Cartwright, N. (1999). *The dappled world. A study of the boundaries of science*. Cambridge, MA: Cambridge University Press.

Chang, H. (2004). *Inventing temperature: Measurement and scientific progress*. Oxford: Oxford University Press.

Channel, D. F. (2017). *A history of technoscience. Erasing the boundaries between science and technology*. Abingdon: Routledge.

Chemcal & Engineering News. (1983). 61 (32).

Clement, R. C. (1965). Last call for birds of prey. *Audubon Magazine, 67*(1), 37.

Echeverria, J. (2003). *La revolucion tecnocientifica*. Madrid: Fondo de Cultura Economica.

Elowitz, M. B., & Leibler, S. (2000). A synthetic oscillatory network of transcriptional regulators. *Nature, 403*(6767), 335–358.

Endy, D. (2005). Foundations for engineering biology. *Nature, 438*(7067), 449–453.

Fadeel, B., Feliu, N., Vogt, C., Abdelmonem, A. M., & Parak, W. J. (2013). Bridge over troubled waters: Understanding the synthetic and biological identities of engineered nanomaterials. *Wiley Interdisciplinary Reviews: Nanomedicine and Nanobiotechnology, 5*(2), 111–129.

Forman, P. (2007). The primacy of science in modernity, of technology in postmodernity and of ideology in the history of technology. *History and Technology, 23*(1/2), 1–152.

Funtowicz, S., & Ravetz, J. (1993). Science for the post-normal age. *Futures, 31*(7), 735–755.

Galison, P. (2017). The pyramid and the ring. A physics indifferent to ontology. In B. Bensaude Vincent, S. Loeve, A. Nordmann, & A. Schwarz (Eds.), *Research objects in their technological setting* (pp. 15–26). Abingdon: Routledge.

Gelfert, A. (2013). Synthetic biology between technoscience and thing knowledge. *Studies in History and Philosophy of Science Part C: Studies in History and Philosophy of Biological and Biomedical Sciences, 44*(2), 141–149.

Gibbons, M., Limoges, C., Nowotny, H., Schwartzman, S., Scott, P., & Trow, M. (1994). *The new production of knowledge - the dynamics of science and research in contemporary societies*. London: Sage.

Gibson, D. G., et al. (2010). Creation of a bacterial cell controlled by a chemically synthesized genome. *Science, 329*(5987), 52–56.

Greco, J., & Groff, R. (Eds.). (2013). *Powers and capacities in philosophy: The new Aristotelianism*. London: Routledge.

Guchet, X. (2011). Les technosciences: un essai de définition. *PhilonSorbonne, 5*, 83–95. https://philonsorbonne.revues.org/348

Hacking, I. (1983). *Representing and intervening*. Cambridge, MA: Cambridge University Press.

Haraway, D. (1997). *Modest_Witness@Second_Millennium.FemaleMan©Meets_OncoMouse™: Feminism and Technoscience*. New York: Routledge.

Haraway, D. (2003). *The companion species manifesto : Dogs, people, and significant otherness*. Chicago: Prickly Paradigm Press.

Harré, R. (2003). The materiality of instruments in a metaphysics of experiments. In H. Radder (Ed.), *The philosophy of scientific experimentation* (pp. 19–39). Pittsburgh: Pittsburgh University Press.

Hottois, G. (1979). *L'inflation du langage dans la philosophie contemporaine*. Bruxelles: éditions de l'Université de Bruxelles.

Hottois, G. (1984). *Le signe et la technique. La philosophie à l'épreuve des techniques*. Paris: Aubier.

Hottois, G. (1996). *Entre symboles et technosciences. Un itinéraire philosophique*. Seyssel: Champ Vallon.

Houkes, W. (2009). The nature of technological knowledge. In A. Meijers (Ed.), *Philosophy of technology and engineering sciences* (pp. 309–350). Amsterdam: Elsevier.

Idhe, D. (1991). *Instrumental realism: The interface between philosophy of science and philosophy of technology*. Bloomington: Indiana University Press

Idhe, D., & Selinger, E. (Eds.). (2003). *Chasing technoscience, a state of the art view of technoscience studies featuring the work of Donna Haraway, Don Idhe, Bruno Latour, Andrew Pickering*. Bloomington: Indiana University Press.

John, D. (2012). *Life as a process*. Oxford: Oxford University Press.

Jones, R. A. (2004). *Soft machines. Nanotechnology and life*. Oxford/New York: Oxford University Press.

Klein, U. (2005). Technoscience avant la lettre. *Perspectives on Science, 13*(1), 226–266.

Kroes, P., & Meijers, A. (Eds.). (2001). *The empirical turn in the philosophy of technology*. Amsterdam: Elsevier.

Lasswell, H. D. (1946). The interrelations of world organization and society. *The Yale Law Journal, 55*(5), 889–909.

Lasswell, H. D. (1948). The prospects of cooperation in a bipolar world. *The University of Chicago Law Review, 15*(4), 877–901.

Lasswell, H. D. (1957). The political science of science. *Scientific Monthly, 84*(1), 34–44.

Latour, B. (1987). *Science in action*. Milton Keynes: Open University Press.

Latour, B. (2013). *An inquiry into modes of existence*. Cambridge, MA: Harvard University Press.

Loeve, S. (2011a). Sensible atoms: A techno-aesthetic approach to representation. *NanoEthics, 5*(2), 203–222.

Loeve, S. (2011b). 'Ceci n'est pas une brouette' Grands et petits récits des nanotechnologies. In O. Thierry & S. Houdart (Eds.), *Humains, non-humains* (pp. 208–220). Paris: La Découverte.

Loeve, S., & Bensaude Vincent, B. (2017). The multiple signatures of carbon. In B. Bensaude Vincent, S. Loeve, A. Nordmann, & A. Schwarz (Eds.), *Research objects in their technological setting* (pp. 185–200). Abington: Routledge.

Lynch, M. (2013). Ontography: Investigating the production of things, deflating ontology. *Social Studies of Science, 43*(3), 444–462.

Lyotard, J.-F. (1979). *La condition postmoderne*. Paris: Minuit. English Edition: Lyotard, J.-F. (1984). *The postmodern condition* (G. Bennington & B. Massumi, Trans.). Minneapolis: University of Minnesota.

Marliere, P. (2009). The farther, the safer: A manifesto for securely navigating synthetic species away from the old living world. *Systems and Synthetic Biology, 3*(1–4), 77–84.

Mazzolai, B. (2014). *A robot that grows like plant roots*, EU project. https://ec.europa.eu/programmes/horizon2020/en/news/barbara-mazzolaiplantoid-project-robot-grows-plant-roots

Nandagopal, N., & Elowitz, M. B. (2011). Synthetic biology: Integrated gene circuits. *Science, 333*, 1244–1248.

Nordmann, A. (2006). Collapse of distance: Epistemic strategies of science and technoscience. *Danish Yearbook of Philosophy, 41*, 7–34.

Nordmann, A. (2010). Philosophy of technoscience in the regime of vigilance. In G. Hodge, D. Bowman, & A. Maynard (Eds.), *International handbook of regulating nanotechnologies* (pp. 25–45). Cheltenham: Edward Elgar.

Nordmann, A., Radder, H., & Schiemann, G. (Eds.). (2011). *Science transformed? Debating claims of an epochal break*. Pittsburgh: Pittsburgh University Press.

O'Malley, M. (2009). Making knowledge in synthetic biology: Design meets kludge. *Biological Theory, 4*(4), 378–389.

O'Malley, M. (2011). Exploration, iterativity and kludging in synthetic biology. *Comptes rendus de l'académie des sciences-Chimie, 14*(4), 406–412.
Pestre, D. (2003). Regimes of knowledge production in society. Towards a more political and social reading. *Minerva, 41*, 245–261.
Pestre, D. (2004). Thirty years of science studies: Knowledge, society and the political. *History and Technology, 20*(4), 351–369.
Pestre, D. (2008). Challenges for the democratic management of technoscience: Governance, participation and the political today. *Science as culture, 17*(2), 101–119.
Pestre, D. (2010). What about participation, governance and politics? Remarks on contemporary techno-science and the field of STS. In M. Epple & C. Zittel (Eds.), *Science as cultural practice: Vol. I: Cultures and politics of research from the early modern period to the age of extremes* (pp. 163–182). Munich: Akademie Verlag.
Raynaud D. (2015). Note critique sur le mot 'technoscience'. *Zilsel*. https://zilsel.hypotheses.org/1938
Raynaud, D. (2016). *Qu'est ce que la technologie?* Paris: Éditions Matériologiques.
Rheinberger, H.-J. (1997). *Toward a history of epistemic things*. Stanford: Stanford University Press.
Riskin, J. (2015). *The restless clock. A history of century-long argument over what make living things tick*. Chicago: The University of Chicago Press.
Roherty, J. M. (1960). Oskar Morgenstern, the question of national defense. *American Political Science Review, 54*(2), 530–531.
Rudd, R. L. (1964). Pesticides and the living landscape. *Natural Resources Journal, 4*(1), 432–434.
Sebbah, F. D. (2010). *Qu'est ce que les technosciences? Une thèse épistémologique ou la fille du diable?* Paris: Les Belles Lettres.
Seris, J.-P. (1994). *La Technique*. Paris: Presses universitaires de France.
Simondon, G. (2005). *L'individuation à la lumière des notions de forme et d'information*. Paris: Millon.
Simondon, G. (1958). *Du Mode d'existence des objets techniques*. Paris: Aubier (2013 edition). English edition: Simondon, G. (2016). *On the mode of existence of technical objects* (C. Malaspina, Trans.). Minneapolis: Univocal Publishing.
USA Today. (2010). *Scientists create 1st bacteria strain from man-made DNA*. 20/05/2010.
Woolgar, S., & Lezaun, J. (Eds.). (2013). Special issue: A turn to ontology in science and technology studies? *Social Studies of Science, 43*(3), 321.

Part III
Revisiting Anthropological Categories

Chapter 12
Technology and Nature

Catherine Larrère and Raphaël Larrère

Abstract It is generally taken as obvious that artificial objects are fabricated, and that the technical process of producing artificial entities is a fabrication. The paradigm of *fabrication* may serve as a reference for a whole series of technical activities: the production of objects and tools, the construction of buildings, of infrastructures, the synthesis of substances which do not exist in nature. It is the art of *making*; it applies equally to the art of the craftsman (unique creations) and to industrial fabrication (the serial production of a number of identical objects).

We aim to show that besides fabrication, there is another model of technical action, that may variously be called "steering," "stewardship" or "husbandry" but for which we will use here the more generic term of "piloting." It resides on using natural forces or living beings, or on orienting natural processes in order to obtain desired results. These are the multiple ways of adjusting to nature as could be done with a partner. These are not the arts of making but of doing-with, of inducing things to happen. To such a model belong agriculture and animal-raising, all the arts of controlled fermentation, as well as therapy.

It is not a question of successive periods in the history of technology: these two models are neither mutually exclusive nor do they follow one after the other. A certain number of recent technologies (nanotechnologies, biotechnologies), which are generally considered as fabrications, can equally well be considered under the heading of piloting. We will attempt to show that the latter characterization is actually preferable; the advantage being that far from separating technologies and nature and making them independent realities, it makes it possible to understand how technologies can act in and with nature.

This chapter has been written in French for this volume and translated by John Stewart, University of Technology Compiègne.

Keywords Fabrication · Making · Making-do · Piloting

C. Larrère (✉)
University of Paris 1 Panthéon-Sorbonne, Paris Cedex 05, France

R. Larrère
Institut National de la Recherche Agronomique (INRA), Paris Cedex 07, France

S. Loeve et al. (eds.), *French Philosophy of Technology*, Philosophy of Engineering and Technology 29,
https://doi.org/10.1007/978-3-319-89518-5_12

On March 11th 2011 Japan was hit by the Tohoku earthquake, and the tsunami which followed submerged the reactors of the Fukushima nuclear power-plant which thus went out of control. After recalling the circumstances of this accident, a science journalist for the newspaper *Le Monde* made the following comment on the aftermath: "In front of a bemused world, there was played out what can be interpreted as a conflict between humans and their machines." (Foucart 2014) This remark is surprising: the nuclear power-plant did not cease to function as a result of a breakdown, but following a natural event, and as a consequence the humans were faced with natural forces: radioactivity, in itself, is not a human creation! Why then was this incident described as a revolt of technical objects, rather than as an incapacity to take into account the natural environment?

Is it that because of the increasing importance of the human techno-sphere, nature has if not disappeared at least retreated far into the background? It is commonplace to say that the development of the "new technologies" – information and communication technologies, biotechnologies, nanotechnologies and above all their convergence[1] – has obliterated the distinction between what is natural and what is artificial. This does not really mean that Nature, considered as a global entity or as a set of processes which occur independently of our will, no longer exists (seismic events, tsunamis, volcanic eruptions, storms and tempests and so on are still there to remind us of its existence); but what it does mean is that, in this extension of the scope of technology into natural domains that were hitherto beyond reach (such as the genome and the "elementary bricks" of living organisms with biotechnology, the molecular level with nanotechnology), technical objects acquire a form of independence; and by this very fact, they threaten to turn against us. It is as though the question has been shifted, away from the relation of human beings to nature towards the relation of human beings with their own technological creations, a relation which can take the form of a confrontation. Fictions abound which tell stories of such confrontations between men and machines. Instead of being intermediates between humans and nature, have technical objects become the competitors of humans?

However, we have to ask whether this supposed conflict is actually real; when the relation to Nature is masked in this way, is it not rather an optical illusion due to the way in which we apprehend technical action? It is generally taken as obvious that artificial objects are fabricated, and that the technical process of producing artificial entities is a fabrication. Now it is true that "fabrication" is one way of portraying technical processes; but it is not the only way. We aim to show that besides fabrication, there is another model of technical action, that may variously be called "steering," "stewardship" or "husbandry" but for which we will use here the more generic term of "piloting." It is not a question of successive periods in the history of technology: these two models are neither mutually exclusive nor do they follow one after the other. A certain number of recent technologies (nanotechnologies, biotechnologies), which are generally considered as fabrications, can equally well be considered

[1] The NBIC convergence (between nano-, bio-, info- technologies and cognitive science) has been a watchword of American nanotechnology programs since 2002.

under the heading of piloting. We will attempt to show that the latter characterization is actually preferable; the advantage being that far from separating technologies and nature and making them independent realities, it makes it possible to understand how technologies can act in and with nature.

12.1 Two Models: Fabrication and Piloting

It has often been said that the Greeks had little interest in technology; some authors even go so far as to say that they had a mental block on this point (Schuhl 1938). To be sure, this commonplace should not be pushed too far, because the Greeks did have machines, engineers and even a form of technical thought (Gille 1980). Nevertheless, as The historian Jean-Pierre Vernant has shown, in their myths and in Greek thought there is no general concept – either for technical activity or for work – which regroups all the various technical activities in a specific category while distinguishing them from other sorts of human activity. The concept of *tekhnè* runs from theatre to rhetoric and includes medicine and pottery. Jean-Pierre Vernant (1965: 511) recalls that, in Homer, the category of *demiourgoi* includes not only artisans (professional workers of metal and wood) but also "the confederations of soothsayers, harbingers, faith-healers and bards." What is dominant is an undifferentiated vision of action, a vision which valorizes effort and application and which aims at success: the latter is viewed as a recompense for the capacity to achieve success in a moving world, by virtue of the energy deployed and because one has managed to seize the right moment (*kairos*). This goes for the relations with things, and for the relations with other men, but also for relations with the gods.

It is maybe on the basis of this primary absence of distinction, from which a differentiation of activities emerges gradually, that we should understand the way in which Plato, in the *Timaeus* (1950), borrows from the vocabulary of human crafts the term of the demiurge (*demiourgos*) to designate the divine architect who fashions or forges a world out of the primeval chaos. As a historian of philosophy expresses it, "the divine demiurge creates and engenders as much as he works," he is at one and the same time a craftsman, a poet and a father (Joly 1974: 277). The divine craftsman or architect breathes life into the work that is produced in this way, he actually animates it. What Plato presents in the *Timaeus* is thus a complete model of creative activity; by means of the Platonic reading of the biblical story of Genesis by Christianity, this model will be transmitted to Christian theology and, from there, to Western philosophy as a whole. A particularly important aspect is the idea (known as *verum factum*, the *maker's argument* or the *argument from design*) that if God knows the world it is because He has made it according to an intelligible model.

In this perspective, to create is to produce according to a model: "every creation where the worker has fixed his view on that which is always identically conserved, using such an object as a model (*eidos*) in order to reproduce its essence and its key properties, such a work will necessarily be beautiful." (Plato, *Timaeus* 28a) This model of divine creation can also serve as a reference model for human acts of fabrication (which can easily be assimilated to a creation) as the imposition of a form on matter. This conception is widespread throughout reflections on technique and technology. It is in this way that Aristotle distinguishes an *artifact* from a *natural entity*: the former, concerning its movements and changes, depends on the purpose imprinted in it by its maker; the latter bears within itself the principle of its movements. Marx says nothing else when he compares the work of a bee and that of a human architect:

> A spider performs operations which resemble those of a weaver, and a bee – by the skill with which it produces the structure of wax cells – would put to shame many an architect. But what distinguishes the worst architect from the most expert bee, is that the architect constructs the cell in his head before constructing it in the hive. The result the architect achieves pre-exists in an ideal form in the mind and imagination of the worker. (Marx 1867: t.I, section 3, chap. VII)

Introduced in this way, and progressively refined, the paradigm of *fabrication* may serve as a reference for a whole series of technical activities: the production of objects and tools, the construction of buildings, of infrastructures, the synthesis of substances which do not exist in nature. It is the art of *making*; it applies equally to the art of the craftsman (unique creations) and to industrial fabrication (the serial production of a number of identical objects). In the art of the craftsman, technical action still appeals to the know-how and the astute intelligence (*mètis*) of the producer. Organized "scientifically," the technical actions of industry pretend – but do not succeed – to dispense with both the know-how and the intelligence of the worker (Dejours 2003). This possibility of passing from traditional crafts to contemporary industry, without really changing the model, is doubtless one of the reasons for the success of the latter.

There is, however, another reference for technical activity, which can also be found in Plato: it is that of the pilot. The pilot, for Plato, is most often the captain of a ship, the one who has "authority over the sailors" whom he "governs" in the same way as he governs his ship (Plato, *Republic*, 342e). Piloting – or governing – is thus a political metaphor. As such, it is related to metaphors of the "pastor," the shepherd or cowherd, who likewise governs his flock (Plato, *Republic*, 343a; *Politics* 267c, 268c). But the art of navigation, the art of animal-rearing, as well as the art of medicine which Plato frequently associates with them, can also be studied for their own sake, as a form of technical activity, a mode of relating to nature which makes it possible to obtain desired results. These other modes of technical activity cannot, however, be identified as forms of fabrication: the pilot does not fabricate his course, he plays with the winds and the currents in order to lead his ship where he wants to go. He does not "make," he *makes-do with.*[2]

Now it is important to realize that a large number of technical activities are comparable to the conduct of a ship, and belong to the same paradigm that we have up

[2]This is the reason why we speak of piloting. We could certainly have taken other figures of *doing-with*, but it seems to us that the navigator, to the extent that he has to accommodate himself to natural forces over which he has no control, is a prime example of this mode of acting, taking into full account the context in which it occurs.

until now termed « piloting ». It is the case of the farmer: like the pilot, the farmer does not himself fabricate the result; he diverts natural processes to his own ends, in order to procure certain benefits. Cultivating amounts to favoring certain plants in the play of inter-specific competition. At the end of the eighteenth century, a physiocrat characterized agriculture in this way, referring to "the infinite multiplication of species in the forces of nature; a multiplication whose only limits are those of the amount of help that humans are prepared to provide by repelling many of the species which dispute the ground and the resources in favor of those he wishes to see predominate." (Grivel 1784)

The same goes for the art of the animal breeder. It is a mistake to consider that domestication is a form of fabrication (Sigaut 1988). The domestication of animals proceeds less from the imposition of human will on animal material than from subtle negotiations between humans' and animals' wills. Human will is not almighty. Some animal species are easily tameable while others are recalcitrant. For instance, we never achieved the domestication of spiders as we did with silkworms. Domestication rests on practices of selection (which thus take opportunistic advantage of the natural processes of heredity) and on certain forms of socialization. In their natural state, domesticated species were almost always sociable species, with all that implies about hierarchies and dominance within groups of social animals (Midgley 1983). The success of domestication depended then on the capacity of humans to take over themselves the role of the dominant animal in the group, and to make use of the other hierarchical relations which subsist within a coherent group of animals. Therefore domestication results from a process of socialization involving humans and the animals who lived around them (domesticated commensals). Human societies have always included animals, they have always been mixed societies. Domestication was only able to succeed because humans entertained exchanges of services, information and emotions with animals from whom they were able to obtain this or that form of collaboration.[3]

We should add to agriculture and animal-raising (whether it is a question of stock animals or animals of labor) all the arts of controlled fermentation: without the manipulation of natural ferments (and thus of diverse natural processes that humans reorient and bend to their own ends) there would be neither bread, nor wine, nor cheese, nor preserves[4]... And, in this overview of techniques which belong to the realm of piloting, we should certainly not omit the therapeutic arts which help the organism to be cured of ailments.

[3] Such exchanges and such relations carry certain obligations, which we have designated by the expression "the domestic contract." What we want to say by this expression is that the relations within these mixed communities – and human communities have always been mixed communities including animals – are not "natural" relations (they are neither automatic, nor inscribed in a pre-existing natural order); on the contrary they are the result of a history, they are fashioned by a certain form of free mutual consent, which has to be renewed from generation to generation, to the form of society which is created in this way (Larrère and Larrère 2001).

[4] The work of Marie-Claire Frederic (2014) offers an impressive panorama of the considerable set of practices which make it possible to preserve foodstuffs without cooking and without additives.

Less omnipresent than fabrication, the paradigm of piloting has nevertheless not been entirely neglected. Jean-Jacques Rousseau gives a presentation of it in the eleventh letter of the fourth part of *La Nouvelle Héloïse*. Julie takes Saint Preux, her former lover, on a visit round her garden. Julie's garden is a wild, luxuriant specimen nature, which leads Saint Preux (who has just come back from an escapade in tropical seas) to believe that he has before his eyes the islands of the New World. Julie teases him: it is only the scruffy old orchard that Saint Preux had so liked to wander in 8 years ago. Faithful to his creed as a good dualist, Saint Preux has difficulty grasping this illusion of a wilderness: this "artificial desert" can only be the result *either* of a natural process, *or* of considerable human labor. Julie reassures him, with a touch of irony: "nature did it all, but under my direction, and there is nothing here that I have not designed." (Rousseau 1761: 388) The embarrassment of Saint Preux demonstrates the originality of the approach: Julie's garden just does not fit in the stark opposition between the natural and the artificial, the wild and the fabricated. It is neither one nor the other, belonging to a sort of "in-between," to a *relation* between natural spontaneity and human intervention. It is the result of an "art of the natural." (Larrère C. 1988; Larrère R. 2013)

Without presenting in detail all that Rousseau has to say about this art, let us note the gathering in the enclosure of a large number of plants, both wild and cultivated, from the surrounding countryside; and the combination of vegetal landscapes. Here, trees have been felled and the stumps removed, leaving a clearing planted with grasses and flowers from the fields. There, the apple-trees have been replaced by "underbrush of rose, raspberry, and currant bushes, patches of lilac, hazel, elderberry, mockorange, broom, trifolium, which decked the earth while giving it a fallow appearance." (Rousseau 1761: 388–389) Elsewhere, in order to simulate a dense forest, Monsieur de Wolmar and his men used the techniques of runners and, in order to render the shade more dense, they associated the trees of the orchard with parasitic plants (clematis, honeysuckle, jasmin, vines, hops and bindweed) "that, trained upon the tree trunks, surround their crowns in the thickest foliage and their feet in shade and coolness." (Rousseau 1761: 389) A stream which ran into the lake has been diverted to meander through the old orchard, forming a set of branching brooklets. Running lively here, dawdling there, the waters gather in a place where, by plantations of bushes and some sown cereals, birds have been attracted and animate the garden by their songs. Finally, let us imagine the multiple interventions which have rendered the enclosure impenetrable to view, abolished all straight lines, nullified any perspective and produced a baroque profusion of vegetal curbs and arabesques. "Common grasses, common shrubs, a few trickles of water flowing without frills, without ducts have been enough to embellish it" (Rousseau 1761: 397) were enough to introduce a note of wilderness into the old orchard. This art of the natural is inspired by nature and deploys natural dynamics. The gardening accomplished by Monsieur de Wolmar and his men leaves to the vegetation itself the task of growing, of reproducing, of multiplying; they contented themselves with orienting and deflecting its spontaneous dynamics. Julie's maxim is not one of abstinence; at most it is one of least effort, because it is not only a question of *doing-with* (natural processes) but also of *inducing to do* (nature acts in your place).

There are thus two quite distinct models, that of fabrication and that of piloting; that of *making* and that of *making-do with* (and *inducing to do*).

But does not the very development of technologies lead to the pervasive extension of fabrication and to the retreat, or disappearance, of piloting? At first sight fabrication and piloting belong to two different domains, the mechanical and the organic. Now is progress in technology not identical with the advance of mechanization? It would seem to be the case in the domain of energy where mechanical traction has replaced animal traction. Modern biotechnologies, and in particular the cloning of mammals, can be considered as the intrusion of the mechanical in the domain of the living: biological filiation, where there is a succession of distinct individuals, is rendered obsolete by cloning which supposedly involves identical replication, in an unlimited number of copies, by a form of fabrication. What is erroneously designated as the disappearance of the distinction between the artificial and the natural (because, as shown by Descartes, "artificial things" do not cease for all that to be natural, they continue to obey the same laws[5]), should rather be understood as a mechanization of the organic, or even as a substitution of the self-maintaining machine for spontaneous living organisms. But is this really the case? Can fabrication really make piloting disappear? If we look closer, nothing could be further from the truth.

12.2 Piloting and Fabrication: What Is Their Relation to Nature?

Piloting and fabrication: these are two idealized prototypes. In reality, one always finds a mixture of the two. Piloting makes use of fabricated objects: the farmer uses tools (the scythe, the cart or a tractor); there is no pilot without a ship which is a pretty complex fabricated object. Conversely, there is no fabrication which does not incorporate processes that are not themselves fabricated and are never completely controlled: it is the case, to take one notable example, of the blacksmith, who relies on physico-chemical processes that he turns to his own profit, often without even knowing what they are (the strength and resistance of the traditional Toledo steel depends on molecular reactions at the nanometric level, reactions which as such were completely unknown in the Middle Ages). The mode of *doing-with* is thus common in the mechanical arts, in the activities of repairing and even in the upkeep of all sorts of materials. This has been well illustrated by Matthew Crawford:

> Fixing things, whether cars or human bodies, is very different from building things from scratch. The mechanic and the doctor deal with failure every day, even if they are experts, whereas the builder does not. This is because the thing they fix are not of their own making, and are therefore never known in a comprehensive and absolute way. The experience of failure tempers the conceit of mastery; the doctor and the mechanic have daily intercourse with the word as something independent. (Crawford 2009: 81)

[5] "All things which are artificial are by that very token natural." (Descartes 1641: IV §203)

As is shown by the examples of the blacksmith and the motor mechanic, the distinction between piloting and fabrication is not reducible to the distinction between the organic and the mechanical. The distinction between piloting and fabrication is not so much between the organic and the mechanical, but rather between the dynamic nature of the processes that piloting calls upon, and the static character of the model or the form that fabrication imposes on matter. There is therefore no good reason for associating piloting with routine or obsolete ways of doing things. Quite the contrary. While "fabrication" accords with a static vision of the duality of form and matter, "piloting" appeals to an apprehension of nature in terms of dynamic processes which is well in accord with contemporary notions.[6] The technologies of piloting have benefited from scientific developments (be it in agronomy, in zootechnology, in forestry, in ecology). The history of wine-making is a good illustration of how piloting and technological sophistication go hand-in-hand: a simple visit to a cellar in the Bordeaux area of France where the finest wines are produced largely suffices to attest to this fact. More often than not, precise control of fermentation requires chemical analysis and sometimes recourse to a computer.

The relation between fabrication and piloting is not one of historical succession, but one of logical precedence, that of a nature that we have not made and which necessarily pre-exists anything which we ourselves may do. This is particularly clear in the case of medicine where, as Canguilhem has shown, the question is not to oppose a medicine which is non-interventionist (close to nature or Hippocratic) as against a medicine which is interventionist and highly technological; it is rather to appreciate that any medical practice necessarily relies on the capacities of *auto-poïesis* and self-reparation of the body. One counts on the body, one acts *in concert* with it. Thus, progress in the field of technical intervention and the recourse to fabricated objects does not lead to the disappearance of nature which remains the unavoidable reference: "one can thus continue, even in the age of industrial pharmacology, of the imperialism of laboratory biology, and the electronic treatment of diagnostic information, to speak of nature in order to designate the initial fact of the existence of auto-regulating living systems." (Canguilhem 2002: 31)

The scheme of fabrication, that of action according to a prior model, that of the house which pre-exists in the mind of the architect before being realized materially, is that of an *independence* of the object. The technical object is the result of a process of fabrication but, because it was conceived beforehand, it takes on an independent existence once it is realized. The accent is placed, notably by Bacon, on the unprecedented character of the objects which are fabricated in this way. They are without any natural counterpart:

> to create a technical object, explains the philosopher and economist Cornelius Castoriadis, is not to alter the present state of nature, as one does by simply moving one's hand; it is to constitute a universal type, to posit an *eidos* which henceforward "exists" in its own right quite independently of any particular empirical examples. (Castoriadis 1978: 302)

[6] A conception to which the disappearance in ecology of reference to natural equilibria bears witness (Botkin 1990; Blandin 2009; Larrère and Larrère 2015).

It is this independence of the fabricated object with respect to its context, whether it be its context of origin (the nature within which it was conceived), its context of fabrication (the concrete empirical process at the end of which it comes into existence), or the context of its subsequent existence (the environment in which it is introduced) which is at the heart of its success. It has enabled the objects produced by Western industry to invade the whole world. This has been taken to such an extent that even techniques which clearly belong to the realm of piloting, such as agriculture, have come to be apprehended on the model of fabrication: the "industrialized" agriculture of the twentieth century has replaced the natural functionalities mobilized in traditional agriculture, by the massive use of products of industrial origin. With this conception of agronomy, nature is no longer a potential partner whom one learns to work with, that one seeks to pilot in order to procure foodstuffs; the tables have been turned and nature has come to be perceived as a threat because it is the source of hostile forces (weedy plants, voracious or pathogenic insects, capricious climates) that will have to be actively countered. It is now necessary to prepare the ground, to sow seeds, to spread fertilizer and herbicides; and for that it will be necessary to mechanize, to provide chemical treatments against illnesses and harmful pests, often to irrigate in order to avoid drought. However in spite of all that, even if it has modified the environment to allow highly productive seeds that have been selected in experimental fields to manifest all their potential, this "industrialized" agriculture is still unable to control *all* the factors of production. In the soil, even though it has been plowed and enriched with chemical fertilizers, there remain insects which feed on fecal matter, worms and fungi which obtain nutrients from dead organic matter or which are symbiotic, and micro-organisms which are not domesticated but which decompose wastes, recycle organic matter and thus nourish plants. Many harvests are pollinated by insects which are beyond the control and maintenance of the farmer.

Thus, it is precisely the root of the success of fabrication that is at the source of problems. Because of the fact that artificial objects do not cease to be natural, they inevitably continue to interact with the natural environment in which they find themselves… in ways that go beyond what was expected of them. This is the realm of unwanted secondary side-effects (various sorts of pollution, but also perturbations of the great bio-geo-chemical cycles, of which climate change is the paradigmatic example) which constitute the current environmental crisis. We have arrived at the situation, described by Hans Jonas, where.

> action takes place in a context where every deployment on a large scale of a given capacity engenders, in spite of the honorable intentions of the agents, a series of secondary effects which are inseparable from the immediate and intended "beneficial" effects, a series which after a cumulative process finally results in harmful and even disastrous consequences which far outweigh the sought-for benefits. (Jonas 1997: 232)

In addition, however "independent" it may be, a fabricated object never functions all on its own, it requires what we may term an associated environment: in order for a train to roll, it requires rails, and these must be adapted to the topography of the place where the railway-line has been built. Similarly, a large number of technical objects that we use daily (domestic electrical appliances, telephones, computers…) only function because of technical mega-systems which aliment them in electricity, or which transmit electro-magnetic waves. Thus fabrication requires conjoint operations of abstraction (from the natural conditions in which the object will be introduced), standardization (so that it can be used across a broad range of contexts), and inter-operability (for it to be compatible with other technical objects and with its "associated milieu"). Now the varied contexts of use of a technical object are not without importance: the success of Peugeot motors in North Africa was due to their high performance in arid climates. There are many anecdotes which illustrate how technical objects have found uses which were in no way foreseen by those who conceived them. For example, certain scissors made in France in the eighteenth century were composed of steel so poorly tempered that they were hardly able to cut; in spite (or because) of this they sold very well in Turkey where their buyers used them to snuff candles (Clicquot de Blervache and Vincent de Gournay 1758: 94). The use and the context (both natural and social) in which they intervene are thus decisive elements in the forms of existence of any technical object.

Always under the threat of failure, the arts of *doing-with* can also have unintended effects on their natural environment. However, in contrast to objects produced in the mode of fabrication, objects in the *doing-with* mode systematically involve taking full account of the context (i.e. the complex natural environment and the equally complex social environment) in which they are inscribed, as a clear condition for having any chance at all of succeeding. Thus they structurally presuppose interventions which take into account the evolution of the piloted system in its relation to its context, whether this evolution is unfavorable (thus requiring corrective action) or, as sometimes happens, fortuitously favorable (in which case it is important to take advantage of the situation). This holds even when the action is controlled at a distance, as is the case for example in contemporary long-distance sailing races: piloting (here in the literal sense of steering a course) remains a form of intervention where knowledge of the context is decisive. This can be seen as a limiting case where the attempt to "fabricate" the course from a terrestrial control post still cannot avoid taking into account the local data in all their singularity.

Piloting is an approach which is attentive, empirical and precautionary, so sensitive to the context of production that it must always be adapted to local conditions and can hardly be reproduced identically. In agriculture, when one does not try and force it into the fabrication mode, the diversification of ways of producing accompanies the spatio-temporal variability of natural and social conditions: this is what makes the difference between high-productivity models which ignore the soil by treating it as a simple substrate, and ecological agriculture which is constantly attentive to the natural functionalities of the soil in all its complexity and seeks to harness them to its advantage (Larrère 2002; Griffon 2013).

Ecological agriculture consists of "shifting from a logic of development based on mastery of local environments, to another logic based on connivance and complicity with the eco-systems: playing *with* and not against the natural variability of ecosystems, deploying the knowledge accumulated by ecological science." (Griffon and Weber 1996: 120) It could hardly be clearer: what is at stake is a return to the "piloting" mode of relating to natural processes. First of all it is necessary to consider every cultivated plot as a complex ecosystem. That supposes no longer considering the soil as a mere substrate, that is plowed to aid plants to take root and that is enriched with fertilizers to "force-feed" their growth. In the agronomic conception of high-productivity agriculture, the life of the soil with its fauna, its myco-flora which recycle wastes, its bacteria which mineralize organic matter and in some cases fix atmospheric nitrogen – in short, the soil as an ecosystem with all its complex interactions – is considered as a black box. The turn to ecological agriculture involves opening this black box, and understanding (in order to take advantage of them) the biological mechanisms which preside over reproduction and fertility, as well as those which contribute to maintain a structure and a content of organic matter favorable to the rooting of cultures and the retention of water. This also involves devoting effort to the selection of seeds adapted to the local environmental conditions. It invites consideration of the other biological components of the agro-system (and the landscape in which it is inserted) not as a reservoir of threats (insects and other sources of devastation that have to be held at bay by an arsenal of pesticides), but rather as a source of auxiliaries able to limit aggression by competition, predation or parasites. Finally, it supposes associating animal-raising with growing crops, and rotating crops including leguminous plants (fodder plants and beans) which fix atmospheric nitrogen. The appropriate rotation of crops on the same plot of land helps to break the reproductive cycles of species which compete with the harvest crops and which cause damage. In this way it is possible to control weeds, parasites and harmful species with a minimal use of chemical sanitary products. An astute piloting of the flux of fertilizing elements and crop rotation thus makes it possible to limit the impact of agriculture on wild fauna and flora.

The true development of our technical capacities, far from abandoning the paradigm of piloting in favor of fabrication, leads us to consider that the greater the importance of context, the more it is preferable to apprehend our actions in terms of piloting rather than fabrication. Each of these ways of acting can be harmful for our environment and for ourselves. But whereas the defects of fabrication accompany its apparent successes – because it is the increase in our powers which render the involuntary consequences of our technical actions ever more dangerous – the failings of piloting come from the inadequacies of our knowledge of the context. Piloting thus renders indispensable a deep knowledge of the nature to which we relate – a form of knowledge which fabrication has the pretention of being able to dispense with…

The point at issue is indeed how we conceive our relation to nature. Doubtless, during the course of its development, reflection about technology has become differentiated from political reflection, separating the relations from things (or with nature) from social relations between humans, each aspect being studied in its own right. But the metaphorical use of the terms employed continues to put into relation the domains even if they are formally separated. Piloting remains a metaphor of political action, ambivalent because it can designate both unshared authority (the captain is sole master on board) and skill in maneuvering. However in general it is a question of relations between humans, as is attested by its use in the vocabulary of management. However it is possible to reverse the metaphor: a way of characterizing human social relations can be transferred to the relations of humans with nature.

This is typically what is done by Bacon when, after having condemned the political ambition of the conquest of humans by other humans, he declares that it is good and praiseworthy when it is a question of the relation to nature, speaking of "empire" and attributing to relations with nature terms that are generally used for praising political action: compared to sorts of political conquests, "if a man endeavor to establish and extend the power and dominion of the human race itself over the universe, his ambition (if ambition it can be called) is without doubt both a more wholesome and a more noble thing than the other two. Now the empire of man over things depends wholly on the arts and sciences. For we cannot command nature except by obeying her." (Bacon 1620 § 129) Still, may we not say that Bacon does recognize a certain form of submission? It is constantly in the political metaphors of command and obedience that Bacon pronounces what is only superficially an inversion of our relation to nature: it is still a question of dominating it. The whole passage is oriented towards the power we can have over nature. In 1992, the Heidelberg appeal, initiated by certain scientists to reject the environmental concerns expressed by the Rio summit meeting, echoed Bacon in affirming that "Humanity has always progressed by putting Nature at its service, and not the other way round." (Lecourt 1993: 172)

Now it is precisely this arrogant relation to nature that Rousseau refuses, whether it be a question of human social relations or of our relations to nature. The "direction" that Julie claims in no way excludes reciprocity: one can see this in the way that she speaks of the birds as her "hosts," playing on the twin sense of the term (we receive them/they receive us). We may therefore ask whether the pre-eminence of the paradigm of fabrication over that of piloting, when presenting technical action, does not stem from its pretentions of domination – mastery and control – that it allows itself to proclaim, rather than from any particular precision in its analysis of technical procedure. This is what we shall examine when we look at the new technologies.

12.3 New Technologies: Fabrication or Piloting?

Towards the end of 2013, the announcement was made, with much pomp and ceremony, that an artificial heart had been successfully implanted into a cardiac patient. Early in 2014, when the patient died, the death was attributed to a "breakdown." But it was not clear, in the way the news was presented, whether it was a breakdown of the artificial heart, or whether it was the result of a failure of the organism, metaphorically treated as a mechanical breakdown. This shows the extent to which the vocabulary of fabrication has invaded medicine, but also the extent to which the distinction between the literal and figurative sense remains salutary: the body is not a machine, at most it can be compared to a machine (a comparison which may be more or less pertinent). It is indeed important to distinguish between the actual employment of machines in technologies (be they medical or other), and the assimilation of living organisms (or other natural inorganic entities) to fabricated mechanisms, an assimilation that certain technological developments renders tempting, but which remains dubious.

The "new technologies," bio- or nano-technologies, are most often presented in terms of fabrication. For biotechnologies, the old Cartesian model of the animal-machine has been brought back into service, renewed on the occasion by cybernetics.[7] Thus, one speaks of "fabricating" animals that are "programmed" to develop desirable characteristics – whether it be for medical research (transgenic mice which develop certain cancers, or Alzheimer's disease); for animal-raising (such as the "enviro-pigs" who fix more phosphate and hence pollute less than normal pigs – Golovan et al. 2001); for the production of drugs or special foodstuffs; or yet again for duplicating, by cloning, the animals with optimal performances...

Nanotechnologists proclaim a "bottom-up" approach, supposedly capable of manipulating matter "atom by atom." Each molecule is envisaged (at best) as a machine, and the comparison is systematically developed. For example: "In 2001, we designed a machine which we dubbed a 'molecular wheelbarrow.' It had two molecular front wheels, with a diameter of 0.7 nanometers, attached to an axle; two legs at the rear, like the legs of a wheelbarrow; and finally two little sleeves at the back to act as handles, where the tip of the STM would push." (Joachim and Plévert 2009: 64) "Making" and "knowing" converge. What confers on the scanning tunneling microscope its decisive importance in nanotechnology, is that it not only makes it possible to observe molecules, it observes them *because* it modifies them, by exploiting their surface energy (Loeve 2009; Guchet 2014: 50).

It is thus in the tradition of the demiurge and the Platonic identification of doing and knowing, recast in modern terms (the *verum factum*, or *maker's argument*), that the physicist Richard Feynman, consecrated as the father of nanotechnology, declared "*What I cannot create, I do not understand*."[8] This formula is often employed to express the ambition of scientists in synthetic biology: if they are able to assemble an artificial biological system which has the properties attributed to a living system, incorporating a rational design which leaves nothing to chance, they consider that they have understood the living system. Synthetic biology has the ambition of being an engineering science, aiming at producing systems conceived in advance, but which are predictable and make it possible to act on the world in a deliberate way. Synthetic biology thus has the pretension of substituting its own design, its own intentional model, for the tinkering, the chance trial-and-error of biological evolution. (Marlière 2009). According to an OECD report, "Synthetic biology may be especially powerful in this respect because it

[7] Even if cybernetics does include, in its very name, a central reference to piloting (*kubernêtês*, pilot), it is based on the postulate of an equivalence between finalized behaviors and mechanisms that can be calculated (analyzed into a finite sequence of operations). Thus cybernetics has oscillated between an art of piloting which exploits the complexity and context-dependency of systems and a reductionist science for which everything can be represented and regenerated from discrete code (Dupuy 2000; see also Dupuy in this volume).

[8] This phrase was found, written in chalk on Feynman's blackboard, at his death in 1988.

frees the design of biological system from the process of natural evolution." (OECD and the Royal Society 2010: 8)[9]

There is, in these proclaimed ambitions, a manifest will to dominate nature and to acquire power over it: one affirms that with nanotechnologies one will be able to "track down matter in its furthest retrenchments," one claims that by mastering the gene one has "cracked the secret of life," one aims at taking over the role of evolution, including the forthcoming of a trans-humanity which will take over from current humanity which is a highly imperfect result of natural evolution. But this glorification of power runs the risk of turning into its opposite (which comes finally to the same thing): the denunciation of total power. The ancient myths which are critical of technology, such as the sorcerer's apprentice, portray a power which escapes from all control, and which is thus terrifying. But whether one glorifies it or fears it, in both cases what one focuses on is the *power* of technology. Eric Drexler has been one of the promoters of nanotechnology; he has also propagated fear of the "grey goo," a magma potentially produced by the reproduction of nanorobots which could threaten to invade and to absorb everything which exists. Here again we find the idea that our creatures may turn against us, that the hour has come when it is no longer question of the combat of man against nature, but of a combat against his own technical creations. Having passed over to the technical object itself, the power escapes from its author.

But what if this turn-around from power to powerlessness happens all the more easily because the proclaimed ambition hides what is really happening? If we look closer, we will see that what is presented as a fabrication is actually much better understood as piloting. This is indeed the case with nanotechnologies. What unifies them is not so much the nanometric scale at which they intervene than what that implies: one reaches a level – that of atoms, molecules and clusters – where the individual properties that emerge are not fully predictable from statistical physics alone (which governs the material behaviors of larger macroscopic scales) or from quantum physics alone (which governs the behaviors of subatomic scales). Of course, there are cases when nanoscientists can predict and control very well the behavior of nano-objets and systems. However, the capacities of control emphasized in nanoscience publications are generally based on "retrodictions," i.e. ex-post predictions for explaining this or that surprising behavior, as fieldwork epistemological studies have shown (Lenhard 2004; Nordmann 2004; Loeve 2011). It is only at the end of a long series of trial-and-errors, semi-empirical simulations, and construction of local and provisional models that nanoscientists become able to pilot nanoscale processes. The kind of control that is achieved in nanotechnology is not synonymous of mastery of the subject over the object; it rather consists in generating specific relationships between the individual nano-object and the specific features of the environment of operation (Riedel et al. 2008: 77). Therefore, nanoscientists do not achieve control by imposing their will on matter or acting

[9] See also Dupuy (2008): "The truly metaphysical aim of this program, whose ambitions have already triggered a technological, industrial and military race at the scale of the planet, is to make man into a demiurge, or more modestly "the engineer of evolutionary processes."

directly on the nanoscale; they do not "take control of matter." They rather *delegate* control to other nanoscale entities that surround the nano-object and constitute its associated milieu (Simondon 1958). Only the constitution of an associated milieu allows stabilizing the behavior of the object and making it partially predictable – at least in the laboratory or in highly-standardized industrial processes in microelectronics, for instance: concerning nanoparticles in whole organisms and ecosystems, the environment is so complex that uncertainty remains irreducible.

Nanotechnology is thus a field where the will of controlling things does not contradict the search for surprising and unexpected behaviors (Lenhard 2006). Both advance hand in hand. In their laboratories, nanoscientists have to control as many parameters as possible in order to arrange the conditions for a surprise to occur – otherwise, they would not even recognize the surprise (Loeve 2009). But if one retains the paradigm of fabrication, where it is a question of producing objects with predictable behaviors on the basis of an *a priori* model, this manner of placing the unexpected at the heart of the approach is highly unsettling. It amounts to making the engineer, in the words of Jean-Pierre Dupuy (2009), "a sorcerer's apprentice by design." However one can see things differently, if one does not consider the technical process as a fabrication of objects without any natural antecedents, but rather as a way of intervening in a pre-existent context where one aims to exploit the potentialities.

The same can be said of biotechnologies (transgenesis and cloning) and synthetic biology. An enquiry as to what actually goes on in the laboratories is enough to reveal that transgenesis and cloning actually involve tinkering that is far from being mastered, where the consequences on the biology and the behavior of the organisms modified in this way are far to be fully understood (Larrère 2006, 2011). While it might be rendered fully predictable regarding the desired outcome (as when a single gene is "silenced" for instance), holistic effects on whole organisms and ecosystems are emergent, predictable only *a posteriori* by trial-and-error. In a similar vein, Bernadette Bensaude Vincent and Dorothée Benoit-Browaeys write that synthetic biology attempts "to make living organisms in order to see what they can do"; it is a question of "exploring possible avenues, with an open mind with respect to what is unforeseen, unexpected, unpredictable." (Bensaude Vincent and Benoît-Browaeys 2011: 55–56) Maureen O'Malley (2010) has highlighted the opportunistic tinkering, the tricks of the trade exhibited by the engineers of synthetic biology to achieve their results. In order to characterize these groping and astute improvizations she uses the term "kludging," originally invented by the hackers of free software.

"What the setup teaches us about a phenomenon comes from what we can 'make the various agents do'," writes Bensaude Vincent (2009: 117) about nanotechnologies and their various experimental setups. We gladly generalize this statement by adding: whether these agents be atoms, electrons or molecules in nanotechnology, or the enzymes, plasmids or elementary building-blocks of living organisms manipulated by biotechnology. In transgenesis, one puts a restriction enzyme to work to isolate a DNA fragment that one wants to transfer, and another enzyme to obtain the polymer chains in sufficient quantities. The insertion of the transgene in the host genome is sometimes achieved by exploiting the natural aptitude of the bacterium *Agrobacterium tumefaciens* to integrate the "genetic construction" – previously inscribed in a plasmid – and to transmit it to the host cells. When cloning mammals, it is the cytoplasm of the egg-cell which "re-programs" the transferred nucleus, a specific process of despecialization which renders it totipotent. The experimental conditions "activate" the egg-cell and put it to work. Without this work of the egg-cell, there would not be any "clones." Finally, the molecular circuits that are synthesized by synthetic biology must necessarily be inserted in "chassis organisms" which, according to the circumstances, will either do what those who conceive the circuits ask of them, or else do something quite different because the synthetic system entertains interactions with its cellular environment and acquires unforeseen functional properties.

These *bottom-up* procedures thus pertain to the domain of *inducing to do*. In the event it is a form of *making-do with* since the result depends on the situated behavior of the agent that one puts to work and it is vital to take careful note of this behavior, to adapt to it or attempt to adjust it. Now, unlike machines or robots (which are programmed to have predictable behaviors – neglecting possible breakdowns) that are put in operation in industrial production-lines, the behavior of these agents is not entirely predictable and comprises a large portion of uncertainty.

Rendering experimental conditions artificial leads to the emergence of properties that are unprecedented in the natural conditions which occur on Earth; it is a question of exploring these potentialities and to select those which present a particular scientific, economic or military interest. The presentation of new technologies as an exploration of natural possibilities, which is frequent (Debru 2003), can thus be much better understood by taking piloting as a reference rather than fabrication.

In this context, the phrase of Feynman ("*what I cannot create, I cannot understand*") can be understood quite differently from the pretention to the powers of a demiurge (playing down the notion of "creating"): I only "understand" after the event, by observing what I have unwittingly brought into existence by manipulating natural processes without having any clear idea in advance of what was going to happen. The objects that nanotechnologies bring into existence are no longer materialized representations, illustrating general laws of nature. (Bensaude Vincent 2004) They are rather complex systems, which are inseparable from the particular experimental setup that has revealed them and which makes it possible to observe their behavior; they contribute to the "natural" context in which they have arisen. In virtue of this, these new technologies (the same goes for synthetic biology) do not produce new independent models; rather, one introduces perturbations in a network of interactions without any *a priori* knowledge as to what the reactions of the system may be. Whereas "modernity" proclaimed the equivalence between the knowledge of things and the capacity to reproduce the process of their production (assimilated to a fabrication), technologies such as nanotechnologies are defined by the fact that we do *not* have any knowledge in advance of the constitutive processes of the systems we are interested in. That radically changes the relation between technology and nature. One can no longer think of technology in terms of domination, with the predictability that is therein implied. We must be prepared to be surprised, and not to make plans to avoid being surprised. It is less a question of "mastering" nature, but rather of entering into a conversation with nature as a partner. Renouncing relations of domination, we can enter into relations of co-operation.

12.4 Conclusion

How are we to explain the fact that the paradigm of fabrication has had so much success, to the point where all technical activity has come to be thought of as fabrication? There is doubtless the will to affirm human creativity, the human capacity to bring into existence what nature alone has not been able to achieve. As Castoriadis (1978: 301) has noted, "in nature there is no equivalent whatsoever of the pulley, the stirrup, the potter's wheel, the steam locomotive or the computer." There is also the will to affirm the domination of man over nature. A theological model, a model of knowledge, a basis for justifying the appropriation of a nature that we transform in our own image: fabrication (assimilated to a creation) is not a modest model. It is a model which is both imperial (Bacon's "the empire of men over things") and imperialist.

However this model has met its limits, at the very measure of its success. Deliberately ignoring the context in which it is inserted, fabrication comes to have relations with nothing other than itself; and men, through this sort of technical object, meet only themselves. As has been shown by Arnold Gehlen, the desire or the nostalgia for a purely autonomous mechanism is an archaic fantasy, maybe even *the* original human fantasy: "a sort of internal sense of the specific constitution of mankind which reacts to that which, in the external world, is analogous to this specific constitution." (Gehlen 1990: 112)

But the automatic nature of technology is only the external projection of the representation that we have of ourselves as autonomous beings, cut off from any relation. Now just as this definition of "autonomy" does not really suit human beings, so being perfectly automatic is not the perfect form of mechanism. As has been shown by Gilbert Simondon (1958), a technical object does not become more and more concrete in becoming increasingly an automaton, but on the contrary by becoming more and more related to other technical objects, and via the intermediary of their associated milieu to the whole of nature. With the concept of fabrication, we do not get out of ourselves and, being unable to conceive of anything that is not us, we end up by meeting ourselves… in the form of so many threats.

The concept of piloting enables us to escape from this sort of solipsism. It is a concept which is much more modest (but also much more astute); it reminds us that we always act in a world that we have not ourselves created. Because it obliges us to take into consideration the context in which we intervene, the conception of piloting, unlike that of fabrication, is not forgetful of nature.

Acknowledgements A preliminary version of this text has been discussed by Bernadette Bensaude Vincent and Pierre de Jouvancourt during the authors' workshop held in Paris Sorbonne on 22–23 June 2015 in view of the preparation of the collective book.

References

Bacon, F. (1620). *Novum organum* (2012th ed.). Bottom of the Hill Publishing.

Bensaude Vincent, B. (2009). *Les vertiges de la technoscience. Façonner le monde atome par atome*. Paris: Editions La Découverte.

Bensaude Vincent, B., & Benoit-Browaeys, D. (2011). *Fabriquer la vie. Où va la biologie de synthèse?* Paris: Editions du Seuil.

Bensaude Vincent, B. (2004). *Se libérer de la matière? Fantasmes autour des nouvelles technologies*. Versailles: INRA éditions.

Blandin, P. (2009). *De la protection de la nature au pilotage de la biodiversité*. Versailles: Editions Quae.

Botkin, D. (1990). *Discordant harmonies. A new ecology for the twenty-first century*. Oxford: Oxford University Press.

Canguilhem, G. (1978). L'idée de nature dans la pensée et la pratique médicales. In G. Canguilhem (Ed.) (2002). *Écrits sur la médecine* (pp. 15–48). Paris: Editions du Seuil.

Castoriadis, C. (1978). Technique. In C. Castoriadis (Ed.), *Les carrefours du labyrinthe 1*. Paris: Editions du Seuil.

Clicquot de Blervache, S., & Vincent de Gournay, J. (1758). *Considérations sur le commerce, et en particulier sur les compagnies, sociétés et maîtrises*. Amsterdam [i.e. Paris]: Pierre Guillyn.

Crawford, M. B. (2009). *Shop class as soul craft. An inquiry into the value of work*. New York: The Penguin Press.

Debru, C. (2003). *Le possible et les biotechnologies*. Paris: Presses Universitaires de France.

Dejours, C. (2003). *L'évaluation à l'épreuve du réel – Critique des fondements de l'évaluation*. Versailles: Editions Quae.

Descartes, R. (1641). *Les Principes de la philosophie* (1953rd ed.). Paris: Editions Gallimard, bibliothèque de La Pléiade.

Dupuy, J.-P. (2000). *The mechanization of the mind: On the origins of cognitive science*. Princeton: Princeton University Press.

Dupuy, J.-P. (2008). Günther Anders, le philosophe de l'âge atomique. Preface to G. Anders. *Hiroshima est partout*. Paris: Editions du Seuil.

Dupuy, J.-P. (2009). *La marque du sacré*. Paris: Carnets Nord.

Foucart, S. (2014). Le nucléaire et la vie. *Le Monde*, 14/03/2014. Retrieved from http://www.lemonde.fr/planete/article/2014/03/14/le-nucleaire-et-la-vie_4383506_3244.html

Frédéric, M.-C. (2014). *Ni cru ni cuit. Histoire et civilisation de l'alimentation fermentée*. Paris: Alma.

Gehlen, A. (1990). La technique dans l'optique anthropologique. In A. Gehlen (Ed.), *Anthropologie et psychologie sociale*. Paris: Presses Universitaires de France.

Gille, B. (1980). *Les mécaniciens grecs. La naissance de la technologie*. Paris: Editions du Seuil.

Golovan, S. P., et al. (2001). Pigs expressing salivary phytase produce low-phosphorus manure. *Nature Biotechnology, 19*(8), 741–745.

Griffon, M. (2013). *Qu'est-ce que l'agriculture écologiquement intensive?* Versailles: Editions Quae.

Griffon, M., & Weber, J. (1996). La révolution doublement verte: économie et institutions. In CIRAD, *Vers une révolution doublement verte*.

Grivel, G. (1784–1788). Article « Économie ». In *Encyclopédie méthodique. Economie politique et diplomatique*. Paris-Liège: Pancoucke, 4 vol. in-4°, t. 2.

Guchet, X. (2014). *Philosophie des nanotechnologies*. Paris: Hermann.

Joachim, C., & Plévert, L. (2009). *Nanosciences: The invisible revolution*. Singapore: World Scientific Publishing.

Joly, H. (1974). *Le renversement platonicien. Logos, episteme, polis*. Paris: Vrin.

Jonas, H. (1997). La technique moderne comme sujet de réflexion éthique. In M. Neuberg (Ed.), *La responsabilité. Questions philosophiques*. Paris: Presses Universitaires de France.

Larrère, C. (1988). Jean-Jacques Rousseau: la forêt, le champ, le jardin. In A. Cadoret (Ed.), *Chassez le naturel…* (pp. 23–61). Paris: Editions EHESS.

Larrère, R. (2002). Agriculture: Artificialisation ou manipulation de la nature ? *Cosmopolitiques, 1*, 158–173.

Larrère, R. (2006). Une éthique pour les êtres hybrides – De la dissémination d'Agrostis au drame de Lucifer. *Multitudes, 24*, 63–73.

Larrère, R. (2011). Le clonage des mammifères. Petite revue des arguments éthiques qui le concernent. In G. Le Dref et al. (Eds.), *Les usages du vivant* (pp. 49–64). Strasbourg: Néothèque.

Larrère, R. (2013). Chi protegge la natura è rousseauiano ? *IRIDE, Philosophia e discussionne publicca, 69*, 339–355.

Larrère, C., & Larrère, R. (2001). L'animal machine à produire: la rupture du contrat domestique. In F. Burgat & R. Dantzer (Eds.), *Les animaux d'élevage ont-ils droit au bien-être?* Versailles: INRA Editions.

Larrère, C., & Larrère, R. (2015). *Penser et agir avec la nature. Une enquête philosophique*. Paris: Editions La Découverte.

Lecourt, D. (1993). *Contre la peur, suivi de Critique de l'appel de Heidelberg*. Paris: Hachette, Pluriel.

Lenhard, J. (2004). Nanoscience and the janus-faced character of simulations. In D. Baird, A. Nordmann, & J. Schummer (Eds.), *Discovering the nanoscale* (pp. 93–100). Amsterdam: IOS Press.

Lenhard, J. (2006). Surprised by a nanowire: Simulation, control, and understanding. *Philosophy of Science, 73*(5), 605–616.

Loeve, S. (2009). *Le concept de technologie à l'échelle des molécules-machines. Philosophie des techniques à l'usage des citoyens du nanomonde*. Ph.D. dissertation, University Paris-Ouest Nanterre.

Loeve, S. (2011). "Ceci n'est pas une brouette." Grands et petits récits des nanotechnologies. In S. Houdart & O. Thierry (Eds.), *Humains, non humains. Comment repeupler les sciences sociales* (pp. 208–220). Paris: La Découverte.

Marliere, P. (2009). The farther, the safer: A manifesto for securely navigating synthetic species away from the old living world. *Systems and Synthetic Biology, 3*(1–4), 77–84.

Marx, K. (1867). *Le Capital* (1985th ed.). Paris: Champs Flammarion.

Midgley, M. (1983). *Animals and why they matter*. Athens: University of Georgia Press.

Nordmann, A. (2004). Molecular disjunctions. In D. Baird, A. Nordmann, & J. Schummer (Eds.), *Discovering the nanoscale* (pp. 51–62). Amsterdam: IOS Press.

O'Malley, M. (2010). Exploration, iterativity and kludging in synthetic biology. *Comptes rendus de l'Académie des sciences, 30*(2C), 3350–3357.

OECD & The Royal Society. (2010). *Symposium on opportunities and challenges in the emerging field of synthetic biology – Synthesis report*.

Platon. (1950). *Timée. Oeuvres Complètes, II*. Paris Editions Gallimard, bibliothèque de la Pléiade.

Riedel, D., Mayne, A., Dujardin, G., Bellec, A., Chiarravalotti, F., Cranney, M., & Lapastis, M. (2008). Probing the movements of a single molecule. *Plein Sud, Spécial Recherche 2008–2009* (Vol. 8, pp. 74–79). Retrieved from http://www.pleinsud.u-psud.fr/specialR2008/en/08_Probing_the_movements_of_a_single_molecule.pdf

Rousseau, J.-J. (1761). *La Nouvelle Héloïse, in Œuvres complètes, II* (1964th ed.). Paris: Gallimard. English edition: Rousseau, J.-J. (1997). *Julie, or the new heloise: Letters of two lovers who live in a small town at the foot of the Alps works* (P. Stewart & J. Vaché, Trans.). Hanover/London: University Press of New England.

Schuhl, P.-M. (1938). *Machinisme et philosophie*. Paris: Presses Universitaires de France.
Sigaut, F. (1988). Critique de la notion de domestication. *L'Homme, 28*(4), S.59–S.71.
Simondon, G. (1958). *Du Mode d'existence des objets techniques*. Paris: Aubier. English Edition: Simondon, G. (2017). *On the mode of existence of technical objects* (C. Malaspina, Trans.). Minneapolis: Univocal Publishing.
Vernant, J.-P. (1965). Remarques sur les formes et les limites de la pensée technique chez les Grecs. In J.-P. Vernant (Ed.), *Mythe et pensée chez les Grecs*. Paris: Librairie François Maspero.

Chapter 13
Leroi-Gourhan: Technical Trends and Human Cognition

Charles Lenay

Abstract The work of Leroi-Gourhan (1911–1985) has had a strong impact on twentieth century French thought. To account for the origin of our human capacities of memory, anticipation and language, Leroi-Gourhan builds on a "Technology" understood as the study of the functional linkage between the organisms and their environment. In continuity with the biological world, without sudden event (by miracle or by chance), it is to explain the gradual separation of social memory by the interplay of technical innovations that will allow free thinking detached from the immediate situation. The fulcrum of this liberation is the *tool*: both a biological fact and a movable organ, it permits the passage from the biological world to the human world.

Keywords Anticipation · Embodied cognition · Externalism · Cognitive science · Hominization · Leroi-Gourhan · Paleoanthropology · Social memory · Technology

The work of André Leroi-Gourhan has had a strong impact on twentieth century French thought. His work covers a vast spectrum, ranging from the history of technology to prehistoric art, prehistory, ethnology, paleontology and anthropology,[1] all

[1] Leroi-Gourhan (1911–1986), after studying Russian and Chinese and a mission in England to the British Museum (1933–4), worked on organizing the Far East and Arctic collections of the Paris *Musée de l'Homme* for its opening in 1937. After a mission to Japan in 1936–1938, during the war he was Assistant Curator at the *Musée Guimet*; at this time he wrote *L'homme et la matière* (*Makind and Matter*), a vast synthesis centered on human technologies, as well as a first thesis in ethnology, *L'Archéologie du Pacifique Nord* (*The Archeology of the North Pacific*), defended in 1946 under the supervision of Marcel Mauss (1873–1950). He then conducted a series of excavations (at Arcy-sur Cure, then at Pincevent) and prepared a second thesis, in paleontology, *Les tracés d'équilibre mécanique du crâne des Vertébrés terrestres* (*Traces of mechanical equilibrium in the skulls of terrestrial vertebrates*). Professor at the Sorbonne University, replacing Marcel Griaule in 1956, he published an essay on *Les religions de la préhistoire* (*The religions of prehistory*) (1964); and then

C. Lenay (✉)
Connaissance, Organisation et Systèmes TECHniques (COSTECH),
Université de technologie de Compiègne (UTC), Compiègne, France
e-mail: charles.lenay@utc.fr

S. Loeve et al. (eds.), *French Philosophy of Technology*, Philosophy of Engineering and Technology 29,
https://doi.org/10.1007/978-3-319-89518-5_13

linked and mutually embedded in an over-riding project: to try and understand the "human phenomenon" at all time-scales, in continuity with the biological world, and through a "Technology,"[2] i.e. the study of functional couplings between the organisms and their environment. The approach of Leroi-Gourhan is of great interest today, and it can offer cognitive science some fruitful and original hypotheses. To Leroi-Gourhan, cognitive faculties are not explained by the brain, but by an evolutionary process in which the brain is one of the consequences. Thus, in his masterwork *Le geste et la parole* (*Gesture and Speech*) (1964–1965), Leroi-Gourhan proposes an explanation, ranging from the biological process of hominization to the freeing of social memory, which accounts for the specificity of our cognitive capacities of anticipation and language. The fulcrum of this liberation is the *tool*: both a biological fact and a movable organ, it permits the passage from the biological world to the human world.[3]

In order to follow Leroi-Gourhan's approach, it is first necessary to grasp the originality of his study of technology in his ethnological works. I will then explain how he approaches paleontology, and provide several elements elucidating his conception of anthropological evolution. Several preliminary comments are in order.

First, it is important to insist on the fact that the work of Leroi-Gourhan is characterized by rigorous empirical investigations and great prudence with respect to philosophical generalizations; yet at the same time it opens up vast theoretical perspectives of great originality. In these pages I will restrict myself to this theoretical dimension, whose originality has not escaped a number of French philosophers (Georges Canguilhem, Gilbert Simondon, Michel Foucault, Gilles Deleuze, Jacques Derrida or more recently Bernard Stiegler). (Guchet 2015)

Second, in this chapter I consider only a part of Leroi-Gourhan's work. I will refer neither to the schools of ethnology and history of technology which have been inspired by his work, nor to his work on cave paintings, nor to his methods which have largely contributed to renewing prehistory (excavations by horizontal strata for historical, statistical and topographic analysis). Many discoveries have been made since the 1960s when Leroi-Gourhan published his main work in anthropological paleontology (for example the discovery of 'Lucy' and the characterization of *Homo habilis*), but there is nothing which seems to put into question the set of theoretical intuitions I shall present.

Finally, let me mention that I adopt a naturalistic reading of Leroi-Gourhan against certain hasty assimilations of his evolutionary perspective with a teleology of technology tinged with spiritualism, even though certain passages in his texts do seem to motivate such interpretations. The feature which marks the singularity and

---

his most important work of synthesis *Le geste et la parole* (*Gesture and speech*) (1964–1965). He subsequently devoted most of his time to excavations, and to a general reflection on the arts of prehistory. He was nominated professor at the Collège de France in 1969.

[2] Which he often writes using a capital "T" to distinguish the discipline from the operations, tools and systems constituting its object of study.

[3] In this Leroi-Gourhan extends an intuition of Marcel Mauss (Mauss 1936).

the interest of Leroi-Gourhan's approach is that by openly assuming a naturalistic posture with respect to the phenomenon of technology, he makes it possible to grasp the irreducibility of the human phenomenon… while at the same time avoiding a sharp break with the biological world.

13.1 Technological Trends and Technological Facts

Leroi-Gourhan began his career by a considerable amount of work in ethnology. In order to arrange the collections for the opening of the *Musée de l'Homme* in 1938, he undertook the construction of a terminology and a system of classification that allowed for the study of technologies from pre-historical times up to the industrial period. This work is presented in the two fascinating volumes of *Evolution and Technology* (Leroi-Gourhan 1943, 1945). The first thing that is striking is that such an undertaking is even possible. By taking into account the types of material, the basic means of action, and the forces which can be mobilized, it turns out that only a limited number of techniques are possible, and so they can be subject to a systematic description. Three central concepts are forged by Leroi-Gourhan in order to perform this scientific research into technology: the concepts of "trend," "degrees of factuality," and "technical milieu." (Leroi-Gourhan 1943: 325)

The term "trend" (*tendance*) does not designate any sort of final causation, but rather the determinism stemming from the limited number of possible modes of coupling between living organisms and matter.[4] According to the laws of geometry and rational mechanics, there are only a limited number of ways in which a given function can be realized. It is normal that roofs should have a double slope, that axes should have a handle, and that arrows should have a balance-point at one-third of their length. This being so, for the technical principles which are thus defined, it is possible to *construct* a series of objects and to speak of "progress," for instance, from the first flint choppers to copper knifes to steel swords (Leroi-Gourhan 1964a: 91).[5]

However, trends are not to be confused with the *facts*, i.e. the concrete local and historical observations of objects and practices. Leroi-Gourhan distinguishes several "degrees of factuality," i.e. for each object observed at different levels of description, starting from its function described in very general terms (which amounts to a materialization of the trend), followed by determinations which are

[4]The "trend" is "a simple abbreviation to characterize in one word the whole set of potentialities which only becomes realities under favorable conditions of the environment, symbolizing the penchant which is followed in the living world by all the needs for survival according to modalities which are increasingly complex." (Leroi-Gourhan 1943: 325)

[5]"In zoology as in ethnology, (…) *everything seems to happen as if* an ideal prototype of a fish or a flint blade evolved according to preset lines from the fish to the amphibian to the reptile to the mammal or bird; from the rough flint chopper to the finely hewn blades, to the knife of copper, to the sword of steel. Let there be no misunderstanding: these lines of evolution are simply the result of an aspect of life, that of the limited and inevitable choice that the milieu offers to living matter." (Leroi-Gourhan 1943: 14) My emphasis.

more and more complete up to the designation of the tool of a precise ethnic group at a given moment of its history. The trends are only abstract principles whose concrete realization is perturbed by multiple external and internal conditions. The external milieu comprises the physical environment as well as the ethnic environment. Contacts between ethnic groups (movements of men, of objects, of practices) can induce the appearance of new techniques. And above all, each ethnic group is characterized by a *technical milieu* which determines the changes it can accept. The adoption of a new technique, either by internal invention, or by reception during contacts with other ethnic groups, depends on the capacity of this technical milieu to reproduce the innovation in question. From this point of view, "between the autonomous invention and the straightforward borrowing from a neighbor, the difference is not very great" (one and the other both result in the creation of the same technical milieu). "In other words one only invents the spinning-wheel, or one only borrows it, if one is in condition to use it." (Leroi-Gourhan 1943: 320) But conversely, one should not read into the facts a descent or an origin without taking into account the existence of universal trends which produce similar technical inventions quite independently in different ethnic groups, separated in space and time.

> The search for improvements in the technique of throwing is in the order of the most natural technical trends, its simultaneous realization at several points of the globe or its diffusion from a unique source are of the order of facts which admit of only one demonstration: putting a sufficient number of instruments of propulsion into concordant geographical and chronological series. (Leroi-Gourhan 1945: 62)

The great difficulty in ethnological studies is thus to unravel what derives from the diffusion of technologies or practices, and what derives from convergent independent inventions (Leroi-Gourhan 1945: 95).

13.2 Functional Palaeontology

When Leroi-Gourhan turns to biology to follow the evolution of the mechanical structures of vertebrate skeletons, he deploys a similar logic. The bodily conformations of each species are considered as technical devices destined to ensure the survival of the organism by functions such as the acquisition of food, movement or defense against predators.[6] The stereotypic nature of an anatomical structure, its constancy or its distribution among species is not determined only by heredity or phylogeny. It is also the product of constraints in the coupling of living organisms and matter with respect to particular functions. Like the trends, these stereotypes can be theoretically described, and accounted for independently of factual phylogenetic considerations concerning the filiation of species. This is demonstrated by the

[6] "Technical action is found in invertebrates as much as in human beings and should not be limited exclusively to the artifacts that are our privilege." (Leroi-Gourhan 1965: 237)

cases of convergence, both in the case of living organisms from various phyla and in that of techniques from various ethnic groups.

> One can show by dragging a plastic mass in water, that any solid whatsoever in displacement in a liquid medium necessarily takes on a particular fusiform shape, and that the tuna fish, the ichthyosaurus, the whale and the ship could not have had any other general plan than the one that is imposed by physics. (Leroi-Gourhan 1945: 337)

In the same way, the general technical principle of a mechanism of prehension, such as the hand, crops up repeatedly in the most diverse lineages, both for the anterior limb of rodents or primates, and for the posterior members of birds.

> The case of birds is of interest because it proves that the possibility of intervention by the "hand" not only exists in a limited number of zoological groups on the direct line of evolution from the crossopterygian fish via monkeys to humans, but is even to some extent independent of any specific anatomical area. (Leroi-Gourhan 1964a: 33)

This functional paleontology makes it possible to account for the courses of evolution that are open to a given species (Leroi-Gourhan 1964a: 31). In the same way that the technical milieu of an ethnic group can only select among a limited repertoire of certain possible changes, so the functional situation of the species only offers certain directions in which selective pressures can operate. The functional situation predates the course of evolution that it generates. Leroi-Gourhan does not go into the details of the biological mechanisms of variation and selection which modify the genetic memory. Here, we will restrict ourselves to the "Darwinist" perspective that he claims elsewhere (Leroi-Gourhan 1982: 18).

If we turn to the development of the nervous system, we observe the same logic. The general structure of the organism determines the range of possible actions in the world. It is only afterwards that this functional situation selects the evolution of a brain which is best able to control the actions that are available.[7] There is thus an "*advance*" of the technical situation over the development of the control system, which combines the operational sequences that the situation allows.

> The progressive enrichment of the nervous system is an evolutionary fact of the same order as the perfectioning of *automatic controls of machines* with respect to the evolution of mechanical organs. (Leroi-Gourhan 1983a: 29, my emphasis)

In the series of mammals, one witnesses the development of the diversity of operations that are accessible. This diversity is already great among the carnivorous animals and primates; and in the phylum of hominids, it will progressively increase further. Cortical development then materializes the necessary increase in the capacity to complexify the relations between perception and action in complex operational sequences. The capacity to think depends strictly on the power to act.

[7] "We cannot cite a single example of a living animal whose nervous system preceded the evolution of the body, but there are many fossils to demonstrate the brain's step-by-step development within a frame acquired long before." (Leroi-Gourhan 1964a, b: 47)

13.3 The Question of Hominization

A general account of the development of the nervous system is not in itself an explanation of hominization. One has still to explain how this is related to the emergence of human experience comprising memory, anticipation and language. It is not so much a question of establishing factual lines of filiation between species, but rather of rendering intelligible the trends explaining human evolution. For that, Leroi-Gourhan takes on and resolves the paradox of a biological *determinism* of technical evolution that in the end accounts for a *liberation* from that determinism.

He first of all spells out the particular sequence of trends which, in the burgeoning patterns of evolution, explains the paleontological succession of functional types which led to the first hominids (Leroi-Gourhan 1964a: 36). To summarize briefly, the trend of animals to mobility leads to a functional type with bilateral symmetry; a fundamental type which itself defines a trend to the development of an anterior field devoted to relations with the environment, which will take various forms including those where the "relational field" is shared between the face and the forelimbs; and this in turn will define a trend to a vertical stance and the freeing of the hand. Among the organisms which are able to grasp, there are still two possible trends. On one side grasping may be only intermittent, limited to certain bodily postures (for example rodents in a sitting posture). On the other side, the grasping may be constant, continuing while the animal is in movement, as with the Australopithecus. This second solution, which is specific to our phylum, leads to a vertical posture having two further corollaries: the free hand, and the short face.

> Freedom of the hand almost necessarily implies a technical activity different from that of apes, and a hand that is free during locomotion, together with a short face and the absence of fangs, commands the use of artificial organs, that is, of implements. (Leroi-Gourhan 1964a: 19)

From a paleontological point of view, the general and sufficient criterion to distinguish our phylum from all other primates is thus present very early in the Australopithecus lineage. Amongst them, Leroi-Gourhan gives the name Zinjanthropus (*Zinjanthropus boisei*, now *Paranthropus boisei*) to the first hominids equipped with a few very simple tools more than two million years ago (today we rather attribute the first tools to *Homo habilis*). It is then shocking to find that these beings, whose general posture is so similar to our own, had such tiny brains.

> This uneasy feeling is due to the fact that the Australanthropians are really not so much humans with monkeys' faces as humans with a braincase that defies humanity. We were prepared to accept anything except to learn that it all began with the feet! (Leroi-Gourhan 1964a: 65)

There is still a long evolutionary path to be trodden to reach the "Neanthropians" of which we *Homo sapiens* are part. In the absence of direct traces of creative intelligence equipped with language, all we can do is to follow the transformations of the material traces of techniques of coupling between living organisms and their environment. In order to evaluate this slow evolution, Leroi-Gourhan proposed to measure, for each stone industry, the number of different tools and the length of cutting edge obtained per kilo of flint.

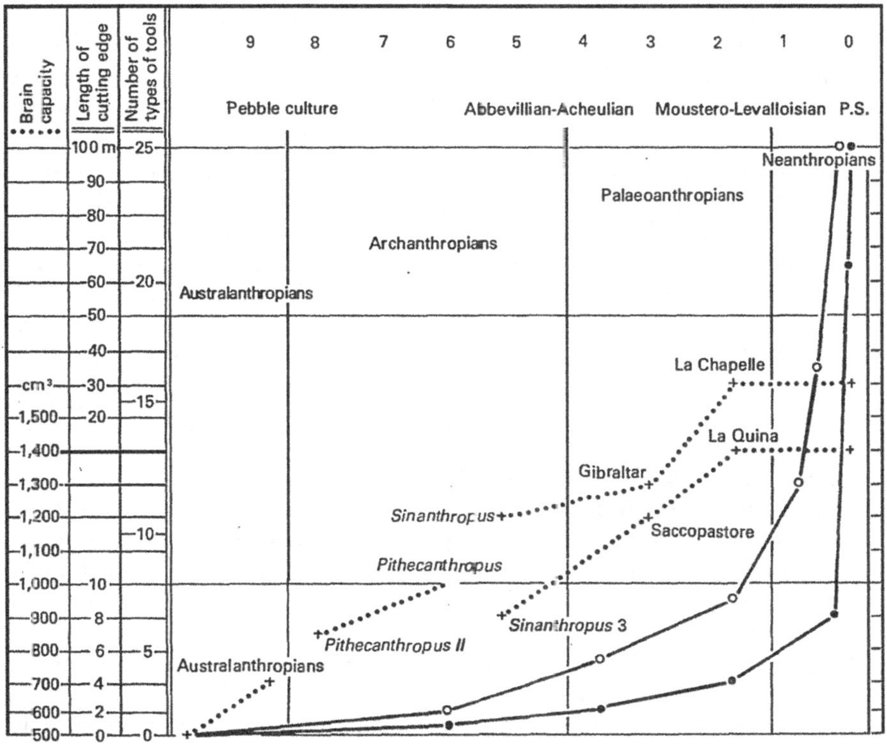

Fig. 13.1 The relationship between increase in brain volume and technical evolution during the Quaternary Period of the Cenozoic era (relative length of blade per kilogram of material and diversity of tool types). Australanthropians (*Zinjanthropes*); Archanthropians (*Homo Erectus*); Palaeoanthropians (*Neanderthal*); Neanthropians (*Homo sapiens*). (Leroi-Gourhan 1964a: 138)

If we compare these curves with that of the volume of the brain cavity (Fig. 13.1), two striking features appear. First, the extremely slow evolution of the stone industry. Technical progress and biological evolution advance at the same slow rate, "a fact that confers a curiously biological character to the prehistory of sharp-edged objects." (Leroi-Gourhan 1964a: 134) Then, with the Neanthropians (Sapiens), there is such an acceleration of technical evolution that it seems to become completely independent of new biological transformations: "from being governed by biological rhythms, human cultural development began to be dominated by social phenomena." (1964a: 141) Nevertheless, throughout this evolution, the general formula of hominid anatomy does not change much. One only observes a lightening of the bony structure of the skull, and a filling by a brain that increases in volume.

Classically, the development of the nervous system is considered the relevant explanatory feature, and one starts from the forms of intelligence and culture in animals to account for the origin of our cognitive faculties and our socio-technical systems (Tomasello et al. 2005). This supposes an evolutionary scenario in which the faculties of learning and social transmission of the first hominids would carry a trend towards their complexification. Hereditary variations in the brain which went toward increased cognitive faculties would constitute in themselves an adaptive advantage. An increasingly complex technical system would follow from an increase in these cognitive abilities.

For Leroi-Gourhan, these "brain-centred" approaches, which one finds in the myth of "a monkey ancestor of man," do not correspond to the facts and are quite unable to account for the specificity of human evolution. Rather, one must recognize that the advance of technical situation over the development of the control system observed in the whole of the living world, is found again here. The development of the nervous system can only represent an adaptive advantage if it is related to a repertoire of possible actions. There is an advance of the general anatomical structure that defines the concrete living conditions of the organisms over the cortical variations which can take advantage of the new possibilities offered by these conditions.

At the start of the process of hominization, the tool was a biological fact which derived from the upright posture and the freeing of the hand. Just like any other organ, it would be an *obligatory* product of zoological ontogenesis, systematically produced, under normal conditions, quite independently and prior to an encounter with a situation in which it can be exploited. The first forms of tools were in a way "exuded" by the organism, and hence could not evolve more rapidly than their biological bearer could.[8] The constancy of their forms over hundreds of millennia proves this.

Now the presence of tools signifies a functional situation for the species in which increasingly complex operational sequences are possible. This complexity defines the conditions of selection for the development of a brain apt to coordinate the behaviors in this new space of possibilities. In turn, this evolution allows for the production of richer technologies... which themselves will induce further developments of the brain. This process leads to a progressive deployment of the "cortical span," i.e. a strong development of the associative zones in the cortex which control actions and their succession in complex operational sequences.

In order for this coupling between tools and the evolution of the brain to function, it is necessary to admit a biological determination of the first tools. This is the essential difference with the behavior of certain monkeys. Since the time of Leroi-Gourhan, our knowledge of primate behavior has been enriched. There can already be tools for infant monkeys. For example, there have been observations in nature of the differentiated ways in which techniques for cracking nuts with a percussive instrument diffuse in various primate populations (Wrangham et al. 2005). However, if we follow Leroi-Gourhan, these premises of cultural transmission remain limited, contained within the repertoire of actions accessible to the species. Even if they bear witness to remarkable cognitive and social capacities, these new behaviors do not

[8] "Australanthropians (...) seem to have possessed their tools in much the same way as an animal has claws (...) as if their brains and their bodies had gradually exuded them." (Leroi-Gourhan 1964a, b: 106)

have the potential to initiated novel directions of biological evolution. They do not impose a constant selective pressure on other characters of the species precisely because they adapt on the spot in function of the circumstances. When chimpanzees assemble a perch by fixing together several sticks in order to catch a bunch of bananas otherwise out of reach, they are only responding to a momentary problem in function of the data present in their environment. Such inventions of monkeys do not determine an evolutionary trend, because the monkeys do not have to adapt to tools that are already there. The human tool poses a *problem* for the ancestors of humans; whereas the perch of the monkey is a *response* to a contingent situation. Thus, paradoxically, it is because the human tool is not so much the product of intelligence, but rather the intelligence that is the product of the tool, that the biological evolution of the human brain is justified.

13.4 The Problem of the Externalization of Memory

If this is so, how is it possible to account for the appearance of a social memory that has the capacity to register technical innovations much more rapidly than the genetic memory of heredity? Two explanatory options are open:

- Either the appearance of this social memory was the secondary consequence of the evolution of the cerebral cortex, which would have passed a somewhat mysterious threshold with the appearance of the Neanthropians (*Homo sapiens*);
- Or else the social memory was the product of a process of externalization that is specific to human technologies.

Certain formulations of Leroi-Gourhan seem to favor the first option.[9] However, as Bernard Stiegler has pointed out, this would mark a serious regression (Stiegler 1994). Instead of a co-invention of mankind and technologies, humanity, in the modern sense of the term, would suddenly appear with the Neanthropians, by the inexplicable means of a mutation. In order to avoid this disappointing backsliding, Bernard Stiegler grants the very first tools of the Zinjanthropes the status of "external memory"; but this amounts to admitting that from this stage on, there is a genuine autonomy of the history of technical differentiation.[10]

Nevertheless, another reading of Leroi-Gourhan is possible, which avoids any sudden appearance of mankind either at the level of the Zinjanthropes or that of the Neanthropians. According to this reading, mankind appeared *progressively*, by means of a *gradual* detachment of social technical memory. This would be more consistent with the general project of explaining human evolution by the play of definite trends. Between the incomprehensible event of a random "accident," and a

[9] "It does seem as though the 'prefrontal event' had marked a radical turning point in our biological evolution as a zoological species governed by the normal laws of species behavior." (Leroi-Gourhan 1964a, b: 137)

[10] Bernard Stiegler speaks of an "epiphylogenetic memory." (Stiegler 1994: 185)

mysterious "predestination," there is a third way, a paradoxical "human solution of the problem of our origin." (1964a: 94) If we admit that "humanity" begins as soon as the upright vertical stance is established, it is a question of maintaining the principle of an "advance of technical situation" over cortical development, so as to account for the progressive genesis of the way "the system that provides human society with the means of permanently preserving the fruits of individual and collective thought came slowly into being." (Leroi-Gourhan 1964a: 187)

13.5 A Human Solution to the Problem of Mankind

The tools that accompanied hominids more than two million years ago are challenging our contemporary human reason. The very simple tools of the Zinjanthropes are the products of a single gesture, the perpendicular shock between two flints. They show no rapid or spectacular progress, no visible differentiation of techniques over hundreds of thousands years. Notwithstanding, that does not preclude from conjecturing that, already at this stage, fortuitous or deliberate variations in an external tool might favor their own reproduction.

Leroi-Gourhan admitted the existence of capacities for individual learning of know-how even in the simplest animals. There is a technical intelligence in all prehensile organisms (a perception of the forms to be grasped and used, a mastery in the combination of actions in action-sequences). Leroi-Gourhan describes "instinct" in the animal world, not as a behavior inscribed in the nervous system, but as the determinate result of a coupling "located at the intersection of the means specific to that individual and the external causes for deploying those means in action sequences." (1965: 221)

As in *situated cognition*, the solving of a problem corresponds to the transformation of the environment, a transformation that includes the participation of the agent just as much as the initial material milieu (Gallagher 2009; Hutchins 1995; Clark 1997).[11] A certain form of memory of learned behaviors must thus have existed in the first hominid societies. These populations must already have been able to transmit behavioral novelties according to different "traditions." Nevertheless, Leroi-Gourhan does not propose to explain the liberation of human history from its biological basis as an effect of the social organization of the first anthropoids. What is lacking are the concrete conditions for learning the reproduction of new techniques. A new tool enriches the range of possible operations, but there is no assurance that it contains precisely the operations for making this new tool (nothing ensures that the new tool can participate recursively in its own reproduction).

Nevertheless, with the movable tools of the first anthropoids, there is already a radical novelty compared with organs that are attached to the organism. By its exter-

[11] We are very close to the notion of "stigmergy" developed at the time by zoologist Pierre-Paul Grassé (1959) who presided in the 1955 the jury of the thesis in natural science of Leroi-Gourhan.

nality and its material permanence, it allows for intergenerational exchanges. The tool is "already there" in the environment of the next generation. By grasping it, the young individual receives from the outside a new power to act, that she has not necessarily produced herself. The coupling between the abilities of the organism and its environment now takes place in a milieu comprising movable tools, exchangeable material inscriptions, which surpass the lifetime of an individual. This new situation bears a *trend* towards an external memory.

With the Archanthropians (*Homo erectus*), there is essentially a second series of gestures: tangential blows, which results in the fabrication of the famous bifacial tools. The progressive complexification of the set of tools is realized in a context where the activity related to a certain tool can be dedicated in part to the fabrication of other tools.[12] The reproduction of a tool can mobilize a social transmission of techniques. The external transmission of tools which can be given and received defines a richer technical milieu. However, the transformation of the technical milieu by learning new tools remains confined in the field of situations of possible fabrications. The simple transmission of an innovation does not directly entrain its reproduction. There is not yet a fully-blown autonomy of the external technical history, but still only a displacement in a fixed field of possibilities.

With the Palaeoanthropians (Neanderthal), the technical milieu becomes still more complex. The operations of fashioning comprise several steps, marked by changes in the tools and the operations (rough fashioning of the original block to give it an appropriate shape; productive chipping and flaking; refashioning the block; pursuit of the productive chipping…). (Pellegrin 1990) The tools are successively grasped and put down, fashioned and used. We may even speak of a sort of technical *syntax*, insofar as the fabrication of the tools proceeds by ordered sequences of operations, and a different arrangement would produce different products.

> Techniques involve both gestures and tools, sequentially organized by means of a "syntax" that imparts both fixity and flexibility to the series of operations involved. (Leroi-Gourhan 1993: 114)

As soon as the conditions of the fabrication of technical objects become recursively themselves transmissible external techniques, the field of possibilities is vastly enriched. There is a genuine external memory when the introduction of a new technique can be the cause, direct or indirect, of its own reproduction. Externalization renders possible a spatial deployment of the syntax of operational action sequences, which in its turn allows for a process of external reproduction of these conditions of learning. The externalization of the movable tool is thus duplicated into an externalization of the conditions of its reproduction.

The creation of new tools and new situations of coupling is no longer the product of a heritable variation, but results from a modification by the organisms of their technical environment. The capacities of reproduction of this social memory

12 "The lump of stone initially intended to become an almond-shaped tool became instead a source of flakes of predetermined shape, and it was these flakes that were eventually used as tools." (Leroi-Gourhan 1964a, b: 100).

participate in the definition of the functional situation of the species. In the game of mirrors between cortex and silex there are now *two* memories which respond to each other, genetical and sociotechnical. To the extent that the possibilities for external reproduction remain limited, this situation can still lead to further biological evolution.

We can understand why the process by which a social memory independent of biological determinism arose was so extremely slow – hundreds of thousands of years! It is because this process was contingent upon a complexification of the specific techniques for the reproduction of tools sufficient for them to be able to progressively encompass an ever increasing diversity of new possibilities.

In a final phase, that of Neanthropians (*Sapiens*), the movement which was thus set in motion accelerates and amplifies. There is no longer time, nor any need, for a selective effect of the technical milieu on biological genetic memory. There is no longer time, since the recording of technical variations in the social memory is infinitely more rapid than that of biological evolution, which must wait for relevant mutations in the genetic memory.[13] There is no longer any need, since the creation and the fixation of innovations can be accomplished directly as a function of their success in this social memory, even if they are useless from the point of view of the biological species. The dynamics of the evolution and differentiation of human productions is thereby profoundly altered. Just as species separate into a diversity of phyla according to their histories inscribed in the genetic memory, so human populations will diversify into different ethnic groups according to their histories inscribed in the social memory.[14]

This explanation of hominization as the product of a trend towards the externalization of social memory allows for an original approach to the evolution of cognitive capacities.

13.6 Intentions and Anticipation

Ever since the first stages of hominization, Leroi-Gourhan admits that the fabrication of tools supposes a form of *technical consciousness* with capacities for prediction and anticipation.[15] These capacities will continue to reinforce themselves progressively, since for the fabrication of tools such as the bifacial flint, there is clearly the aim of a stereotype in spite of the infinite variations in the initial form. Now, as we have seen, Leroi-Gourhan maintains at the same time that at this stage

[13] "In *Homo sapiens* technicity is no longer geared to cell development but seems to exteriorize itself completely – to lead, as it were, a life of its own." (Leroi-Gourhan 1964a, b: 139)

[14] "If it is true to say that the species is the characteristic form of animal grouping and the ethnic group of human grouping, then a particular form of memory must correspond to each body of traditions." (Leroi-Gourhan 1964a, b: 221).

[15] "The Australanthropian making a chopper already foresaw the finished tool because the pebble chosen had to be of suitable shape." (Leroi-Gourhan 1964a, b: 97, personal translation).

technical objects cannot transform themselves independently of a biological evolution.

> Thus the first anthropoids' technicity [implies a state] of technical consciousness to which, however, we must not apply our own yardstick. It is undoubtedly less of a risk to see human technicity as a simple zoological fact than it would be to credit *Zinjanthropus* with a system of creative thought. The countless millennia during which his industry remained unchanged – conditioned, as it were, by the shape of his skull – disproves the latter hypothesis. (1964a: 92)

Leroi-Gourhan thus invites us to delve into the strange realms of a technical consciousness capable of certain sorts of learning but incapable of innovation; which has the power to aim at certain archetypes amidst the myriad diversity of perceptual situations, but remains destitute of the capacity for free creation... An effort of this magnitude to try and imagine the obscurity of the most archaic forms of thought seems to us nevertheless absolutely necessary if we wish to grasp the "stages in which the link between the zoological and the sociological has become progressively more tenuous." We are invited to admit that, at the very beginnings of humanity, there was the capacity to aim at a goal without there being the capacity to discover new goals.[16] If one tries to elaborate a conception of intentionality which can fit this specification, we must first of all reject the idea of an intentional behavior guided by the representation of a perceived model. That would be to give ourselves what it is our task to discover, since the perception of a novel form is not sufficient to set in motion the learning of a way of making it. We must rather look for a conception of intentionality in a technical consciousness that is directly anchored in the living world.

There was a time when the stability and the constancy of the forms within any given species induced biologists to invoke the notion of a "final cause" which, like a causally effective intentionality, would operate to direct the processes of ontogenesis. By doing so, the scheme of the conscious productions of a craftsman was projected into biological explanations. Nowadays, however, a complex causality regulated by genetic memory is considered sufficient to account for the appearance of final causation. If there is still some resistance to recognize a similar process in the case of the fabrication of the first tools, it is merely because their ontogenesis is external: it mobilizes the organs of perception (choice of materials, adjustment of the gestures) and of action (finding the materials, controlled gestures).[17] This makes it difficult not to attribute to the makers of these tools the same conscious intentions that we would have for the same work.

> There is little reason to distinguish between the Palaeoanthropian technician's attitude and that of any technician of a more recent age – at any rate in strict terms of technical intelligence. (Leroi-Gourhan 1964a: 102)

[16] At least in the domain of the production of stone tools that we can observe, because in the case of woodworking there are few if any observable remains.

[17] "It is logical that the standards of natural organs should be applied to such artificial organs: They must exhibit constantly recurring forms, their nature must be fixed." (Leroi-Gourhan 1964a, b: 91)

If we follow Leroi-Gourhan in his search for a *continuity* between biological causality up to human cognition for which the capacity to aim at a goal has to be admitted, a reversal of the terminology is possible. Rather than renouncing the idea of an intended form from the moment when an explanation in terms of memory is available, one could say that the aim derives from a memory construed as a capacity to produce the same forms in a diversity of material situations. However, as long as this memory is genetic, even if one admits a form of consciousness of the intended forms, this consciousness is limited to a choice within a repertoire which is biologically fixed and limited.

Any form of memory involves a temporal lag, a retention programing future action, in other words an anticipation. For Leroi-Gourhan, an essential characteristic of human tools is that their production occurs in a situation that is independent of the context of use: "the operations involved in making a tool anticipate the occasions for its use and the tool is preserved to be used on later occasions." (1964a: 114) The production of a tool involves a preparation for a situation, which is absent and merely possible, where the tool will be used. In this way Leroi-Gourhan attributes to the first hominids a *concrete* capacity of anticipation in the fabrication of their tools. The organism "foresees" without *choosing* what it foresees. It is only to the extent that this anticipation contributes to the survival and the reproduction of the organisms in question (and hence to the reproduction of the biological memory) that it is conserved. The biologically specified tool *concretely* anticipates its future use, in the same way that an organ produced in the course of ontogenesis anticipates its future use, just as the nest of a bird anticipates its function of protecting the future nestlings. The anticipation is *concrete* by its biological determinism, and by the biologically functional nature of the situation that is anticipated.[18] However, with the externalization of the tool, there is already an externalization of the anticipation. With progressive enrichment of this external memory, new forms can be aimed at. Their number and complexity increase ever further as this memory becomes autonomous.

At the stage of the Archanthropians, when certain tools served the production of other tools, we find ourselves in a situation where the initial tools anticipate the situation of the fabrication of subsequent tools. The operations follow on from each other in action sequences which become highly complex, which "implied a good deal of foresight on the part of the individual performing the sequence of technical operations." (Leroi-Gourhan 1964a: 97) A system of embedded anticipations comes into play: the initial forms anticipate a number of subsequent forms, which are themselves produced with a view to future use. Nevertheless, the meaning of the possible anticipations derives from their origin in a biological memory of feasible operations, and corresponds to their adaptive utility.

> Operating behavior remains completely rooted in lived experience, for projection can only take place once operations have been freed from their materiality and transformed into sequences of symbols. (Leroi-Gourhan 1964a: 226)

[18] The "concrete" character of the anticipation on which I insist here is not justified by the concrete character of the tool or the organ bearing the anticipation. We will see that a "symbolic anticipation" can be based on substrates that are just as concrete.

This freeing is only attained at the succeeding stage. As we have seen it is with the Paleanthropians that a veritable external social memory develops. Tooling up reaches the stage of becoming a means of producing new tools. It is in this reflexivity that it acquires immense capacities, and in particular the power to reproduce innovations. The variation and the reproduction of anticipations can occur following their success in social interactions, independently of the concrete character of the situation. This is what Leroi-Gourhan calls "liberation from lived experience." (1964a: 33) In this way, an external memory opens the way to creative anticipation, a full anticipation that is no longer derived from a biological determinism and a utilitarian content. Leroi-Gourhan speaks of *externalizing a symbolic representation*, which is the beginning of an intelligence that is no longer *strictly technical*, that is to say the possibility of thinking, of reflecting about the future, in the absence of concrete actions in the environment. Just as reproduction in biological memory specifies the aim of constant forms in the behavior of living organisms, so reproduction at the level of social memory makes it possible to specify the aim of constant forms in technical behavior. The anticipations henceforth available and produced are now the fruits of a social history, and they are related to the development of language.

13.7 The Development of Language

In the absence of material traces we are reduced to hypotheses. Leroi-Gourhan proposes to imagine that language develops in the same way technical artifacts do, in the same movement and following the same basic logic.[19] The proximity of the brain areas involved can be read as a result of their functional proximity.[20] During the initial stages (Zinjanthropes then Archanthropians), although there were probably already exchanges of auditory and gestural signals, these exchanges were limited, as were the anticipations, to the context of actions in concrete situations.[21] However, following through on the analogy between language and techniques, in the same way that tools came to be made and rendered available independently of the situation of use that they anticipated, the "verbal forms" came to be reproduced and available *before* their use in concrete situations With the complexification of

[19] "Technics and language are not two distinct typically human facts but a single mental phenomenon neurologically based on contiguous areas and expressed jointly by the body and by sounds." (Leroi-Gourhan 1965: 403)

[20] "This leads us to conclude, not only that language is as characteristic of humans as are tools, but also that both are the expression of the same intrinsically human property, just as the chimpanzee's 30 different vocal signals are the precise mental counterpart of its use of several sticks to pull down a banana hanging overhead – in other words, no more a language than fitting the sticks together is, properly speaking, a technique." (Leroi-Gourhan 1964a, b: 114)

[21] "The purpose of verbal figures – words and syntax – is, like the purpose of tools and manual gestures, their equivalents, to provide an effective hold on the world of relationships and of matter." (Leroi-Gourhan 1965: 365)

the techniques and a veritable syntax of action sequences, one can very well imagine that there was an equivalent structuring of language, even if it was still limited "to the expression of concrete situations."

> If language did indeed spring from the same source as technics, we are entitled to visualize language too in the form of operating sequences limited to the expression of concrete situations, at first concurrently with them and later involving the deliberate preservation and reproduction of verbal sequences going beyond immediate situations. (Leroi-Gourhan 1964a: 116)

It is only when an external memory became fully autonomous that a truly symbolic language was able to make its appearance. For Leroi-Gourhan, the faculty of symbolization consists of producing a distance between the human and the milieu (both internal and external) in which it is immersed: "a detachment, which expresses itself in the separation between tool and hand and between word and object." (1964a: 235) When the forms thus produced (tools, vocal or gestural expressions) are no longer linked to biological memory and utility, they come to depend only on their reproduction in social interactions. Language can then be applied to "areas beyond that of purely vital technical motor function," and so "used for *post facto* transmission of the action symbols in the form of narration." (Leroi-Gourhan 1964a: 115) It is then that one observes the development of activities of figuration,[22] which can take on an esthetical-religious character as with signs of the anticipation of death (sepultures) and a taste for the unusual (fossils, pyrites). (Leroi-Gourhan 1964a: 107; 1964b)

With language, there is also the development of "reflective thought" at the level of the individual (Leroi-Gourhan 1964a: 195). The system of social memory allows, up to a certain point, a personal liberation with respect to biology and to the social dimension itself (Leroi-Gourhan 1965: 227). By appropriating the reproducible forms that are available, each individual can construct on her own account a specific memory, the last degree in ethnic differentiation. The mastery of the concepts born by the social memory allows her to construct her own anticipations.

13.8 Conclusion

The perspectives opened up by Leroi-Gourhan in his work on paleo-anthropology are potentially fruitful for inspiring novel lines of research in cognitive science, in particular for the "enactive, embodied, embedded" approaches which refuse the facility of simply giving themselves the capacities of a representational computational system. Taking "technology as anthropologically constitutive" (Stiegler 1994)

---

[22] "Figurative behavior cannot be dissociated from language: It forms part of the same human aptitude, that of reflecting reality in verbal or gestural symbols or in material form as figures. Just as the emergence of language is connected with that of hand tools, figurative representation cannot be separated from the common source from which all making and all representation spring." (Leroi-Gourhan 1965: 363)

makes it possible to propose an explanation for the passage from the instinctual capacities of the biological realm (which Leroi-Gourhan conceptualizes in terms of a coupling between the organism and its milieu), to the symbolic capacities of creation, reflexivity and free anticipation, that are commonly associated with an internal representational system, but which here are realized secondarily by the individual reappropriation of external symbols embedded in the social memory.

"Technology" is not only the object of Leroi-Gourhan but the very method he systematically follows, and which could be referred to as "*the principle of concrete operations*": Accounting first for the functions and properties of the coupling between living organisms and their environment *before* examining the possible transformations that they imply for the organism (either taken up by selection in the hereditary memory, or taken up by reproduction in the technical milieu of social memory, or yet again taken up by learning in the cerebral system of the individual). Following this principle, the existence of reproducible external forms is the precondition for any learning of their internal reproduction. Every "symbolic representation," whether individual or collective, presupposes that a process for the reproduction of concrete external forms (tools or symbols) has *already* been set up. In this perspective, cognitive activity and individual memory are not encased within the organism, but are rather constitutively related to material technical inscriptions that are *external* (Lenay 2012).

If we admit that there is an "advance of technical situation," it is necessary to systematically take into account the repertoires of concrete operations in order to define the conditions of possibility for mental operations. Rather than explaining extended cognition as the external deployment of cognitive capacities that are already there, it is a question of understanding how the technical environment is the very condition which makes these capacities possible. Individual thought does not happen in the brain alone, but *with* a brain as it is coupled – via the rest of the body and a set of tools – to the technical and social milieu. This perspective may help to evaluate the contemporary transformations resulting from the development of digital technologies of collective memory.[23]

In the second volume of *Gesture and speech* (not analyzed here), Leroi-Gourhan pursues by an analysis of the techniques of social memory, systems of writing and figuration. He thus prolongs his reflection towards the future, and anticipates many of the developments of cybernetics. Thus, by his conception of technical systems, he helps us to escape from the reduction of cognitive activities to a simple question of information processing; he offers us instead a vision of cognitive activities as stemming from a dialogue between *life* and *matter*.

Acknowledgments I wish to thank Xavier Guchet, Pierre Steiner, Sacha Loeve and Gunnar Declerck for their highly constructive comments during the writing of this text.

[23] "Saying that we are currently being overwhelmed by technical innovations is thus a false problem: technical systems are *always* 'overwhelming', this is quite normal; the real worry is probably elsewhere," Said Leroi-Gourhan in *Le fil du temps* (1983b: 87).

References

Clark, A. (1997). *Being there: Putting brain, body, and world together again*. Cambridge, MA: MIT Press.

Gallagher, S. (2009). Philosophical antecedents to situated cognition. In P. Robbins & M. Aydede (Eds.), *Cambridge handbook of situated cognition* (pp. 35–51). Cambridge: Cambridge University Press.

Grassé, P.-P. (1959). La reconstruction du nid et les coordinations interindividuelles chez Bellicositermes natalensis et Cubitermes sp. la théorie de la stigmergie: Essai d'interprétation du comportement des termites constructeurs. *Insectes Sociaux, 6*(1), 41–80.

Guchet, X. (2015). André Leroi-Gourhan et la philosophie. In P. Soulier (Ed.), *André Leroi-Gourhan "L'homme tout simplement"* (pp. 117–133). Paris: De Boccard.

Hutchins, E. (1995). *Cognition in the wild*. Cambridge, MA: MIT Press.

Lenay, C. (2012). Separability and technical constitution. *Foundations of Science (FOS), 17*(4), 379–384.

Leroi-Gourhan, A. (1943). *Evolution et techniques, Tome I: L'homme et la matière*. Paris: Albin Michel.

Leroi-Gourhan, A. (1945). *Evolution et Technique, Tome II: Milieu et techniques*. Paris: Albin Michel.

Leroi-Gourhan, A. (1964a). *Le geste et la parole. Tome I: Technique et langage*. Paris: Albin Michel. English edition: Leroi-Gourhan, A. (1993). *Gesture and Speech* (Vols. I & II, A. B. Berger, Trans.). Cambridge, MA: MIT Press.

Leroi-Gourhan, A. (1964b). *Les Religions de la Préhistoire*. Paris: Presses Universitaires de France.

Leroi-Gourhan, A. (1965). *Le geste et la parole. Tome II: La mémoire et les rythmes*. Paris: Albin Michel. English edition: Leroi-Gourhan, A. (1993). *Gesture and Speech* (Vols. I & II, A. B. Berger, Trans.). Cambridge, MA: MIT Press.

Leroi-Gourhan, A. (1982). *Les racines du monde*. Paris: Belfond.

Leroi-Gourhan, A. (1983a). *Mécanique vivante*. Paris: Fayard.

Leroi-Gourhan, A. (1983b). *Le fil du temps*. Paris: Fayard.

Mauss, M. (1936). Les techniques du corps. *Journal de psychologie, 32*(3–4), 365–386. English edition: Mauss, M. (1973). Techniques of the body. *Economy and Society*, *2*(1), 70–88.

Pellegrin, J. (1990). Prehistoric lithic technology: Some aspects of research. *Archeological Review from Cambridge, 9*(1), 116–125.

Stiegler, B. (1994). *La technique et le temps, 1. La Faute d'Épiméthée*. Paris: Galilée. English edition: Stiegler, B. (1998). *Technics and time, 1. The fault of epimetheus* (G. Collins & R. Beardsworth, Trans.). Cambridge, MA: Stanford University Press.

Tomasello, M., Carpenter, M., Call, J., Behne, T., & Moll, H. (2005). Understanding and sharing intentions: The origins of cultural cognition. *Behavioral and Brain Sciences, 28*(5), 675–691.

Wrangham, R. W., McGrew, W. C., deWaal, F. B., & Heltne, P. G. (2005). *Chimpanzee cultures*. Cambridge, MA: Harvard University Press.

Chapter 14
The Anthropocene: Where Are We Going?

John Stewart

Abstract The dominant current in the contemporary environmental movement fails to make the connection between the preservation of the environment and the survival of humans. The fashionable concept of the "Anthropocene" is not fully adequate to get to grips with the full gravity of the situation. Contemporary human society, based on a neo-liberal market economy, is "locked in" to a productivist mode of existence, so that it will be extremely difficult to abandon the goal of "growth" and to achieve a sustainable relationship with the eco-system on which human existence depends. The "TAC" thesis, that "Technology is Anthropologically Constitutive," has a dark side: technology may be anthropologically destructive.

Keywords Anthropocene · Ecological crisis · Extinction of human species · Technology as anthropologically constitutive

The concept of the "Anthropocene" is generating a certain amount of interest at the present time. The general idea is that our current epoch is characterized by the fact that human activity has become *the* major influence on the state of the planet, at a global geological scale. This concept is controversial for a number of reasons. To begin with, the uncertainty of the concept is illustrated by the inconclusive debate as to the date of entry into this period: is it after 1945? The nineteenth century industrial revolution? The Renaissance and the birth of modern science? Or does it go back to the Neolithic? But most of all, the question is to know whether this concept provides an adequate framework for correctly identifying the *political* issues, and hence possible *actions*. Another, related issue that is also generating a great deal of discussion is the "ecological crisis." A manifest instance of this is the much-publicized issue of climate change and global warming. Serious as this is, however, it is not clear that the concept of the Anthropocene, as such, adequately focuses on

J. Stewart (✉)
Connaissance, Organisation et Systèmes TECHniques (COSTECH), Université de technologie de Compiègne (UTC), Compiègne, France

S. Loeve et al. (eds.), *French Philosophy of Technology*, Philosophy of Engineering and Technology 29,
https://doi.org/10.1007/978-3-319-89518-5_14

the full gravity of the situation. To put it bluntly, it is within the realms of possibility that what is at stake in this so-called "Age of the Human" is the actual *extinction* of the species *Homo sapiens*. Let us look at this question in more detail.

14.1 Anthropocene or the End of Anthropos?

The basic point, which is well established in ecological science, is that every biological population depends on a species-specific eco-system. Moreover, every such eco-system has a finite, limited carrying-capacity. If the population expands beyond a certain point, so that the eco-system is exploited beyond its carrying-capacity, the result is that the eco-system is destroyed… and the species goes extinct. This much is general ecology, and as I say is widely recognized. The question is: how does this apply to the human species? In an extremely important article published in *Nature*, Barnosky et al. (2012) estimate that the environmental impact of human activity is currently at 150% of the carrying capacity of our eco-system. If this were to continue the result would be the destruction of the human eco-system, leading to the extinction of our species in a matter of centuries from now. Even more worrying, the eco-system on which human life depends is a complex system. It is a characteristic of such systems that the damage may become irreversible well before the worst effects become manifest; Barnosky et al. estimate that we may only have 10–20 years to turn things around. As yet, there are no signs of this happening; on the contrary, far from decreasing or even stabilizing, growth is continuing – and remains a deliberate aim of governments the world over.

With this in mind, I will now rather briefly examine some of the (critical) literature on the Anthropocene.

In a collection of texts directed by Emilie Hache (2014), Isabelle Stengers remarks that the dominant current in the Environmental movement fails to make the connection between the preservation of the environment and the survival of humans. Instead of that, the movement has focused its attention on an abstract idea of "nature and environment" as an entity separate from human beings. This categorical separation between nature and culture, which is prevalent in dominant discourse on the environment, has led critics to declare that the environmental movement "cares more about whales and owls than about poor people." Can the Anthropocene do better?

In November 2015, at a two-day conference in Paris specifically devoted to the Anthropocene (Larrère and Beau 2018), Bensaude-Vincent and Loeve (2018) remarked that the concept of Anthropocene, which does not question the postulate of a unique, linear temporality, perpetuates and strengthens a simplistic opposition between techno-optimists wedded to a belief in "progress," and techno-pessimists who content themselves with looking nostalgically backwards; this opposition can only lead to a political impasse. The concept of Antropocene has a stupefying and globalizing effect: it casts the Earth as an object of governance in our hands while at the same time highlighting our powerlessness to act. At the same Paris conference, Stefan Aykut (2018) remarked that the COP21 conference,[1] in spite of the media attention it has received, is incapable of providing any real governance – the reason being that these negotiations are hermetically *separated* from international negotiations on trade and commerce which go in the opposite direction by seeking to promote unqualified *growth*. The same governments that participate in the COP negotiations simultaneously *subsidize* increased energy consumption. The incoherency is heightened when we add in the efforts of powerful lobbies on the part of multinational corporations and international finance, which seek to nullify any restraints based on environmental considerations. As to the ecologist Pascal Durand, he recalled the speech given by Chilean President Allende to the UN in 1972, shortly before he was killed in the USA-backed military coup. Allende stigmatized the power of multinational corporations, which have no accountability (other than to their share-holders); this makes any notion of "democracy" fallacious when it comes to economic decisions. What is at stake is to find a means of re-establishing the primacy of *common interest*, as opposed to the force of private capital. In line with this, Philippe Descola also (2018) noted the stranglehold of private appropriation (which dates back historically at least to the Enclosures); and called for a return to the notion of "Common Goods." This is reminiscent of Winstanley, the seventeenth century English "Digger" who opposed the Enclosures by peaceful direct action, and who is reputed to have said that much of the evil in the world stems from "this dreadful business of buying and selling." There is considerable convergence here on the importance of the political dimension; with a reminder that these issues do not date from yesterday but have a considerable history.

In an independent contribution to the debate, Fressoz (2012) also emphasizes the historical dimension. Painting the picture of a past which was blithely technophilic, contrasting it with a hypothetical awakening of contemporary ecology, leads to a political impasse. The men who accomplished and experienced the industrial revolution were perfectly aware of the immense risks they were producing. But they decided, knowingly, to go ahead anyway. Already in the eighteenth and nineteenth centuries, there was a whole series of oppositions and controversies. Fressoz cites the examples of vaccination which aroused considerable opposition at the time; the advent of industrial chemistry involving a complete overturn of the forms of environmental regulation in force up until then; the transformation of the environment into an object of financial transactions, putting a "price" on pollution, both of which made it possible to "buy off" opposition; technical norms of security (for example concerning gas) which, far from restricting the potentially dangerous innovations, actually made them possible by specifying the conditions under which they were acceptable; a similar remark holds for the legislation on "occupational risks" in the workplace which actually legitimated the controversial innovations. Fressoz speaks of a set of "petty modern dis-inhibitions"; a set of "coups," which conferred *a posteriori* "legitimacy" to what were actually straight power-plays. Fressoz raises the question as to whether we are not now witnessing a recurrence of technological

[1]The COP21 conference was held in Paris from 30 November to 12 December 2015; it lead to an agreement between the 195 States involved to reduce their emissions of greenhouse gases. This so-called "Paris agreement" came into effect on 4 November 2016.

imperialism, the imposition of technological action as the sole form of legitimate life. This is how we are well on the road to the abyss. He emphasizes that it has its roots in the past: the scientific and political production of a certain form of modernizing thoughtlessness.

In another important contribution, Bontems (2016) refers to the work of Gilbert Simondon and, while remaining fully alert to the dangers of the present situation, also offers some grounds for hope. Simondon, with considerable foresight, identified the *flow of energy* as a key issue. There is an ethical stake here: it is a question of *reforming our technological relation to the world.* If humans persist in treating machines as slaves, they run to perdition. Bontems cites the calculations of McNeill according to which our current energy consumption is the equivalent of 20 energetic slaves per person. This is the place to recall the classical consideration that the institution of slavery is to be condemned not merely because it is not very nice for the poor slaves; more profoundly, it demeans the "masters." Bontems cites André Gorz: the solution lies in aiming not at the accumulation of material riches, but at free time with equal access for all. Bernard Stiegler, himself an heir of Simondon, echoes this theme in a recent publication (Stiegler 2015) when he poses the question of the consequences of the generalization of automatic processes of production. Bontems identifies the key issue as that of the relation between "progress" and "power." "Progress" is only meaningful if it involves *slowing down* the exploitation of available energies and the local (terrestrial) increase of entropy—a proposition recently echoed by Stiegler with his "Neganthropology." (2017) This amounts to radically rethinking our inherited notions of "progress." If the corresponding reorganization is not accomplished in time, a generalized degradation will ensue, and the system could even totally collapse. The difficulty of the present situation is that "growth" (ever more material riches) is flatly opposed to (simple) diminution. What is needed is neither the one nor the other, but a veritable *transformation*: the quest for progress *liberated* from the quest for power.

There is thus a considerable literature on the subject. In addition to the works cited above, a more complete bibliography on the Anthropocene would run to dozens of articles and books. However, a perusal of this literature strengthens rather than weakens a rather strange impression: in all this literature, the possibility (or even the probability) of an *extinction* of the human species in the course of the next century or so is barely mentioned. It is somewhat like the attitude to sexuality in Victorian England: everyone knows that it exists, but it is "not done" to mention it in polite circles. One way of approaching the question is to consider the relationship between individual and collective intelligence. In cognitive science, much is made (and rightly so) of "swarm intelligence": in the case of the social insects (bees and ants), the intelligence of the colony as a whole is far greater than that of the individuals. In the case of contemporary humans, however, the relation is the other way round: our collective intelligence is dramatically *less* than that of individuals.[2] A

[2] Even in the case of humans, collective intelligence may well be greater than the intelligence of individuals, as in the case of Condorcet's jury theorem. But this is not systematically so; and in the present case an argument can be made that the collective intelligence is indeed less than that of the individuals.

certain number of human *individuals* are clearly conscious of the extreme gravity of the ecological crisis: way beyond the bothersome problems of global warming, the real threat is that of the extinction of our species. The dinosaurs went extinct because of a cosmic event (a meteorite hitting the Earth) which they had no means of foreseeing and which they were powerless to prevent. In our case, things are different: the potential "extinction event" is of our own making; we can and do know about it; and – in principle – we ought to be able to do something to prevent it. A reduction of 50% of our current scale of activities would suffice; and this would only mean going back 50 years or so which is manifestly not impossible if there was a genuine political will to do it. However, such measures are not on *any* current political agenda.

14.2 Anthropocene and the Anthropologically Constitutive Character of Technology

At this point, in the hope of moving beyond the impasse, I would like to introduce the work of the "Compiègne school" and the idea that "Technology is Anthropologically Constitutive" (the "TAC" thesis). The University of Technology of Compiègne is characterized by the fact that all the students are in training as engineers; yet owing to the vision of its founder, Guy Deniélou, the University also has a strong Department including philosophy with the human and social sciences. The aim is not to divert the students from their calling as future engineers; on the contrary, it is to enlarge their awareness of the full human significance of technology. When an engineer is inventing a new piece of technology, (s)he is not simply designing a material object (although this is important, and requires considerable expertise and competence). She is also fashioning the *social life* of the people who will use this new technology – their relationships to the world, to themselves and to others. In order to deploy this perspective, the Department of Technology and Human Science (TSH) has taken an interest in Cognitive Science, and in particular in the notion of *Enaction* (Stewart et al. 2010). Originally coined by Francisco Varela, "enaction" is a metaphor from theatre which emphasizes the fact that all living organisms "bring forth" the "lived world" in which they exist. The paradigm case is the "world of the tick" as evoked by Jakob von Uexküll (1909). To a first approximation, every biological species enacts a single characteristic *Umwelt* (lived-world). The exception is the species *Homo sapiens*: members of this species are capable of enacting an immense variety of lived-worlds. This is essentially due to the fact that humans are capable of making and using tools; to which is to be added the fact that, through language, humans *talk about* inventing and using tools, i.e. they invest in techno-logy. This emphasis on the key role of Technology for human cognition is an extension of the Varelian concept of Enaction that Varela himself did not develop. As I announced, this is the core of what has come to be known as the "Compiegne school," which has as its centre-piece the notion that "Technology is Anthropologically Constitutive." (Stiegler 1998, 2009)

The "TAC" thesis is nicely illustrated by Leroi-Gourhan's aphorism "Mankind began with the feet." (Leroi-Gourhan 1964; Lenay, Chap. 13 in this volume) When our hominid ancestors first stood up on their hind feet, this liberated the hands for making and using tools, and the face for articulating speech; the two together giving rise to "techno-logy." This marked the separation between Man and our nearest living biological relatives, the great apes; the large brains of which we are so proud only came later, arguably to cope with the complexities resulting from the development of technology. The TAC thesis is further illustrated by the fact that each epoch is characterized by the dominant technology of the period: flint choppers for the Stone Age; the cities of the great civilizations of Antiquity (Egypt, Babylonia, the Incas…); the "domestication of the savage mind" and the "entry into history" (Goody 1977) with the invention of alphabetic writing and coined money by the Ancient Greeks; the Renaissance (the blast furnace for iron working, the Archimedes screw, the printing press…); the nineteenth century Industrial Revolution (machines and factories). Marx summed up the idea: "the water-mill goes with feudal society; the steam engine with industrial society…"

What is to be noted is that over the course of this long series, the *pace* of technological (and social) change has increased and indeed accelerated. In the initial period when there were a multitude of hominid species – the period leading up to the Stone Age and the emergence of *Homo sapiens* – the time-scale was still that of the millions of years of biological evolution; Leroi-Gourhan (1965) notes that "early man exuded his tools in much the same way as he grew teeth and finger-nails." The great civilizations of Antiquity lasted for thousands of years. "Modern" times are measured in centuries; and nowadays, with the revolution in information and communication technologies, a lapse of 10 years seems a long time. An important factor in this acceleration is the *synergy* between technological innovation and capitalism. Marx predicted that capitalism would "wither away and die" because of the tendency for profit rates to fall due to market competition. What he did not foresee is that through permanent technological innovation, profits could be maintained "at the margin" as long as patent rights held and before the latest innovation became generalized. In contemporary society, economic growth has become a social necessity; an economic recession leads the ills of mass unemployment. The result is that we are "locked in" to a productivist mode of existence.

So what does the "TAC" thesis bring to our analysis of the Anthropocene in a state of ecological crisis? We have already mentioned the conundrum that in the present case, collective human intelligence appears to be *less* than the intelligence of individuals. The fact that we are "locked in" to a productivist mode of existence is probably a key here. We might "know" that continued growth is bad for us; but we cannot seriously envisage giving it up. Latour (2014) makes a revealing analogy with the question of trying to stop smoking. It is one thing to "know" that smoking cigarettes causes cancer; it is quite another to actually stop smoking. One has to actually feel the pain in the flesh in order to measure what "knowing" really means. Coming back to the climate issue, how many additional institutions, how many bureaucracies would be necessary in order to feel personally responsible for something as abstract and distant as the chemical composition of the atmosphere? Latour remarks that it is no accident that the very same lobbies, which nourish the climate-skeptics worked so long to break the connection between cigarettes and the state of your lungs. He recalls the formula attributed to Lao Tseu: "To know and not to act, is not to know."

In a related vein, we may consider the phenomenon of addiction. Moore (forthcoming) recounts an experiment in which laboratory rats, shut up in isolation in a cage, repeatedly self-administered cocaine and heroin to the point of oblivion. However, when the rats were let out of the cage into a more natural environment and provided with other rats and play activities to keep them occupied, the drugs become dramatically less attractive (Alexander 2008). Alexander argues that contemporary Western society is characterized by a similar failure of "psychosocial integration," in which shopping – and increasingly also religious fanaticism – serves as a "pseudo-solution" to the experience of dislocation, filling the void left open by the sacrifice of community and meaningful employment in favor of the creation of capitalistic wealth. This theory casts addiction as an understandable and moreover 'adaptive' response to the demoralizing, alienating effects of community breakdown, or the 'poverty of spirit' that comes about when "society systematically curtails psychosocial integration in all of its members." (Alexander 2008)

Finally, on a slightly different tack, there is a crucial difference between the insect colonies mentioned above, and the current human population. These colonies are genuinely unified entities; "super-organisms" almost comparable to a multicellular plant or animal. By contrast, "mankind" is anything but a unified entity. As Latour points out, the problem is that in order to face up to the challenge of the ecological crisis, there is literally *no-one* who can be considered as responsible. Why? Because there is no way to *unify* the *anthropos* as a generic entity to the point of making it responsible for everything that will play out on this new global scene. Aborigines, and Native American peoples, who themselves have life-styles that are entirely respectful of the environment, are victims of their encounter with Western society; they are powerless, unable to save their own skins let alone "save the planet." Likewise, the impoverished millions in Bangladesh – whose *per capita* environmental impact is well below a sustainable norm – are also powerless victims unable to contribute to alleviating the ecological crisis. On the other side, the individuals who occupy managerial positions in multinational corporations or financial institutions are also (however strange that may seem) quite powerless: if they were to even attempt to act differently, they would simply be rapidly replaced. Latour sums up by noting that ecological questions are not there to peacefully assemble the various parties concerned; they divide even more effectively than all the political passions of the past. If Gaïa could speak, She would say like Jesus: "Think not that I am come to send peace on earth: I came not to send peace, but a sword." (Matthew 10, 34) Or even more dramatically, as in the apocryphal book of Thomas: "Hear thou, Thomas, the things which must come to pass in the last times: there shall be famine and war and earthquakes in divers places, snow and ice and great drought shall there be and many dissensions among the peoples. … For there shall be great disturbance throughout all the people, and death. At that time shall be very great rising of the sea, so that no man shall tell news to any man." With the ecological crisis in mind, this is uncannily prescient. What is to be noted in particular is that "natural" disasters ("earthquakes, snow and ice and great drought") are combined with social disarray ("many dissensions among the peoples. … so that no man shall tell news to any man"). We can't say that we haven't been warned… although the fact that the book of Thomas was relegated to an apocryphal status indicates that already in Biblical times, we are not very receptive to the bad news.

14.3 Conclusion

Where does this leave us with respect to the theme of the Anthropocene? On the face of it, the "TAC" thesis seems to be entirely positive and benign, a matter for unalloyed optimism. What history has shown – in particular our recent history leading to the ecological crisis – is that matters are far more complex. Technology can be the instrument for the flowering of mankind; but it also harbors the potential for its complete destruction. Technology is anthropologically constitutive as much as it is anthropologically destructive. A French saying has it that "Science without conscience is the ruin of the soul." We may adapt and extend this by saying that "Technology without responsibility is the route to an apocalypse," the problem being that technology is involved in shaping – enacting – this responsibility, as well as the blindness to it. The task of inculcating an awareness of this to the upcoming generation, in particular to young engineers, is a noble but daunting one.

References

Alexander, B. K. (2008). *The globalization of addiction: A study in the poverty of spirit*. Oxford: Oxford University Press.

Aykut, S. C. (2018). Le climat et l'anthropocène. Cadrage, agentivité et politique climatique mondiale après Paris. In C. Larrère & R. Beau (Eds.), *Penser l'Anthropocène* (pp. 499–522). Paris: SciencesPo Les Presses.

Barnosky, A. D., et al. (2012). Approaching a state shift in Earth's biosphere. *Nature, 486*(7401), 52–58.

Bensaude Vincent, B., & Loeve, S. (2018). Penser carbone. In C. Larrère & R. Beau (Eds.), *Penser l'Anthropocène* (pp. 375–389). Paris: SciencesPo Les Presses.

Bontems, V. (2016). L'énergétique de Simondon: progrès versus puissance. In V. Bontems (Ed.), *Gilbert Simondon ou l'invention du futur* (pp. 45–56). Paris: Klincksieck.

Descola, P. (2018). Humain, trop humain? In C. Larrère & R. Beau (Eds.), *Penser l'Anthropocène* (pp. 19–35). Paris: SciencesPo Les Presses.

Fressoz, J.-B. (2012). *L'apocalypse joyeuse: Une histoire du risque technologique*. Paris: Editions du Seuil.

Goody, J. R. (1977). *The domestication of the savage mind*. Cambridge: Cambridge University Press.

Hache, E. (Ed.). (2014). *De l'univers clos au monde infini*. Paris: Editions Dehors.

Larrère, C., & Beau, R. (Eds.). (2018). *Penser l'Anthropocène*. Paris: SciencesPo Les Presses.

Latour, B. (2014). L'Anthropocène et la destruction de l'image du Globe. In E. Hache (Ed.), *De l'univers clos au monde infini* (pp. 27–54). Paris: Editions Dehors.
Leroi-Gourhan, A. (1964). *Le geste et la parole. Tome I : Technique et langage*. Paris: Albin Michel. English Edition: Leroi-Gourhan, A. (1993). *Gesture and speech* (Vols. I & II, A. B. Berger, Trans.). Cambridge, MA: MIT Press.
Leroi-Gourhan, A. (1965). *Le geste et la parole. Tome II : La mémoire et les rythmes*. Paris: Albin Michel. English Edition: Leroi-Gourhan, A. (1993). *Gesture and speech* (Vols. I & II, A. B. Berger, Trans.). Cambridge, MA: MIT Press.
Moore, G. (forthcoming). The pharmacology of addiction. In J. Foster & P. Gratton (Eds.), *The continental philosophy of science*. London: Bloomsbury.
Stewart, J., Gapenne, O., & Di Paolo, E. (Eds.). (2010). *Enaction: Toward a new paradigm for cognitive science*. Boston: MIT Press.
Stiegler, B. (1998). *Technics and time, 1. The fault of epimetheus* (G. Collins & R. Beardsworth, Trans.). Stanford: Stanford University Press.
Stiegler, B. (2009). *Pour une nouvelle critique de l'économie politique*. Paris: Galilée.
Stiegler, B. (2015). Demain, le temps de l'automatisation. *Intellectica, 63*, 151–160.
Stiegler, B. (2017). Escaping the Anthropocene. In M. Magatti (Ed.), *The crisis conundrum. How to reconcile economy and society* (pp. 149–163). Dordrecht: Springer.
von Uexküll, J. (1909). *Umwelt und Innenwelt der Tiere*. Berlin: Springer. English Edition: von Uexküll J. (1957). *A Stroll through the world of animals and men* (C. H. Schiller, Trans.). New York: Indiana University of Pennsylvania.

Chapter 15
Toward an Object-Oriented Philosophy of Technology

Xavier Guchet

Abstract This chapter focuses on the contemporary "philosophy of technical artifacts" in the Dutch context to open up a discusion. It demonstrates how French philosophy of technology may enrich current debates on ontological and normative issues related to artifacts. In the French tradition, "thing," "artifact" and "object" are not equivalent terms. Furthermore, French philosophers and anthropologists have paved the way for a "biological philosophy of technology." They considered technology in a close relationship to biological life. Insofar as contemporary philosophers have to pay attention to puzzling *bio-objects* and unprecedented arrangements of technology and biology, such as GMO, clones, molecular bio-machines, biomarkers for precision medicine, big data, bio-repositories etc., the paper claims the relevance of both an *object*-oriented and a biological philosophy of technology for overcoming some limits of the "artifactual turn" in the philosophy of technology.

Keywords Aesthetics · Artifact · Bio-object · Normativity · Object · Postphenomenology · Technology · Values · Vitality

One of the most inspiring developments in contemporary philosophy of technology for the last 20 years has consisted in carefully examining technical artifacts *per se*, with great attention to their materiality and the way they embody norms and values. In opposition to classical philosophers of technology, who mainly considered artifacts as such a matter of poor philosophical interest (see for instance Heidegger 1954), scholars have dramatically challenged this reluctance, and strongly claimed an artifact-centered approach to contemporary technology (Dipert 1995; Kroes and Meijers 2000; Achterhuis 2001). In the last 15 years, this "philosophy of technical artifacts" has mainly taken two paths: firstly, a postphenomenological approach inspired by American philosopher Don Ihde and now carried out by Peter-Paul

X. Guchet (✉)
Connaissance, Organisation et Systèmes TECHniques (COSTECH), Université de technologie de Compiègne (UTC), Compiègne, France
e-mail: xavier.guchet@utc.fr

S. Loeve et al. (eds.), *French Philosophy of Technology*, Philosophy of Engineering and Technology 29,
https://doi.org/10.1007/978-3-319-89518-5_15

Verbeek (Verbeek 2005); secondly, a metaphysical one, inspired from analytical philosophy and known under the label "The Dual Nature of Technical Artifacts." (Kroes and Meijers 2006) Both approaches intend to better combine three perspectives that are at stake in the so-called "empirical turn" in the philosophy of technology (Brey 2010): the society-oriented viewpoint, the engineering-oriented one and the "applied technology ethics" one. Scholars of both postphenomenological and metaphysical currents have recently strenghtened their joining together and reinforced their dialogue, in particular through a shared focus on a controversial subject: the "intrinsinc morality of artifacts." (Kroes and Verbeek 2014)

This paper intends to argue that French philosophy of technology can fruitfully feed this "artifactual turn," as it has developed in the Dutch context for the past 20 years. More precisely, it focuses on French authors who have paved the way for an object-centered philosophy of technology. Are technological *object* and technological *artifact* equivalent terms however? Contrary to most of the contributors to the current philosophy of artifacts, who consider equivalent the words "artifact," "thing" and "object," this chapter answers no, and claims the potential interest of a specifically *object*-centered approach (see especially Simondon 1958) to overcome some limits of the "artifactual turn" in the philosophy of technology.

While *artifacts* are above all man-made, intentional entities, French authors focus much more on technological *objects* as beings-in-the-world among other entities: beyond the intentions of their makers, technological objects prove to have an activity of their own in the natural and the social world. They link to other objects – be they natural or artifactual – in an unpredicted way, and may give rise to processes that escape our control – a very common statement undoubtedly, at least since Hannah Arendt's claim that technology (especially nuclear) prompts new processes inside Nature. As Jean-Pierre Dupuy, following Arendt, pointed out more than 10 years ago, nanotechnology primarily consists in making technical devices of unpredictable behavior (see Dupuy, Chap. 9, in this volume). They are less *artifacts* fitting human intentions than *objects* which have somewhat unwanted effects on natural entities. In other words, technical objects such as nano-objects may have an "existence" of their own – a "mode of existence," as French philosopher Gilbert Simondon says. The word "existence" may be puzzling here, but it must be literally understood: technical objects "ex-ist" insofar as they spread out, and must be critically assessed, beyond the intentions of their makers.

To be sure, the current literature does not reduce artifacts to human intentions. On the one hand, artifacts have an intentional nature but also a material one. They are physical systems and careful attention must be paid to their materiality. But do scholars of the Dual Nature program consider the two "natures" of equal importance? One can doubt it (Vaccari 2013). On the other hand, artifacts are "mediations" that shape our moral equipment beyond any intention we may have (Verbeek 2010). Postphenomenology, however, focuses on the Human-world relationships, and leaves quite unexplored the unpredicted relationships that may arise between technical and non-Human natural entities.

The approach tackled here is in agreement with Verbeek's claim to dismissing any transcendentalist view of technology,[1] but it also overcomes the focus on human-world relationships. It agrees with the Dual Nature program insofar as it focuses on technical objects *per se*, but it dismisses any static metaphysical description in favor of an ontogenetic, dynamic one. So this essay provides a critical reading of the philosophy of artifacts and claims the fruitfulness of an *object*-oriented approach to technology, with reference both to ontological and axiological issues.

From an ontological viewpoint, an *objectual* approach focuses on processes rather than "natures" – a claim which proves to be of great interest today, especially with respect to the large amount of technical entities in which the making *process* is more important than the *intention* of the maker (in bio- or nanotechnology for instance).

From an axiological viewpoint, it is commonly admitted that technical artifacts embed non-technical norms, among which ethical ones. In the existing literature devoted to technological design however, these non-technical norms are mostly refered to *society*, and more precisely to conflicting expectations and values that shape it. As a matter of fact, very poor attention is usually paid on *biology* as a possible source of norms for design. The Dual Nature program clearly focuses on man as an intentional agent, with no reference to man as a living being. To be sure life is of great concern in Verbeek's postphenomenology, however it is much more understood in the sense of the *lived* (the lived world, the lived experience etc.) than in the sense of the *living*.

This chapter intends to overcome such a dismissal – a challenge that is incidentally at stake in recent developments within the philosophy of design (see in particular the emergence of a new paradigm of design, namely the Life-Based Design, grounded on a holistic concept of life including its biological dimensions) (Leikas 2009). It argues that a strong reference to what French philosopher Georges Canguilhem called the "vital normativity" (Canguilhem 1943), and more broadly a reference to the "biological philosophy of technology" developed by French scholars such as paleoanthropologist André Leroi-Gourhan (1964, 1965), can help enrich the normative debates in the current philosophy of artifacts.

A "biological philosophy of technology" removes technology from any intellectualist framework and roots it in the process of life itself. In the French context, concepts such as "exteriorization" (of biological functions into technical objects) or "technical *milieu*" were keys to renew the understanding of technology. Let us consider whether the concept of technical *objectivity* (instead of *artifactuality*) together with a biological approach to technology could bring about a robust analytical framework capable of enriching the current philosophy of technical artifacts.

[1]One could ask however if a non-transcendantalist approach to the Human-world relationship, as Verbeek points out, still belongs to a phenomenological framework.

15.1 Critical Insights into the Philosophy of Technical Artifacts

For at least the last 30 years, philosophers of technology have faced an exciting situation and a great challenge. Strange artifacts are continuously made and studied in the laboratory, such as clones, genetically modified organisms, synthetic bacteria or molecular machines, blurring the boundaries between the natural and the artificial or the living and the technical (Bensaude Vincent and Newman 2007). How can philosophy help clarify this ontological "continued creation" in the realm of technology, so to speak, and which are the relevant concepts for this purpose? How to handle ethical issues related to this ontological profusion? As emphasized by many scholars today, such questions are not only speculative but also practical: can philosophers help engineers and scientists design "good," ethically-robust technical beings? (Verbeek 2010).

More and more scholars share the conviction that ontological examination of technical beings should come first and shape ethical assessment. In this context, some have proposed to engage philosophy of technology in a "thingly turn" – a term coined by Verbeek (2005). Such a thingly turn was meant to intensify the "empirical turn" initiated by American philosophers of technology in the 1980s: from now on, technical beings *per se*, their materiality and their design should be the focus of philosophical investigations. As pointed out in the introduction, this new empirical trend in the philosophy of technology has highlighted the concept of "artifact" and has taken two major directions.

15.1.1 The Dual Nature Program

The Dual Nature program emphasizes that technical artifacts have a *function*ing (how do they work?) and a *function*ality (what are they supposed to do?). Both aspects are to be carefully scrutinized (Vermaas and Houkes 2006). As Pieter Vermaas pointed out however (Vermaas 2006), while artifacts in themselves have a Dual Nature, their components can be considered material structures with no reference to intentionality. Last, the way technical artifacts achieve their functioning closely depends on their physical features, while "social artifacts," such as banknotes for instance, achieve their goals with less dependence on their materiality (Kroes and Meijers 2006).

Peter Kroes – a leading figure of the Dual Nature program – provided useful clarifications on what is required for a thing to be a "technical artifact." (Kroes 2012) Technical artifacts are a sub-domain of the artificial world, which means first that all man-made i.e. artificial things are not *de facto* technical artifacts – for instance, Americium atoms are human-made insofar as they do not spontaneously occur in Nature (a complex apparatus is required for that), yet they are not technical artifacts owing to the fact that their properties are not intentionally designed.

Scientists "create" phenomena in their laboratories, but they do so in the weak meaning of the term "creation": they just create the conditions for the phenomena to appear, but they do not intentionally shape their properties. Conversely, in a stronger sense designers of technical artifacts intentionally shape the properties of what they create.

Moreover, a natural thing that is used for a specific task (for instance, a flint stone used as a knife) should not be considered a technical artifact unless it was intentionally carved to achieve this task – at best, it should be considered a technical *object*. Thus, technical artifacts are defined as "human-made physical constructions that fulfil practical function" or "useful physical objects" that "embody human intelligent design." Technical artifacts result from both mental and physical human works.

Last but not least, functional properties are not intrinsic properties of artifacts, but relational ones. They refer both to human intentions and to physical properties that achieve the desired functions. For instance, an acetylsalicylic acid molecule is not a technical artifact *per se*, but it becomes one as far as it is refered to the history of a drug that was intentionally designed for achieving a specific purpose: Aspirin (acetylsalicylic acid is the active component of Aspirin). To be sure, Kroes does not mean that functional properties can be added to things *afterwards* – once again, considering an existing thing as a functional one, in a particular context, does not suffice to turn it out to be a technical artifact (cf. the flint stone used as a knife). Technical artifacts have properties that are intentionally designed, but functional properties are not intrinsic properties in the way physical ones are. For instance, a clockwork is a physical structure that we conspicuously refer to a specific desired function. We do so at the macroscale level because we are familiar with it. At the molecular or nanoscale level, this is not so obvious and the functional property of the acetylsalicylic molecule is less immediately "legible" in its physical structure. However, Kroes argues, there is no fundamental difference between both macro- and nanoscales: in both cases things that are designed with intentional properties must be called "technical artifacts."

The Dual Nature program has been critically examined (Mitcham 2002; Vaccari 2013). For instance, Vaccari argues that despite their claims, the contributors to the program mainly focus on the intentional nature of artifacts, leaving the physico-chemical one quite underinvestigated. In particular, Vaccari notes, the program underestimates the fact that matter has an "agency" of its own, due to its structuration at the nanoscale which determines to a certain extent the function ascription. In short, Vaccari's criticism is about the residual hylemorphism of the Dual Nature program: matter has no agency *per se* and is in-formed by external agents. As Vaccari consequently suggests, the Dual Nature program may be unsuitable to give accurate insights into current research fields, especially bio-nano-technology, which do not fit the hylemorphic paradigm (Loeve 2009; Guchet 2014).

Actually the concepts of the Dual Nature program may be challenged by bio-nano-technology, especially with regard to the fact that *intentionally* designed nano-objects are not always the outcome of a *function ascription*. Several years ago, a French team of physicists (Mayne et al. 2004) adsorbed a biphenyl molecule on a silicium surface and addressed it by means of a probe microscope: an electric current was applied to one part of the molecule and the molecule pivoted. After applying a current once again, the molecule turned back to its first position. So the experiment gave rise to a bistable molecular configuration – a property that was not at all anticipated by experimentalists. The bistability of the molecule was neither a desired function that was intentionally ascribed to it with regard to a "use plan," nor a function ascribed to a component entering an artifact. Furthermore, experimentalists considered the molecule a "molecular machine" insofar as its motion could be controlled. Last, it was suggested that this device could be viewed as a "molecular switch" to be used in electronic circuits. Thus, to conclude: (1) a useful function (molecular switching) could be ascribed to this device, but afterwards only, insofar as the device was not intentionally designed to achieve this function. Consequently, in Kroes' terms this device should not be considered a technical artifact, although it is undoubtedly a human-made one. Kroes would probably argue that the biphenyl molecule is the same kind of thing as the aforementioned flint stone; (2) to be sure, unlike the flint stone, something was intentionally designed in the biphenyl molecule experiment – the molecule itself (which has to be synthesized), the silicium surface which has to be prepared with great accuracy, the scanning tunneling microscope which is required to provide the electric current. But Kroes would probably say that we are dealing here with an experiment that "creates" its phenomenon (a specific molecular dynamic) in the weak sense of the term "creation": experimentalists create the appropriate conditions for the bistability to appear, but they do not create the bistability itself, i.e. the property of the device. As a consequence, it still cannot be considered a technical artifact; (3) however, this statement may give rise to confusion. In the Dual Nature framework, functional properties are useful tasks that technical artifacts achieve with regard to human intentions (e.g., being used as a molecular switch). Nonetheless, the experimental device demonstrates a property that is irreducible, neither to useful functions, nor to physical properties. The bistability is not an intentionally designed function, but it is neither a property that takes place among physical properties of the device. Beyond physical and intentional properties, there is thus a third kind of properties – namely "*operational*" ones.

The molecular dynamic is not a physical property of the biphenyl molecule adsorbed on the silicium surface, and it is not yet a useful function. Kroes would rightly argue that it is a natural phenomenon the occurrence of which is due to appropriate experimental conditions, but he would then conflate two kinds of phenomenon-that-appears-in-appropriate-conditions: first, a phenomenon like the Hall effect, which occurrence actually requires an experimental set, ending up in evidencing a universal law of Nature; second, a phenomenon like the bistability of the biphenyl molecule, a property which is attached to the particular experimental set (changing the shape of the surface modifies the behavior of the molecule) and which appears to be the specific operation performed by the whole device molecule-surface-microscope – on the contrary, the Hall effect cannot be considered an "operation" performed by the instrumental set. The biphenyl molecule experiment did not end up in evidencing a new natural law of Nature or an unknown natural phenomenon, but in validating a technological protocol – whatever the unpredictability of the phenomenon in the initial design of the experiment: the bistability appears to be the technological operation performed by THIS specific device. To be more precise, the experimentalists involved in the biphenyl experiment may act as scientists and focus on the natural phenomenon evidenced by it: e.g., the behavior of electrons when they cross a unique molecule. However, they also act as technologists and focus on the whole set molecule-surface-microscope not qua a scientific instrument evidencing a natural phenomenon, but qua a technological device that performs a specific operation: a reversible *pirouette* of the adsorbed molecule. To be sure, the molecule itself is not a technical artifact *per se*: the whole device molecule-surface-scanning tunneling microscope is; as A. Mayne and colleagues pointed out, "the surface is part of the machine."

As it happens, even at the macroscopic level *useful functions* and *technical operations* should not be confused. For instance, a wind mill is a sequence of mechanical "operations" that provides a machine (the millstone) with driving force. The useful "function" of the wind mill is to provide human beings with flour.

Furthermore, for the last 15 years, a great amount of literature has highlighted how deeply nano-objects are value-laden, or should be value-laden. Anthropologist Christopher Kelty for instance (Kelty 2009) has claimed that non physico-chemical properties of nanoparticles, such as safety, should from now on be considered at the same level as physico-chemical ones – so they have to be designed as such in the engineering process. The "engineering view" has to be broadened and brought beyond a purely mechanistic and structural approach to components.

The normative side of the question is at stake too. Contributors to the Dual Nature program mainly consider the normative dimension of artifacts in a functional perspective. The "goodness" of an artifact relates to the way it perfoms its intended functions. In a book devoted to clarify normative issues as part of the Dual Nature program (de Vries et al. 2013), Krist Vaesen highlights three major critical points related to this function-centered approach to technological normativity: first, a function may be perfomed by different artifacts – the choice between various possibilities should then be based on non-functional criteria; secondly, the use of an artifact may have some unpredicted and unwanted consequences (a well-known issue in nanotechnology); thirdly, an artifact may affect other entities in the world. So a technical artifact should be considered not only with regard to its intentional function, but also as an entity of its own, taking place in the world among other entities and interacting with them. This shift from an artifactual to a "being-in-the-world" concept of technical objects may be of great interest in particular when such objects are living ones, such as genetically engineered bacteria or animals in industrial farming.

15.1.2 Postphenomenology and the Concept of Technical "Mediation"

To be sure, Verbeek's postphenomenological approach seems to achieve this shift. Far from considering technical artifacts with reference to the function ascription criterion, Verbeek analyses them as beings *per se* and especially as "moral agents" in their own right (Verbeek 2011). Verbeek's strong and demanding claim is, first, that technical "mediations" should be considered in their ontological productivity – far from being mere "instruments" at the disposal of the Human, they contribute to shape what the Human is and what the world is – secondly, that philosophy must help engineers design ethically "good" mediations, and thirdly that technological design should favor a better "attachment" of humans to their technical artifacts (Verbeek 2005), by means of a deeper and broader commitment of sensibility in the human-technology relationship.

Verbeek seeks to clarify how technical "things" both broaden our repertoire of actions on the material world, and dramatically contribute to shape our moral equipment. This stimulating approach however, which to some extent echoes Bergson's (1932), still focuses on the human-world relationships. To be sure, Verbeek promotes non-human entities as fully moral agents, among them technical artifacts. Nevertheless, postphenomenology is about "mediations" in which humans are always involved. Yet, numerous technical beings made in the laboratory today require a more extensive, non human-refered, concept of relation. Let us get back to the bistable biphenyl molecule. To become a *machine*, the molecule involves several kinds of relations, among which some are non human-refered: the relation between the molecule and the silicium surface; the relation between the two parts of the molecule (two benzenic rings); the relation between the ensemble molecule-surface and the STM tip. Most of these relations operate under the order of magnitude where a human experiences phenomenologically the world. Verbeek's concept of "mediation" may be too restricted for investigating the whole range of relations prompted by technology – a remark that may be addressed to the Dual Nature program as well, insofar as the concept of useful functional properties does not cover the whole range of relations that underly technological operations.

From a normative standpoint, contemporary postphenomenology rightly raises the question of what "morally good technical mediations" may be. Philosophers have to "accompany" engineers in designing such "good" mediations (Verbeek 2010). But what criteria will define the goodness of a mediation? Verbeek's "technology accompaniment" relies on the capacity we have to anticipate the effects of technical artifacts on the world and on our moral equipment. However, despite the fact that this belief has been extensively criticized, it does not solve the major problem: even if we were able to predict all the consequences of an artifact which is under design, some of these consequences may not be morally qualifiable in themselves, i.e. objectively, but only with regard to a set of potentially conflicting values. Hard impacts such as toxicological effects can actually be considered "bad" effects as such and to some extent Kelty's *safety by design* fits Verbeek's technology accompaniment. But what about "soft impacts" such as the effects of technology on the human condition? Here conflicts of values may be at stake. A technology accompaniment primarily requires a public arena where such conflicts can be formulated, and above all, it requires to explicit the values that underly any design choice. To design "good" technical mediations means to embody a set of chosen norms in the artifact, but these norms are favored in accordance to values: what are these values and how to address them? This point is not clear enough.

Finally, Verbeek's proposal to reinforce our sensible "attachment" to technical artifacts and a "techno-aesthetic" commitment with them (a term already coined by Simondon in the early 1980s) is undoubtedly very stimulating. At first glance, it should go hand in hand with careful insights into the vital dimension of human-technology relationships – a demand however that seems dismissed in Verbeek's postphenomenology and needs further investigations.

Let us now turn see how French philosophy of technology can help address these challenges by looking at authors such as Leroi-Gourhan, Canguilhem, Simondon and Dagognet. To a large extent, this French philosophical tradition is rooted in Bergson's philosophy of life and, consequently, the following section of the paper starts with an insight into Bergsonian concepts related to technology.

15.2 Technical Objects, Norms and Values: French Perspectives

15.2.1 Bergson's Philosophy of Technology: A Short Overview

In his renowned essay *Creative Evolution* (1907), Bergson refers to a very common topic at that time, at least since the publication of Ernst Kapp's book *Grundlinien einer Philosophie der Tecknik* in 1877, whose main ideas were introduced in France by the end of the nineteenth century by sociologist Alfred Espinas: according to Kapp, tools are "organ projections," they extend the organic body. In Bergson's terminology, technology is what conveys the "vital impulse" (*l'élan vital*) alongside the line of evolution to mankind. In 1932, 25 years after the publication of *Creative Evolution*, Bergson meaningfully reconsiders the relationships between technology and life (Bergson 1932). While he still defines technology as a body extension, he now claims that industrial machinism can destroy life, at least if a "supplement of soul" (at first glance, an improvement of our ethical equipment) does not accompany its development. This threat however is not external to life itself. Bergson rather argues that life is ambivalent: the "vital impulse" faces no external impediments, it has *per se* a tendency to fall back as matter. Matter is not extraneous to life, it is a fallout of life. To be sure, in *Creative Evolution* mankind is the livingform through which the vital impulse overcomes obstacles, and continues to create new forms. Conversely, in *The Two Sources of Morality and Religion* (1932) mankind is rather viewed as the livingform in which the ambivalence of life reaches its maximal intensity. According to Bergson, humanity continues the vital impulse by creating values in the universe, which is a "machine for the making of gods." However industrial machinism can destroy our capacity to create values, i.e. it can stop the vital impulse that has until now managed to make his way throughout mankind. To be more precise, humanity values things (such as a frenetic consumption of any kind of commodities) that may prompt a collapse of the vital impulse. Bergson means that we may cherish values which are likely to threaten life itself, stopping its creative dynamic. A gap may dramatically divide life as a value-creative process on the one hand, and the values we may cherish as human beings on the other hand. In this regard, Bergson's "supplement of soul" does not convey the need for a checklist of moral principles or guidelines, with the aim of framing the development of technology. It rather means that what we cherish must be "valued" with regard to life itself.

So what does this "supplement of soul" precisely consist in? Bergson argues that industrial machinism has significantly extended our body, insofar as we can *do* and *perceive* more with our machines. To be sure, Bergson closely links action and perception in an original way. In *Matter and memory* (Bergson 1896), he reversed the commonly admitted relationships between both terms: action does not come after perception; it comes first and perception is adjusted to it. We perceive exactly what is of interest for acting. So, as far as it has extended the repertoire of our actions (making it "[reach] as far as the stars," Bergson says), industrial machinism claims a readjustment of our perception. Our perception should be proportioned to our extended technological action which "reaches the stars," and which from now on, due to recent developments in science and technology, reaches also individual molecules and atoms. Bergson's "supplement of soul" refers to this need for a modified perception. Before relating to our moral equipment, it relates to our sensibility. It is worth noting that for Bergson and Verbeek as well, the adjustment of sensibility regarding technological action is not a natural process: it is rather a matter of ethical involment. We should be morally committed to actively reshaping our perception.

So Bergson was concerned with issues that are still at stake in contemporary postphenomenology, such as the need for a technology accompaniment, and the reshaping of our sensibility. Furthermore, Bergson significantly emphasized that moral issues related to technological human-world "mediations" should be addressed with regard to life. Admittedly, Bergson did not exactly meet the thorny challenges that philosophers of technology face today. More precisely, Bergson does not provide a suitable concept of technical artifact, capable of covering the ontological strangeness of bio-technological objects such as clones, GMO, synthetic bacteria, nano-bio-devices and so on. Furthermore, Bergson focuses on human life and its internal ambivalence, but he has no strong concern for how non-human living beings may be affected by technology. Nevertheless, he opened a fruitful way to French philosophers of technology who actually seized upon these challenges, namely Canguilhem and Simondon. Let us focus on the latter, who was pervaded by Canguilhem's concept of "vital normativity," i.e. the idea that living beings create their own values: their *milieu* is not reducible to a set of physico-chemical processes and properties; it is above all framed by biological valences or values.

15.2.2 *Simondon's Approach to the "Mode of Existence of Technical Objects": Beyond the Artifactual Perspective*

What makes Simondon a pivotal author for the general argument of this paper is the following strong claim: to carefully address normative issues concerning technology, it is essential to closely intertwine both an *object*-centered and a *biologically*-based approach to it. The paper now examines Simondon's concept of technical object. More precisely, it emphasizes the analytical significance and the normative content of this concept.

Two characteristics of Simondon's concept of technical object are of great interest for contributing to enrich the philosophy of artifact.

First, technical objects must be described with regard to their *genesis*. As de Vries rightly acknowledges (de Vries 2008), the Dual Nature program highlights a *static* concept of technical artifact. Artifacts have two *natures* and their intertwining is of ontological significance, whatever the *process* of their production may be. On the contrary, Simondon considers the *process* as ontologically prior to the final result.

De Vries confronts both the Dual Nature program and Simondon's concepts in a pragmatist perspective: which concepts better allow to provide a robust analytical framework for accurately describing contemporary technology? De Vries supports his Dutch colleagues although, in the same pragmatist way, arguments might be found in favor of Simondon.

Biotechnology provides suitable examples for this test. From a static viewpoint, a genetically modified animal and a "naturally" born one, so to speak, cannot always be distinguished: no special feature unavoidably brings the former to our attention. To be sure, both can be considered artifacts in the sense of the Dual Nature program, as both result in a Human intentions/biological matter intertwining (to be sure, the "naturally" born veal does not arise from a spontaneous Nature: a long history of artificial selection produced it). In the framework of the Dual Nature program, no ontological difference divides the genetically modified and the "naturally born" animals. Conversely, Simondon's processual approach makes it possible to distinguish them with ontological relevance. Artificial selection consists in intervening at a macroscopic level, where interesting phenotypical features are identified and favored by breeders. Genetics consists in directly intervening on the cell components, at a microscopic level, where molecular mechanisms operate. Orders of magnitude are different and this is of major consequence for ontological claims. Although both animals embody human intentions into biological structures and consequently fit the Dual Nature program framework, they do not hang together *technological* norms and *biological* norms in the same way. Artificial selection undoubtedly modifies the phenotypical features of the species, however it does no violence to their "vital normativity:" artificially selecting phenotypes rather consists in exploiting inherent capacities of species. To be sure, crossbreeding gives rise to new living beings, with no equivalent in Nature. Furthermore, industrial farming deeply, and most often horribly, alters the "vital normativity" of living beings. Consequently, artificial selection with no targeted intervention at the molecular level may result in a violence done to the "vital normativity" of living beings. In these cases however, technological intervention primarily consists in disminishing and even destroying the inherent capacities and, consequently, the "vital norms" of animals, not in creating new ones. On the contrary, biotechnology reshapes living beings at a molecular level and, in doing so, forces them to adopt new "vital norms" that are of interest to us – as evidenced for example by goats that have been genetically modified with the aim of producing silk. Undoubtedly, the production of silk is not an inherent capacity of goats....

Secondly, Simondon defines the "mode of existence" of technical objects with regard to *relational* features – a statement that undoubtedly requires some clarification. To what extent are technical objects, as Simondon claims, relational beings? To be sure, relations are also at the core of the aforementioned approaches in the contemporary philosophy of technical artifact. Claiming the Dual Nature of artifacts makes it necessary to explain what kind of relation links both natures, and defining them as "mediations," as Verbeek does, makes the relation between both mediated terms questionable. It is worth noting that in both approaches, artifacts are not defined in a substantial way, prior to relations – as if relations were deprived of any ontological consistency. On the contrary, relations ontologically make artifacts what they are. However, in the current literature, relational aspects of artifacts are most often defined with regard to Human beings. Anthonie Meijers for instance (a major contributor to the Dual Nature program) distinguishes "intrinsic," "context-dependant" and "relational" properties of artifacts (Kroes and Meijers 2000). He defines the latter with reference to engineers' or users' intentions. While physico-chemical properties are intrinsic ones, relational properties refer to social commitments with artifacts. In Verbeek's view, as claimed above, technical mediations shape the man-world relationship: the concept of "mediation" as such does not extend to object-object relationships.

On the contrary, Simondon strongly claims that technical objects have an "existence" of their own, beyond human intentions. The human-made *product*, wathever it may be, becomes an *object* as soon as it detaches itself from its designer and producer. Consequently, technical objects should not be considered arte-facts anymore, but "beings-in-the-world" in relation with other entities (including non human entities) – a statement that undoubtedly echoes contemporary philosophy of artifacts. What makes Simondon's approach original and unprecedented however, with no equivalent in the current philosophy of technical artifacts, is an astonishing claim: technical objects are ontologically independent from human intentions or human beings. No reference to human features is required for accurately highlighting the ontology of technology. A very counterintuitive and challenging option, indeed. In particular, relationships between users and technical objects are not of ontological relevance. To be sure, Simondon does not deny the importance of psychosocial commitments with technology. However, such aspects are external to ontological insights into technology.

The first part of *On the mode of existence of technical objects* is precisely devoted to clarify what could be a relational definition of technology, with no reference at all to humans. Relations shape technical objects at a physico-chemical level, both with regard to internal features (how do components of a technical object interact with each other?) and to the coupling of objects with their "associated *milieus.*" (*milieux associés*) Due to its functioning, a technical object has effects on its environment. For instance, it may heat it. If such effects are essential to the operations, the object is said "concrete" (*concret*) and defined in terms of "recurrent causality" (*relation de causalité récurrente*) with an "associated milieu."

Although Simondon, as de Vries emphasizes, does not exactly provide an analytical model for understanding how materiality and morality are linked in the design process, he may fruitfully contribute to current reflections about ontological issues related to technology. Insofar as it pays careful attention to genesis, orders of magnitude, and relations, Simondon's concept of technical object may prove relevant for questioning the ontological differences between technical beings, with reference to their mode of production, to the level at which technical interventions operate, and to the kind of relations they have with other entities in the world.[2]

Contemporary philosophers of technical artifacts most often take their examples in the realm of macro-objects. At this scale, the Dual Nature program and/or the postphenomenological approach may prove superior. To be sure, Simondon claimed the inconsistency of the hylemorphic schema at the microscale and at the macroscale as such. A moulded brick for instance does not fit the arte-factual framework. The paper however does not intend to disqualify any artifactual approach to technology, but rather to hold that today, most of the technological developments that prove challenging occur at the nanoscale. Due to the fact that resulting objects such as molecular machines or GMOs cannot be reduced to *arte-factuality*, as emphasized above, alternative concepts are needed.

So, Simondonian concepts seem of great interest for addressing ontological diversity in contemporary technology. What about their interest for addressing ethical issues related to technological developments?

15.3 Claims for the Normative Consistency of an Object-Oriented Approach to Technology

15.3.1 Ethical Technology Assessment: The Hard Issue

One could object however that such a definition of technical objects as "beings-in-the-world," with an "existence" of their own, may dangerously undermine any ethical questioning of technology. If human intentions and values are removed from ontological descriptions, does an ethical assessment of technology remain possible?

[2] As a matter of fact, Simondon's concept of object finds an echo in Graham Harman's metaphysics. As a co-founder of the Speculative Realism, Harman (2002) opened the way for revisiting the realm of objects with no reference to humans. Even though Simondon cannot be considered a new metaphysician, his concept of object and Harman's are akin. In getting closer to Harman's views, contemporary philosophers of artifacts would certainly find stimulating perspectives.

Simondon's concepts may prove suitable for an ontological description of technology, but they seemingly limit technology assessment to a purely functional point of view, with no reference at all to human values – for instance, Simondon considers technological development relevant and value-laden insofar as it increases what he calls the "internal resonance" of technical objects (i.e. a strong functional interrelation between all components), and to the extent that it reinforces the coupling of the object and its "associated milieu." These norms are functional and at first glance, Simondon's approach may be of very poor interest today, insofar as philosophers of technology have to face challenging *ethical* issues related to technological developments, and to propose accurate concepts to assess these issues. The way Simondon discussed technological norms in his book seems to leave no space for ethical questioning – a situation that could open the way to absurd statements, such as for instance an enthusiastic evaluation of functionally improved gas chambers or electric chairs – both examples refer to Fred A. Leuchter who defined himself as a "humanist" insofar as he considered a moral mission to improve electric chairs in prisons, with the aim to avoid unnecessary suffering for condemned prisoners. His "humanism" led him to take an interest in gas chambers, with the same goal, and finally to support Holocaust deniers – an ethically unacceptable shift, undoubtedly.

This objection against Simondon's object-oriented philosophy of technology would yet be misleading. Simondon was influenced by Canguilhem (he supervised his complementary thesis) who strongly claimed that technology primarily consists in creating values (Canguilhem 1938). Technology does more than extending life and "projecting" biological organs into artifacts: it contributes to shape the *milieu* of human life with reference to values. Furthermore, the *milieu* is a social and a cultural one: according to Canguilhem, man is the living being who realigns his biological norms with reference to cultural value (Canguilhem 1943). Consequently, functional norms are not autonomous and Simondon, following Canguilhem, clearly assumes that technology is both rooted in life and culturally shaped. While in the 1950s, at least in France, technology and culture were often contrasted, Simondon dramatically challenged this common view. The first part of *On the mode of existence of technical objects* undoubtedly focuses on functional norms, but the second part of the book sheds a cultural light on technology.

A problem remains unsolved however. As long as technology assessment still refers to cultural values, it seems that no further step has been taken beyond existing philosophies of artifacts. First, most of contemporary philosophers of technology now admit that technology is value-laden. They have found inspiration in Science and Technology Studies (STS) since the end of the 1970s, and accordingly they have given up "instrumental" and "neutral" statements about technology with no difficulty. In this renewed context, Simondon's views may have lost their former originality. Secondly, as claimed above, a critical reading of Verbeek's stimulating proposal resulted in a challenging issue: what are the criteria for defining the "goodness" of a technical mediation? Answering that culture provides such criteria is not statisfactory however.

Let us go back to the case of genetic engineering. As claimed above, the reference to "vital normativity" proved to be relevant for assessing technology on the basis of ontological investigations, with no reference at all to particular cultural values. Ethical assessments turned out to be derived from ontological inquiries, focusing on the "modes of production" of biotechnological beings such as transgenic animals. One would object however that such assessments are still made under the control of cultural values. For instance, some may denounce industrial farming as far as it brutalizes the "vital normativity" of animals, while free-range farming will be considered respectful of this normativity. On the contrary, others like vegans may consider farming as a violence against nature, at least in our historical context. So the reference to "vital norms" for technology assessment is still shaped by conflicting cultural and political values. And as soon as culture becomes the right answer for technology assessment, ethics and ontology divorce again. Ethical assessment, one would conclude, refers to conflicting values at stake in a particular society, with no regard to ontological differences in the "modes of existence" of technical beings *per se*.

On the one hand, the value of life, even in non-human living beings, depends on what we politically value as human beings. On the other hand, as Bergson emphasized, the values we create as human beings can exhaust life. The "supplement of soul" consists in contrasting our values (what we cherish) with life as a creative process. It primarily means that life is *valuable* as such, beyond our current human values. To be sure, both statements could hardly been made compatible – an issue that makes unsure the relevance of an orented-object and biologically-grounded approach to technology today.

Consequently, contributors to a philosophy of artifacts would rightly declare unconvincing an object-oriented and non artifactual approach to technology. On the one hand, they may admit that the reference to "vital normativity" beyond human intentions and values is relevant for ontological and ethical assessment, at least if living beings are a matter for technological intervention. But now, in any case, human intentions, plans and values appear to be the underlying framework for any technological assessment – a statement that philosophers of artifacts never refuted. Simondon may be relevant for reinforcing ontological investigations on artifacts, but he seems unable to enrich the current philosophy of artifacts from a normative standpoint.

15.3.2 Life, Norms and Aesthetics

To dodge this situation, let us get back to Canguilhem's statement that social norms reshape biological ones. It would be misleading to conclude that for Canguilhem, biological norms are completely deactivated in favor of social norms. In the case of medicine, for instance in the consulting room, social norms conveyed by scientific gaze and medical technology may be in conflict with the "vital normativity" of the patient. Canguilhem rather means that in human life, both kinds of norms cannot be distinguished. There is no way to divide what counts as biological and what counts as social in our organism. It does not mean that biology dissolves in society, and that the latter delimits a new field that would be independent from biology: it rather means that AS A LIVING BEING, the human lives in a milieu where valences are non biological. The human remains a living being who may suffer from the conflict between vital norms and social norms. So Canguilhem makes "vital normativity" a strong reference for assessing the norms and values we create as Human beings – including the ones that are embodied in technology.

As a matter of fact, Simondon gave a broader extension to his concept of technical object in the third part of *The Mode of Existence of Technical Objects* (but also in other texts), especially with reference to what he called an *esthetic* commitment to technology. What *esthetic* means here is clarified in a text titled "Psychosociology of technicity," written in 1960–1961 (Simondon 2014). Simondon presents an almost unbelievable case of technological intervention on man, by means of surgery (a case he found in a newspaper at that time). A young farmer was suffering from a disease in endocrine glands. He had to undergo surgery, but the operation was very expensive. To pay for it, the farmer planned to sell one of his eyes… Simondon explains that such a surgical operation is *monstrous*, insofar as it consists in technically intervening on the living body with no care at all for its vital normativity (which means here, for Simondon, the capacity of the organism as a whole to keep its functional coherence, as opposed to the aforementioned situation in which the restoration of one part of the body – the endocrine glands – results from the functional destruction of another part – the eyes). Simondon calls *esthetic* the capacity of assessing technology with regard to the integrity, the whole coherence of the subject of intervention – be it a living being or a landscape for instance (indeed Simondon also calls *esthetic* the "insertion" of technology inside a natural framework, with regard to preexisting valences in it). This capacity originates from our life as living beings. As Simondon explains, esthetics has biological roots, it relates to our "tropisms," it makes us keep a strong link between our life as we create new values in the world, and our life as we still belong to the living world. We create values of social, political and ethical relevance, but these values are *valuable* only with reference to the "vital normativity" of human and non-human living beings. For Simondon, esthetics is not only the design of "good mediations" with reference to desired values: it is above all the design of technical mediations that make a case for our biological lives. As mentioned in the introduction of the paper, this focus on biological life as a normative reference is of major interest for developing new trends in the philosophy of design today (Life-Based Design), challenging the dismissal of biological aspects of life in most of the existing methods for design.

To be sure, the object-oriented approach leaves a thorny problem unsolved: does it actually avoid to ravive the old metaphysical divide between the object and the subject – a divide that scholars unanimously consider misleading today? Undoubtedly, contemporary philosophies of artifacts challenge the classical definition of the human subject as facing neutral objects. As they reconfigure both epistemical and ethical equipments of subjects, technical objects are not neutral and ontologically external to subjects; reciprocally, human subjects are not ontologically isolated and self-referring, they are shaped by their relationships to objects. Consequently, as far as it strongly links technology to human beings, the concept of technical *artifact* seems a suitable candidate for overcoming the risk of reviving the subject-object dualism. Nevertheless, in the French object-oriented approach to technology, the subject facing technical objects is not the Cartesian self-referring "*cogito*": it is rooted in biological processes. In Canguilhem's but also Simondon's view, the relationship between human beings and technology is the outcome of an "individuation" occuring inside a vital system. Technical objects are the ontological partners ~~vis-à-vis~~ of a subject who remains rooted in biological life. As a consequence, technical objects must be assessed with regard to "vital norms." Both statements are closely intertwined.

15.3.3 Exteriorization and Bio-objectification

It is worth noting, finally, that a term was historically coined to help better articulate, on the one hand the account of technical beings as "beings-in-the-world" (and not only artifacts), and on the other hand a strong reference to vital norms in technology assessments: "exteriorization." In a too narrow meaning of the term, "exteriorization" can be defined as the extension of bodily capacities into artifacts, which can be considered "external organs." However this definition *a minima* leaves out the most important: "exteriorization" means above all that in mankind, biology has from the beginning spread out of the body. We exist outside the limits of our body, not only as Persons with strong moral and spiritual values beyond our biological life, but also AS LIVING BEINGS. In André Leroi-Gourhan's account of human evolution for instance – a major representative of the "biological philosophy of technology" current (Leroi-Gourhan 1964, 1965) – biological and biomechanical features were not firstly formed and then, only after, "projected" outside the body in technical and social artifacts: originally, the biological and biomechanical constitution of the human occured outside the body. We BIOLOGICALLY live outside the limits of our body, and that's what the term "exteriorization" means.

What are the forms of "exteriorization" today? To be sure, in the medical field for instance, various kinds of "bio-objects" (Holmberg et al. 2011; Metzler and Webster 2011) recently proliferated and generated new arrangements of bodies and technologies outside the organism: biological samples stored in biobanks and circulated along networks of healthcare actors (laboratories, big pharmas…); digital data related to physiopathological processes; artificial organs operating vital functions outside the body; biostatistic and bioinformatic models of complex molecular interaction networks; molecular biomarkers for improving drug discovery, diagnosis and/or therapy etc. All these "bio-objectification" processes display original ways to "exteriorize" biological entities and functions into technology. Bio-objects are not stable artifacts, they rather depend on a continual process of redefinition, due to their circulation between heterogeneous actors. As a matter of fact, contemporary medical technologies do not "exteriorize" human biological life in the same way silex tools did. While the latter fit Kapp's "organ projection" thesis – tools extend life by non organic means –, the former give an unprecedented "mode of existence" to biological entities, outside the organism, by redefining their ontological status and linking them to *de novo* entities of biological significance, such as digital data. So new technologies claim a rethinking of the "exteriorization" process today – a very challenging issue undoubtedly, which finds more help in ontogenetic concepts such as "bio-objects" and "bio-objectification," than in static concepts such as "artifacts." A molecular biomarker for instance is not *designed* and cannot be considered an "artifact." However, through a "bio-objectification" process involving high-throughput technologies, biostatistics and bioinformatics, it becomes a technical object which may prove of great interest for medical action (leading to more accurate diagnosis and prognosis; targeting intracellular signaling pathways etc.). Beyond the dualism of the natural and the artificial, molecular biomarkers exemplify the way non-designed "bio-objects" draw new biological geographies outside the human body, opening unprecedented paths to the "exteriorization" process.

To be sure, all forms of "exteriorization" are not equivalent, both from ontological and axiological viewpoints. Beyond the "organ projection into artifacts" thesis, Thierry Hoquet has recently proposed a typology of ontological significance, highlighting differences in contemporary exteriorized arrangements between bodies and technologies (Hoquet 2011). From an axiological perspective, some "bio-objectification" processes may brutalize the "vital normativity" of living beings (be they human or non-human), due to the fact that what is valuable to us may be destructive of life as such – a divorce that has been dramatically highlighted by Bergson and, after him, by Leroi-Gourhan, Canguilhem or Simondon.

15.4 Concluding Remarks

Almost three decades ago, Jean-Yves Goffi interestingly noted that contemporary philosophies of technology were mostly split into three main domains (Goffi 1988): (1) a phenomenological and ontological investigation on technology (what is a technological object and how to describe it?); (2) an anthropology of technology, focusing on how humanity and technology relate; (3) and finally; an ethical evaluation of technology. Unlike their predecessors, contemporary philosophers of technology, Goffi claimed, rarely combine the three approaches. It is worth noting that Goffi's diagnosis seems still valid. Undoubtedly, both trends in the current philosophy of artifacts, namely The Dual Nature program and the postphenomenological approach, seem to be in search of a systematic view combining an ontology, an investigation on the relationships between humans and technologies, and an ethical evaluation of technology. It is not sure however that these three domains of the philosophy of technology prompt equal concern among scholars. Indeed, the Dual Nature program focuses on metaphysical issues posed by artifacts, and, to a certain extent, it carries out normative challenges – but it rather dismisses anthropological concerns. Conversely, the postphenomanological inquiry has a great interest in both ethical and anthropological issues (Verbeek strongly and convincingly links ethical concerns to a phenomenological understanding of humans-technology relationships), however it leaves ontological questions quite underinvestigated. As Simondon already critically noted, the careful examination of technological functionings is beyond the scope of phenomenological investigations. This paper intented to demonstrate how French philosophy of technology may overcome this divide and fruitfully contribute to the current philosophy of artifacts, insofar as it precisely provides concepts and methods for better combining ontological, anthropological and normative approaches. More precisely, the paper highlighted the relevance of both a conceptual, mainly simondonian framework for analyzing the "mode of existence" of technological objects, and a biological-grounded analysis of humans-technology relationships, paving the way for further investigations of how technology "exteriorizes" humans today.

Acknowledgements A preliminary version of this text has been discussed by Daniel Cérézuelle and Steven Dorrestijn during the authors' workshop held in Paris Sorbonne on 22–23 June 2015 in view of the preparation of the collective book.

References

Achterhuis, H. (Ed.). (2001). *American philosophy of technology: The empirical turn*. Bloomington: Indiana University Press.

Bensaude Vincent, B., & Newman, W. R. (Eds.). (2007). *The natural and the artificial: An evolving polarity*. Cambridge, MA: The MIT Press.

Bergson, H. (1896). *Matière et mémoire*. Paris: Alcan. English Edition: Bergson, H. (1911). *Matter and memory* (N. M. Paul, & W. S. Palmer, Trans.). London: Swan Sonnenschein and Co.

Bergson, H. (1907). *L'évolution créatrice*. Paris: Alcan. English Edition: Bergson, H. (1911). *Creative evolution* (A. Mitchell, Trans.). Lanham: University Press of America.

Bergson, H. (1932). *Les deux sources de la morale et de la religion*. Paris: Presses Universitaires de France. English Edition: Bergson, H. (1935). *The two sources of morality and religion* (R. A. Andrà, C. Brereton, & W. H. Carter, Trans.). New York: Henry Holt and Co., Inc.

Brey, P. (2010). Philosophy of technology after the empirical turn. *Techné, 14*, 1.

Canguilhem, G. (1943). *Le Normal et le pathologique*. Paris: Presses Universitaires de France. English Edition: Canguilhem, G. (1991). *The normal and the pathological* (C. R. Fawcett, Trans.). New York: Zone Books.

Canguilhem, G. (1980). Machine et organisme. In G. Canguilhem (Ed.), *La Connaissance de la vie*. Paris: Vrin. English Edition: Canguilhem, G. (2008). Machine and organism. In Canguilhem, G. *Knowledge of life* (S. Geroulanos, & D. Ginzburg, Trans., pp. 75–97). New York: Fordham University Press.

Canguilhem, G. (2011 [1938]). Activité technique et création. In *Écrits philosophiques et politiques (1926–1939)* (pp. 499–511). Paris: Vrin.

de Vries, M. J. (2008). Gilbert Simondon and the dual nature of technical artifacts. *Techné: Research in Philosophy and Technology, 12*(1), 23–35.

de Vries, M. J. (2013). Norms in technology. In S. O. Hanson & A. W. M. Meijers (Eds.), *Philosophy of engineering and technology* (Vol. 9). New York: Springer.

Dipert, R. (1995). Some issues in the theory of artifacts. *The Monist, 78*(2), 119–135.

Goffi, J.-Y. (1988). *La Philosophie de la technique*. Paris: Presses Universitaires de France, Coll. « Que sais-je ? ».

Guchet, X. (2014). *Philosophie des nanotechnologies*. Paris: Hermann.

Harman, G. (2002). *Tool-being: Heidegger and the metaphysics of objects*. Chicago: Open Court.

Heidegger, M. (1954). Die Frage nach der Technik. In M. Heidegger (Ed.), *Vorträge und Aufsätze*. Pfullingen: Verlag Günther Neske. English Edition: Heidegger, M. (1977). *The Question Concerning Technology and Other Essays* (W. Lovitt, Trans.). New York\London: Garlang Publishing, Inc.

Holmberg, T., et al. (2011). Bio-objects and the bio-objectification process. *Croatian Medical Journal, 52*(6), 740–742.

Hoquet, T. (2011). *Cyborg philosophie. Penser contre les dualismes*. Paris: Editions du Seuil.

Kelty, C. (2009). Beyond implications and applications: The story of safety by design. *NanoEthics, 3*, 79–96.

Kroes, P. (2012). Nano-artefacts and the distinction between de the natural and the artificial. In C. Kermisch & M.-G. Pinsart (Eds.), *Nanotechnologies: Toward a shift in the scale of ethics*. Bruxelles: E.M.E. & InterCommunications.

Kroes, P., & Meijers, A. (Eds.). (2000). *The empirical turn in the philosophy of technology*. Oxford: Elsevier Science Ltd.

Kroes, P., & Meijers, A. (2006). The dual nature of technical artifacts. *Studies in History and Philosophy of Science, 37*, 1.

Kroes, P., & Verbeek, P.-P. E. (2014). *The moral status of technical artifacts*. New York: Springer, Philosophy of Engineering and Technology.

Leikas, J. (2009). *Life-based design. A holistic approach to designing human-technology interaction*. VTT Publications. http://www.vtt.fi/Documents/2010_P726.pdf

Leroi-Gourhan, A. (1964, 1965). *Le Geste et la parole*, deux tomes. Paris: Albin Michel. English Edition: Leroi-Gourhan, A. (1993). *Gesture and Speech* (A. Bostock Berger, Trans.). Cambridge, MA: The MIT Press.

Loeve, S. (2009). *Le concept de technologie à l'échelle des molécules-machines*. Philosophie des techniques à l'usage des citoyens du nanomonde, PhD dissertation.

Mayne, A.-J., et al. (2004). Chemisorbed bistable molecule: Biphenyl on Si(100)-2×1. *Phys Rev B, 69*, 045409.

Metzler, I., & Webster, A. (2011). Bio-objects and their boundaries: Governing matter at the intersection of society, politics and science. *Croatian Medical Journal, 52*(5), 648–650.

Mitcham, K. (2002). Do artifacts have dual natures? Two points of commentary on the delft project. *Techné: Research in Philosophy and Technology, 6*(2).

Simondon, G. (2012 [1958]). *Du mode d'existence des objets techniques*. Paris: Presses Universitaires de France. English Edition: Simondon, G. (2016). *On the mode of existence of technical objects* (C. Malaspina, Trans.). Minneapolis: Univocal Publishing LLC.

Simondon, G. (2014). *Sur la technique (1953–1983)*. Paris: Presses Universitaires de France.

Vaccari, A. (2013). Artifact dualism, materiality, and the hard problem of ontology: Some critical remarks on the dual nature of technical artifacts program. *Philosophy and Technology, 26*(1), 7–29.

Verbeek, P.-P. (2005). *What things do. Philosophical reflections on technology, agency and design*. University Park: Pennsylvania State University Press.

Verbeek, P. P. (2010). Accompanying technology: Philosophy of technology after the ethical turn. *Techné: Research in Philosophy and Technology, 14*(1), 49–54.

Verbeek, P.-P. (2011). *Moralizing technology. Understanding and designing the morality of things*. Chicago: The University of Chicago Press.

Vermaas, P. E. (2006). The physical connection: Engineering function ascriptions to technical artifacts and their components. *Studies in History and Philosophy of Science, 37*, 62.

Vermaas, P. E., & Houkes, W. (2006). Technical functions. A drawbridge between the intentional and structural natures of technical artifacts. *Studies in History and Philosophy of Science, 37*, 5.

Chapter 16
Cyborgs, Between Organology and Phenomenology: Two Perspectives on Artifacts and Life

Thierry Hoquet

Abstract Cyborgs are usually loosely defined as the adjunction of a mechanical device to an organism. These hybrid entities have triggered several philosophical comments on their nature: are they harmonious wholes (organism-like) or rather unstable couplings? This paper situates cyborgs between two rival traditions of interpretation: organology and phenomenology. In the "organological" tradition, tools are considered as biological organs, as mere extensions of life, or vital strategies evolved by the organisms striving to survive. In the "phenomenological" tradition, tools are approached from the standpoint of the user, as a significant means available to interact with the organism's environment. This paper argues against a broad conception of cyborgs for which all entities combining organic and mechanical components can be called "cyborgs." I distinguish "organorgs" (or tool-using organisms) from an authentic "cyborg" perspective. While the organorg can easily change its tools, pick different instruments and instantly dispose of them, a true cyborg has its tools literally grafted to its organic parts so that the mechanical parts cannot properly be said to be "used." The cyborg standpoint opens up new perspectives on prostheses and the enhancement of human bodies.

Keywords Cyborgs · Technological evolutionism · Human enhancement · Organology · Prosthesis · Tools · Machines

T. Hoquet (✉)
Institut de Recherches Philosophiques (IREPH) (EA 373), Université Paris Nanterre, Nanterre Cedex, France
e-mail: t.hoquet@parisnanterre.fr

S. Loeve et al. (eds.), *French Philosophy of Technology*, Philosophy of Engineering and Technology 29,
https://doi.org/10.1007/978-3-319-89518-5_16

The concept of "cyborg" originated in an article by Manfred E. Clynes and Nathan S. Kline, first published in the magazine *Astronautics* in 1960, as a shorthand for "cybernetic organism." (Clynes and Kline 1960) According to Clynes and Kline, cyborgs are a mechanical device appended to an organism, so that this hybrid of organism and machine is enabled to live in an environment to which it is not adapted beforehand. Ideally, the cyborg condition allows any organism (say: a rat) to live in outer space: the mechanical device regulates the variables of the environment, so that the organism doesn't have to pay attention to the variations of milieu. However, the "cyborg" left the experimental lab and became a character in fictions – such as *RoboCop*, 1987, or *Ghost in the shell*, 1995 – and a "pop" concept in philosophical discourses – such as Donna Haraway's *Cyborg Manifesto*, 1985 – who contributed to fashioning the conceptual scene surrounding cyborgs, and beyond, to shaping our ideas on machines and organisms.

A creature like the cyborg follows a twofold logic. On the one hand, the bodily needs of the creature are fully satisfied through mechanisms and automatic regulations, determined by cybernetic loops of retroaction. In a Cartesian fashion, one could claim that the mind is freed to deal with intellectual puzzles; with all its wants provided for, the individual has the leisure of seeking answers to questions prompted by the new milieu. On the other hand, the cyborg is a hypnotized creature, mentally controlled by the administrator of the mechanical parts: with pharmaceuticals injected in the body, the whole being is potentially anaesthetized and its ideas can be manipulated through proper injections of chemicals (psychotropic substances, painkillers, stimulants…).[1]

Let us consider a banal case of cyborg: it may be said, for instance, that an individual living with an artificial cardiac pacemaker can be called a "cyborg" as its flesh is maintained in life thanks to a grafting to a mechanical device. However, the insertion of a new machine in the body is an intrusion, by which the whole being is submitted and even annexed to a regime of external regulations. The organism is suddenly deprived of its autonomy and becomes dependent from technology (obsolescence of the pacemaker for instance). This loss of autonomy may happen anytime one uses tools or machines. For instance, an individual wearing glasses or even clothes may be considered a sort of cyborg: thanks to technical utensils, it is better fitted to its environment. However, the concept of "cyborg" should not be extended too far so as to include all cases of tool-organism connections, unless it will lose its radicalness. The present chapter aims at rejecting a broad or loose definition of the cyborg as just another name for humans as "tool-using organisms," or, as Andy Clark (2003) put it, humans as "natural-born cyborgs."[2]

Analyzing the nature of the cyborg, the present chapter investigates the intersection between techniques and organisms. The Greek myth of Epimetheus and Prometheus offers here a good introduction. Epimetheus was in charge of providing

[1] Ian Hacking (1998: 209–210) attributes the former view to Clynes, as "astonishingly" close to the Cartesian dualism, and the latter to Kline.

[2] For Clark (2003), cyborgs are not post-human entities, but merely humans, as human beings are "tools all the ways down."

each animal with a positive feature that would be its natural endowment: the voracious lion with claws, the frail rabbit with velocity, the slow turtle and snail with a protective shell, etc. But in the process of distribution, Epimetheus forgot the human kind, which was left with no natural qualities. Prometheus came then into play and he entrusted the humans with the attributes of *fire* and *arts*, which he stole from the Gods—a crime for which the guilty Titan would be soon punished.[3] In the Greek myth, Prometheus' techniques, in the guise of fire and arts (*technai*), embody a sort of artificial counter-creation, while Epimetheus' natural endowments represent the providence of the organism, fully equipped from the start with all that it will ever need. The myth also provides us with a conceptual framework where techniques are a means to make up for the defects or lacks of the natural body. All animals are divided in two groups: those whose bodies are *naturally gifted*, and whose vital functions are performed by organs; and those whose bodies are lacking, and have to be extended by tools. The German tradition of "philosophical anthropology" (Max Scheler, Helmuth Plessner, Arnold Gehlen) refers to this theme as the "natural artificiality" of humans, in order to stress the uniqueness of the "situation of the human kind in the world." (Scheler 1947) Bernard Stiegler in France has also developed the idea of a coeval relationship between "anthropo-genesis" and "techno-genesis." (Stiegler 1994: 47)[4] But it is also striking that, beyond the opposition between ontological plenitude and existential lack, the two poles of this dichotomy, organs/artifacts can also be equated with others terms, such as beasts/humans, or nature/technology. The cyborg also blurs these boundaries as its tools tend to become organs, to be integrated in its general body-plan.

Thinking the cyborg requires that we find a way out of the organs/artifacts dichotomy. A first approach to thinking the cyborg could be borrowed from René Descartes' technological approach to organs: all organs, Descartes claimed, are eventually understandable in terms of mechanisms, and the machine simile explains organisms. Pumps and bellows offer good analogies for representing the functioning of the heart, weights and pulleys for modelling the contraction of the muscles… But in this view, the cyborg is just another machine (just like any animals) and the singularity of its condition is lost.

A different take on the cyborg was offered when Georges Canguilhem, in the 1940s turned the problem inside out: instead of technologizing organs, Canguilhem suggested that technics should be viewed as vital strategies. He presented techniques in general (tools, machines, artifacts) as vital strategies—a reasoning that is called a "general organology." In the organological perspective, technology is a production of organisms as living beings, by which they deal with their environment (*milieu*). The machine/organism dualism is challenged as machines are merely external organs. But technology also affects organisms in the way they perceive

[3] See Plato, *Protagoras*, 320d–322a.

[4] For a presentation of Stiegler's philosophy of technology as "a reinvention of philosophical anthropology," see Michael Lewis (2013). Starting with his two volumes on "*Symbolic Misery*," Stiegler also makes abundant use of the concept of "general organology" that will be discussed in this chapter (Stiegler 2004, 2005).

their environments, construct meanings and achieve their goals in the world. Here, I borrow concepts from the phenomenological tradition, such as "*corps propre*" or "*discernement*," in order to investigate the autonomy of the cyborg with regard to its artifactual implements.

The dual nature of cyborgs requires a twofold treatment, switching from an organological perspective (technology as a production of living entities) to a phenomenological one (technology as involved in meaningful behaviors). The organological perspective focuses on the natural relationship of the organisms to their artifacts. The phenomenological perspective sets itself to the task of elucidating the autonomy of the organisms regarding the world of technological artifacts. A thorough philosophical treatment of the concept of cyborg requires a combination of those two distinct perspectives.

16.1 The Organological Tradition

In the organological tradition, technology is a production of life and living beings in general, not exclusive to humans. Historically, "organology" refers to a philosophical thread that goes back to early German Romanticism, plunges its roots into Leibnizian thought, and received significant elaboration by Novalis in particular (1872–1801) (Weatherby 2016). It aimed at bridging the gulf that Kantian philosophy had instituted between thinking and being, freedom and nature, knowledge and practice, by way of a *Naturphilosophie* in which the metaphysical *organon* (the tools of thinking) joined the physiological *organa* of the perceiving and acting body. In the nineteenth century the organological tradition met the industrial revolution and fueled a new technological imagination of life as well as a biological view of technology. It is only at this stage of its genealogy that I approach the organological tradition, my account of which being focused on its interest for of a biological philosophy of technology. In this respect, the reference to Charles Darwin is central as it shows that technological features are widely spread among animal species.

16.1.1 Organism-Produced Devices (Darwin)

In 1871, Darwin dedicated a short section of the second chapter of his *Descent of man* ("Comparison of the mental powers of man and the lower animals") to "*Tools and weapons used by animals*." (Darwin 1871: 51–53)[5] Darwin used different terms to designate technical augmentations. On the one hand, he refered to "tools," *e.g.* stones used by monkeys in order to break nuts, or branches broken off by elephants to drive away the flies; or a blanket of straw used by a young female orang for

[5] In the second edition (1874), chapter II becomes chapter III (pp. 81–83). In both editions, this section figures in the table of content, but does not appear in the text itself.

covering herself, when she thought she was going to be whipped. Besides "tools," both in the title of the section and the structure of his development, Darwin mentioned "weapons"— as if instruments of war, meant to hurt others, were a category of their own.[6] He also used the general term "implement," which encompasses all augmentation or complementation devices.[7] Stones and sticks can be viewed as mere "implements," not prescribing any determinate use: they are susceptible of different uses, as tools or as weapons, according to various circumstances. In this section, Darwin confined himself to an enumeration of natural history remarks, documenting the various uses of implements in various species. Darwin did not provide his readers with a complete overview of animal techniques.

Quite strikingly, Darwin did not consider here the devices known as "animal architecture": termitarium, bee cells, nests or burrows.[8] The architectural performances of the hymenopter insects have been dealt with in the *Origin of species*, as early as 1859, where the first edition dwells in details on *cell-making instinct* in bees and other hymenopter species. In the particular context of the *Descent*, Darwin had especially in view the transitional forms by which certain human capacities are foreshadowed in neighboring species. Darwin focused mostly on cases of monkeys and on anthropomorphic apes (chimpanzee, orang, baboons). The only exceptions were some brief mentions of dogs or elephants. Darwin described the habits of primates as a proxy for understanding the origins of human technics, in other terms, "the first steps towards some of the simpler arts; namely rude architecture and dress, as they arose amongst the early progenitors of man." (Darwin 1871: 53)

The emphasis on primates is linked to Darwin's engagement in debates on prehistoric humans. For the bishop of Dublin, Richard Whately, the primitive humans were deprived of the instincts by which other animals build nests or make other constructions: humans had first to be instructed in what they had to do—the Christian version of the myth of Epimetheus and Prometheus. George Campbell (eighth Duke of Argyll) offered a different view on the topic: humans, he claimed, are animals, and as such they were gifted with instincts, although of a different nature than those of brute beasts (Campbell 1869). Whence new instincts peculiar to humans have been developed, so that (and this passage from Campbell is quoted by Darwin) "the fashioning [of] an implement for a special purpose is absolutely peculiar to man," these instincts forming "an immeasurable gulf between Man and the brutes." (Campbell 1869, cited in Darwin 1871: 52) Campbell hence supported a paradoxical thesis, hybrid of biological continuity (humans, like all animals, have instincts) and discontinuity (human instincts are different from those of other animals).

[6] As André Leroi-Gourhan remarks, "weapons have always flattered the taste of travellers and one gets millions of items on which classificatory preoccupations have been set to work." (Leroi-Gourhan 1945: 13) The first section of the book *Milieu et techniques* is devoted to weapons as "*techniques d'acquisition.*" (Leroi-Gourhan 1945: 13–68)

[7] The term also figures in the work of George Campbell, 8th Duke of Argyll, especially in his *Primeval man; an examination of some recent speculations* (1869).

[8] There's only one brief allusion to birds nests, mentioned to support the claim that nonhuman animals have "the idea of property." (Darwin 1871: 52)

Darwin rejected both of these theses and supports John Lubbock's argument of the accidental origin of human artifacts, especially of breaking flints in order to produce fire (Lubbock 1865: 473). However, he borrowed his materials freely from the same corpus of anecdotes. Thus, Darwin added in the 1874 edition a reference to elephants shaking tree branches as fly swatters, an instance also mentioned by Campbell but used by Darwin to support a different argument. On the biological origins of technics, Darwin supported an evolutionary stance, implying gradual and continuous change, against Campbell's view of "technical instincts," directly dependent of superior mental powers. Technology is not so much analyzed for itself, but only within a development on primitive forms of mind (reason, language), such as documented by natural history.

Studying the biological roots of techniques, Darwin's naturalist approach shows that technics are not reducible to applied sciences: they must be situated within the general context of evolution by natural selection and survival strategies or what Darwin calls (after 1869), *survival of the fittest.* In the *Descent*, Darwin explained that natural selection can refine various devices or *features* such as organs and instincts, but also "implements" (tools or weapons).

16.1.2 From Implements to Projections (Kapp, Espinas)

Showing that animals use instruments, Darwin's theory paves the way for various paleoanthropological aspects for whom tools are natural extensions of organs, or ways of extending the power of living beings upon nature. The German Ernst Kapp (1808–1896) for instance, understood early tools under the paradigm of "organ projection." (Kapp 1877)[9] Organ projection suggests that technics are a kind of unconscious externalization, organs being increased and extended beyond the organic body. The concept of "projection" also establishes an analogy between the organism and the mechanical devices. A parallel between organs and tools is inscribed in the etymology itself: the Greek term *organon* both describes a part of the body and a tool. Aristotle called the hand a "tool of tools."[10] For Kapp, the hand is an "innate tool," the "model for all mechanical tools": it takes a fundamental part to the fabrication of those material reproductions, namely, the tools. If tools are merely "projected" organs, they are like duplicates of our native organs (those with which we were born). A bowl or a cup for instance is nothing more than the analogical projection of the hands united so as to retain water. Primitive tools and instruments can be viewed as external duplications of the hands, detached from the main body and becoming modifiable. Kapp's approach is for the most part limited to rudimentary tools that can be understood as projection of the hands or of the extremities of the limbs. The comparisons he made between respiratory metabolism and the steam

[9] It is only in 2007 that a French translation of Kapp's seminal book was provided by Grégoire Chamayou with the title *Principes d'une philosophie de la technique*.

[10] *Parts of Animals*, IV, 10, 687a6–687 b 25; *On the soul*, III, 8, 432 a 1.

engines or telegraphic lines and the nervous system proved more adventurous. However, his concept of "organ projection" will be later extended to all machines by the French Alfred Espinas, for whom a machine is "a system which recomposes the articulations." (Espinas 1897)[11]

16.1.3 General Organology (Bergson, Ruyer, Leroi-Gourhan)

Henri Bergson's *L'Évolution créatrice* (1907) rephrases the organs/artifacts opposition as a difference between *internal* and *external* organs. Interestingly enough, Bergson related both kind of organs to two different directions of evolution: *instinct*, which produces glands or internal organs, and *intelligence* which creates tools or external organs. Instinct and intelligence work in contrary directions. Instinct fosters a complete adhesion to the evolution of life; it is extremely efficient but offers no degree of freedom, operating in a blind and fumbling fashion. Intelligence, on the contrary, is embodied in external artifacts. Bergson understood mechanical invention as a biological function, an aspect of the organization of matter by life. Tools (and by extension, machines) are primordially instruments for organic survival, an indirect offspring of vital activity, which implies that there always is an organism at their origin. Bergson went as far as including a technical aspect in the biological definition of humans: not the wise (*Homo sapiens*) but the craftsman or the toolmaker (*Homo faber*), the artificer whose action is to fabricate, create or forge.

> If we could rid ourselves of all pride, if, to define our species, we kept strictly to what the historic and prehistoric periods show us to be the constant characteristic of man and of intelligence, we should say not *Homo sapiens*, but *Homo faber*. In short, *intelligence, considered in what seems to be its original feature, is the faculty of manufacturing artificial objects, especially tools to make tools, and of indefinitely varying the manufacture* (Bergson 1907: chapter 2) .

But Bergson's view of technology is quite different from Kapp's projection of organs; as organs themselves are not a positive reality from whence tools could be derived: organic organs are the "negative" of the organism thriving to insert itself into matter; they are not perfect means adapted to an end, but imperfect and provisory devices in view of enhancing action and improving the efficiency of life.

For Georges Canguilhem, Bergson's *Creative Evolution* is "a kind of general organology (*une sorte d'organologie générale*)" (Canguilhem 1952: 125): a general theory of tools and instruments, which includes bodily organs and other modes of inserting action in the concrete world (intelligence, vision, language). Those various modes are on a par, as they are all strictly speaking *adaptations*: they became useful in improving the living's chance to survive. Another French philosopher, Raymond Ruyer, also took up the theme of machines as new organs for the human species.

[11] Section "*Projection inconsciente des uns et des autres*." On Espinas, see Jean-Yves Goffi (Chap. 19, this volume).

External tools, he said, are a way for the organism to avoid the risk of metamorphosis: inventing tools is like producing ideas, which are "mutations which affect only what's inscribed on the 'cerebral blackboard,'" and which do not represent any biological danger: a bad idea does not entail a cerebral hemorrhage, or does not threaten survival in the way a deleterious mutation (a "mistake" in evolution) would do (Ruyer 1946: 46–47). A bad idea can easily be erased and replaced, and different ideas might be tried, which increase the rhythm of human inventions, especially in comparison with the slowness of organic inventions. For Ruyer, human technology "opens not 'the reign of intelligence,' as is commonly said, but the reign of external circuits and of inventions accelerated by cerebral processes."

> Not only have humans invented means of protection, of locomotion, of attack and defense, artificial eyes, teeth and skins, but their intellectual life and even their instinctive life tend to get outside of them. Humans already have their memory in libraries, they will soon have their thoughts in propaganda offices: even the individual instinct of reproduction now runs through the bizarre circuit of a state grant [*i.e.* family allowance]. (Ruyer 1946:47)

For Ruyer, the technical world extends well beyond tools, but to all artifacts, including cities, factories, and collective organizations.

> His brain reigns over his cities and factories. His familiar instruments are represented in the centres of 'praxies' of his brain in the same way as his organs. The ratio 'brain's weight/weight of the body', of which Man is so proud, must therefore, in good justice, be replaced by the ratio 'weight of the brain' / 'weight of the body + weight of the equipment', ratio thus considerably weaker than the microcephalic reptiles of the Secondary era." (Ruyer 1946: 47)

Glossing on Bergson's formulas of *The Two Sources of Morality and Religion* (Bergson 1932), Ruyer concluded: "the real human body goes well beyond one's own body." Significantly, Stiegler comes quite close to Ruyer when he describes his own "general organology" as a perspective encompassing physiological organs (including the brain), technical organs, and social organizations, and concerned with their (not a priori harmonious) interrelations (Stiegler 2005: 221).[12]

The paleoanthropologist André Leroi-Gourhan also emphasized the fact that humans could not be literally "overwhelmed by technologies" ("*l'homme dépassé*

---

[12] However, after briefly alluding to the romantic tradition (Novalis), Stiegler situates his general organology in another tradition, that of "mechanology" (Simondon): "Novalis spoke in his time about 'organology.' His ambition was to organically explain the mechanical. There is also a concept of general organology in Simondon, who distinguishes between technical elements, individuals, and ensembles, and for which he proposes a 'mechanology': a science studying the ontogenesis of technical beings, especially those that undergo a 'process of concretization,' a particular kind of what he calls more generally a 'process of individuation.' As the technical elements are elementary components that can fit into different individuals, Simondon assimilates them to the organs of a living organism, and this is why he speaks of general organology as a branch of his mechanology. (…) My concept of general organology would be rather the equivalent of Simondon's mechanology, but in which the living is itself included in the set of transductive relations [*i.e.*, in which the terms are constituted by the relation] that connect the various types of artificial and living organs (including the brain) to social organizations in which they evolve and transform themselves. (…) Such transformations constitute processes of psychic and collective individuation in three branches: the psychic individuation, the social individuation, and the technical system." (Stiegler 2005: 221–222)

par les techniques") as they are the creators of technology as much as the reversal.[13] "Technologies are intrinsically overwhelming" ("*les techniques sont normalement dépassantes*"), as they evolve, develop or progress, he declared, but this is eventually a false problem and one that should not be a cause of too much concern or anxiety (Leroi-Gourhan 1983: 125). As early as 1945, Leroi-Gourhan emphasized a strong analogy between change of technology and change of organic forms (the latter being documented by paleontological records): both produce evolutionary convergences as evolution (be it of organisms or of artifacts) responds to the laws of the universe, creating some kind of "physical harmony" (*harmonie physique*) and leading to a form of "technical determinism" (*déterminisme technique*). He thus applies to technology the theme of evolutionary convergences priory developed by Bergson in *Creative evolution*. For Leroi-Gourhan

> A large part of the technical tendency is closely tied to the construction of the universe itself, to the point that it is equally normal for roofs to have a V-shaped slope, for hatchets to have handles, and for arrows to be balanced at a point located at the third of their length, just as it is normal for gastropods to have a twisting shell, whatever geological era they lived in (...). Simple physical harmony suffices to explain why two identical objects having the same use can appear in two populations having no mutual trade. Next to the biological convergence, there is a *technical convergence*. (Leroi-Gourhan 1945:359)

Leroi-Gourhan later stressed the need for "an authentic biology of technology," calling to consider the "social body" as "an organism independent of the zoological one – an organism animated by humans but so full of unforeseeable effects that its intimate structure is completely beyond the means of inquiry applied to individuals." (Leroi-Gourhan 1964: 146)

From Bergson on, the organological tradition recalls that in the first sense, the Greek word "*organon*" designates an instrument. In other terms, tools or machines are organs, and *vice versa*, an organ is a tool or a machine.

16.1.4 "The Biological Philosophy of Technology" (Canguilhem)

In his 1947 paper on "Machine and Organism," Canguilhem described "general organology" as inverting the Cartesian perspective that machines are good models for organisms (Canguilhem 1952: 101, 125 n. 58).[14] Instead, he suggested that we should think of technology as an outcome of vital activity. For Canguilhem,

---

[13] On Leroi-Gourhan see Charles Lenay (Chap. 13, this volume). For a recent application of Leroi-Gourhan's ideas to recent technologies, see for instance Emanuele Clarizio's on Google Glass (Clarizio forthcoming). Clarizio's paper stresses the singular features of Leroi-Gourhan's organology as strict correlation between human evolution and technical evolution: for instance, bipedalism as liberation of the hand opened new technical possibilities.

[14] Canguilhem's "Machine et organisme" is the text of a lecture given at the Collège Philosophique in 1946–1947, and published in *La Connaissance de la vie* (Canguilhem 1952).

machines are organs, on a par with lever, hands or brain as means to perform functions for the organism. Hence, Canguilhem offered a "biological philosophy of technology," that takes the project of a general organology seriously, *i.e.* the fact that *organon* means both "organ" and "tool." (Canguilhem 1952: 123)

In this perspective, "organ" designates any device that allows a better insertion of an entity into its environment and thus ultimately contributes to its survival. An organ can be either native (born with the entity), or acquired in the course of life and, in the latter case, either grafted, secured immovably, or simply borrowed provisionally. The anatomy describes native and solidary organs, those that could be said "internal" or "natural," while technology designates acquired or temporary organs, "external" or "artificial." But both kinds basically share the same nature: all are organs, even if some of them are rather called "tools."

By that, one sees that Canguilhem refused to take machines as an epistemological model of organization. Canguilhem claimed that the mechanistic view of the body is no less anthropomorphic than a teleological conception of the physical world. To counter the mechanistic trends of post-Cartesian biological thinking, Canguilhem described technology as a universal biological phenomenon, more than a human intellectual activity. Technologies for Canguilhem are not the mere result of sciences or their application, but are anchored in Bergson's concept of *Homo Faber*. Technologies are a vital strategy implemented by numerous species. Simply, many animals have internalized devices (organs or glands), while intelligent humans produced externalized devices (tools, machines). Finally, technology is a primary mode of action by which humans respond creatively to the challenges of their environment. (Schmidgen 2006) Finally, tools and machines are organs of the human body.

While Descartes had naturalized machines claiming their operations follow the laws of nature,[15] Canguilhem vitalized machines, by seeing them as the seamless continuation of a vital individual center. Canguilhem was well aware of the beginnings of cybernetics, automata and servo-mechanisms: he refered to them at the end of "Machine et organisme." However, he considered that cybernetics does not change the direction of the human/machine relationship: humans remain the source or origin of mechanisms, machines being a fact of culture, part of human history.[16] If Canguilhem put machines under the control of humans, human history is understood as part of the mainstream of life. Machines comply with the laws of nature and are even devised so as to take full advantage of these laws. Machines are natural in the sense that organisms produce them to achieve certain purposes. But machines are by no means independent entities, which would have a life of their own. Canguilhem made this point in different ways: by emphasizing that a machine has neither the principle of its dynamism, nor that of its construction; by insisting that for that very reason, machines are not natural entities in the same way or in the same sense than organisms. In other words, Canguilhem deflated the Cartesian machines and deprived them from any kind of radicalism, putting them into a strict dependence to life (or human) history.

---

[15] Descartes, Principes de la philosophie, IV, § 203.

[16] "The construction of servomechanisms or electronic automatons displaces the relationship of man to machine but does not alter its sense." (Canguilhem 1952: 88)

16.2 The Cyborg as a Center of Perspective

The organological perspective makes sense of the cyborg as it considers it as a combination of external and internal organs. But in some sense, the concept of the cyborg is weakened in the organological perspective as all humans become "cyborgs" insofar as they use tools. Ian Hacking offers interesting reflections on this subject in an article provocatively entitled "Canguilhem amid the cyborgs." (1998)

16.2.1 From Tools to Cyborgs: Blurring the Boundaries (Hacking)

Hacking takes his starting point in one of the most important results highlighted by Canguilhem and Bergson: the thesis that "tools and machines are extensions of the body; they are part of life" (Hacking 1998: 205); machines are "extensions of life, of vitality, of living"; they intensify life, expand it, but can also threaten it. But where Canguilhem aimed at ruling out the idea of assimilation of the body to a machine, Hacking seems to distort this result and to inflect it in the sense of a lack of differentiation of another kind: as if the machines could be assimilated to organisms. It is in this sense that Hacking cites both Descartes (form whom machines are both "artificial and natural") and Donna Haraway's "Cyborg Manifesto." Hacking's idea is to disturb "the distinction between machine and organism." For this purpose, Hacking refers to the passage where Canguilhem claimed that "a tool or a machine is an organ, and organs are tools or machines." (Canguilhem 1952: 87) Hence, Hacking invites us not make too sharp a distinction between the natural and the artificial.

However, according to Hacking, philosophical arguments do not suffice to convince us that we should erase the boundary between organs and machines, or life and technology. Studies in developmental psychology, Hacking claims, show how children of a very early age are able to make the difference between living beings and artifacts—as if the human brain had a module to perform this operation as early as 3 years old. (Gelman et al. 1995) How can we deepen our sense of analogy between machines and organs, despite our difficulties (at least psychological) to take that reconciliation seriously?

Our concept of "machine," Hacking stresses, also calls for further reexamination. The philosophy of technology should distinguish between single instruments or tools (such as hammers) and more complex machines (such as the spinning jenny, key to the industry of weaving in the second half of the eighteenth century). Clearly, such machines do not fit nicely into the broad concept of artifactual implements that we have found in Darwin. For example, if one can say that tools are "external" or "projected" organs, what of articulated and complex machines? What about the vacuum cleaner or the microwave oven? Can they still be called "organs" and of what kind? It might be claimed that freestanding machines are a specific feature of modern times that calls for a renewed analysis of artifactual entities (Pickering 1995). If some machines like an electric drill or a chainsaw may still be called tools, most machines are characterized by their freestanding nature: although "fuelled" by humans, they are, most of the time, running on their own, with no direct link to the human hand. Such machines aren't even typical of modernity, some of them predating the first industrial revolution: let us think for instance of the Strasbourg clock (the first model dates from the fourteenth century and is alluded to by John Locke), mills (alluded to by Leibniz), or the trebuchet (a kind of siege engine used in the Middle Ages).

Considering not so much tools, but freestanding machines, we must rethink the relationship of technology to life, as it was understood by the organological tradition, from Darwin to Canguilhem. Tools are extensions of the body, always available but external. The matter is different when the artifacts are autonomous machines: can the spinning jenny still be called an organ? Or better: an organism? And if tools are organs, *i.e.* extensions of the body, then what of machines? A factory for instance somewhat couples machines and organisms (humans), all co-working in order to produce a definite output. But can a factory truly be called a cyborg, as it mixes organic and inorganic components in order to perform functions? "Cyborgs" refer to a mode of composition between machine/organism, that leads us to a reappraisal of Canguilhem's analyses.

16.2.2 A Phenomenological Distinction: Cyborg/Organorg

In the following section, I claim that the concept of "cyborg" should not be extended too far unless it loses its conceptual accuracy. Not all organic entities combined with artificial components are "cyborgs." A human wearing clothes or glasses is not a cyborg, no more than a factory can be called one. Hence, I suggest the neologism "organorg" to describe the tool-using organism, the classical organological view of the tool-using organism. A human reading or cycling is an organorg, not a cyborg. Glasses or clothes for instance, don't do anything on their own.

Conversely, an authentically "cyborg" perspective implies an autonomous functioning of machines. Hence, the organorg/cyborg distinction is in line with other important distinctions: tools/machines, dependant/autonomous. And it is a phenomenological distinction in that from organorg to cyborg, the center of perception shifts from the organic body equipped with its external tool to the freestanding machine incorporating itself into the living reign.[17]

[17] In a way, Stiegler also combines the organological perspective with a phenomenological one as his first mention of "organology" in *Symbolic Misery* (2004) came along in a discussion concerning the anthropological implications of the "machinic turn of sensibility," *i.e.* the handling of the shared sensory experience by machines. At this occasion he recalls that "organology" also refers to the science of musical instruments (*organa*), a discipline interested in both the technical and the sensory properties of musical instruments in their relations with the living body of the musician (Bonnerave 2004). In fact, Stiegler's use of "general organology" was originally an extension of

Cyborgs are associated with special artifacts that tend to be grafted on the organism, and with cybernetics the science of automata and of servomechanisms, which perform self-regulation in a system thanks to feedback processes. While Canguilhem distinguished between tools (simple devices) and machines (complex mechanisms), he considered them all univocally as organs, and refused to attribute any kind of autonomy or freestanding status to machines. For Canguilhem, machines were organs and not entitled to any claim to being full-standing organisms: he considered fictions of independent machines competing with organisms, as mere fancies of imagination. Canguilhem tended to interpret all cyborgs as organorgs: as organisms merely extended with tools such as an osmotic pump regulated by feedback loops. Canguilhem's view of cyborg was highly deflationist: cyborgs were just extended with tools that did not modify or perturb their identity.

Samuel Butler developed a different perspective in "Darwin among the machines." (Butler 1863a)[18] In this half-speculative, half-serious piece, Butler presented instruments as extensions of our bodily organs, but he also envisioned machines as a new kingdom of creation, in addition to the traditional mineral, vegetal, animal and human ones (depicted by Linnaeus and his fellow naturalists). Machines are thus inserted in a general network of ecological interdependences, in what Linnaeus would have called the "policy of nature." Thus, Samuel Butler considered machines not as mere organs, but as real organisms, as they participate to our ecosystems and compete with living beings for resources. Butler also evoked a kind of Hegelian dialectic of master and servant,—one which is still central to contemporary fictions like *Terminator* or *The Matrix,* where humans and machines struggle to determine who is legitimate in defining reality. Butler also imagined a philosopher, aptly called "Lunaticus." Lunaticus is aware, or rather obsessed with the danger represented by machines in general: we think machines are useful to us, but we end up devoting our whole lives to caring and curing them; to the point that Lunaticus resolves to quit civilization and live without the help or artifacts (Butler 1863b). Having decided to fight machines in the faint hope of saving humanity, Lunaticus abandons his house and settles with his family in the woods: they sleep on beds of ferns, give up scissors, mattresses and other tools such as baby diapers. Even eggs are suspicious to him: aren't they surrounded by calcareous shells, which may well pertain to the tool category? The point of no return is reached when the visionary and authoritarian *Pater familias* asks his daughter to renounce on hair-

this restricted meaning to "the aesthetical history of humanity" in general, apprehended as "a series of successive disadjustments between the three major organizations forming the aesthetic force and capability of the human: the body with its physiological organization, the artificial organs (techniques, objects, tools, instruments, works of art), and the social organizations resulting from the articulation of artifacts and bodies". On this basis, we need, he wrote "to imagine a general organology that would study the joint history of these three dimensions of human aesthetics along with the tensions, inventions, and potentials they generate." (Stiegler 2004: 23) However, this organology appears to be dealing mostly with the organorg level and fails to consider what is really at stakes with cyborgs.

[18] Hacking's title – "Canguilhem amid the cyborgs" (1998) – may be a far echo to it.

clips, mirrors and other saucy items. Here, the family breaks, and Lunaticus wakes up on an empty bed of ferns, his wife and children having vanished overnight.

Lunaticus' mad undertaking helps us understand the dangers of adopting a broad and fuzzy concept of technology. Artifacts correspond to various levels, from isolated artifacts that can be pointed at, to the general technical system more global, leading to more refined analyses. When one claims that technology threatens the human kind, as Lunaticus obviously thinks, what kind of technology does one have especially in mind?

Awareness of those various levels has to grow and replace general talk on "technology"— otherwise one falls into the pitfalls of Lunaticus, who targets not only houses and hairpins, but even eggshells in his general hatred of machines; but also in the conservative aporia depicted by René Barjavel in his novel, *Ashes, Ashes* (French: *Ravage*, 1943), where technophobia leads to a patriarchal backlash: describing a world where the use of any artifacts is punished by death penalty. If artifacts are strategies of the living beings or external organs, what can be included in this category? Renouncing to all tools and all artifacts seems an impossible task.

So it appears that there is a clear phenomenological distinction between "organorg" (or the tool-using organism) and the cyborg (the organism wired and grafted to its tools). However, it may be practically difficult to distinguish between both concepts on several practical borderline situations.

Undeniably, the "organicity" of the relation between artifacts and organs brings to mind some reflections on the concept of function, but also, in the wake of Maurice Merleau-Ponty, some considerations on the limits of the organic body and the "*corps propre*" (one's own body) as a permanent condition of experience. Organic hairs for instance are part of the objective body, while artifactual glasses belong to the "*corps propre*": hence, the frontiers of the body, between organs and artifacts, must be reexamined. This is clearly shown by Kevin Warwick's experiments: when he performs the remote control of a robotic hand, or when a chip inserted in his arm commands the opening/closing of doors or the turning on/off of lights.[19] The cyborg condition is clearly an extension of the *corps propre*.

Eventually, all this is a matter of "discerning," a concept suggested by Max Marcuzzi to describe "the understanding of possibilities incorporable to the body." (Marcuzzi 1996: 206) This process of discerning concerns, for instance, food that can be potentially "incorporated" to my body, or larger social spaces such as the hospital or even the pharmaceutical industry, which concern my own body as potential instruments of its survival. Hence, the subject's *discerning* of the world designates their tendency to incorporate to their field of action or horizon of possibilities, a variety of different artifacts, included large social systems or technological networks that can, eventually, be subservient or instrumental to their own ends.

---

[19] Especially in Warwick (2002: 61): "my own definition is that a cyborg is something that is part-animal, part-machine, and whose capabilities are *extended beyond normal limits*. This is much more general than other definitions and includes creatures other than humans. It allows for mental upgrades as well as physical upgrades and *allows the extension to go beyond the normal limits of either the animal or the machine*." (emphasis added)

16.2.3 Humans and the Meaning of Prosthesis

The present analysis of the organic conception of technology, and the cyborg/organorg distinction, eventually meets the concept of prosthesis. In this exemplary case, artifact and organism are combined to build one single functional unit. The meaning of prosthesis change once they are considered from the perspective of human enhancement or "anthropotechny"—a term coined by Jérôme Goffette (2006). Plastic surgery, physical and mental doping, etc.—all these new practices aim at "enhancing," "perfecting," or "modeling" the humans, and relate to the quest for improved individual performance, well-being and even happiness. These new demands addressed to physicians oppose the traditional canons of medicine, whose practice was usually regulated by the reference to the category of "normal." For medicine (*contra* enhancement), the physician's action is required by vital needs to fight against the disease and restore the body in its normal state. Instead, enhancing interventions reflect an existential need or desire that is not triggered by any natural norm. Traditional medical interventions are requested by the expectation of a vital benefit, whereas enhancing ones should entail an existential benefit.[20] The distinction between ordinary and enhancing medicines is an invitation to re-interpret the case of cyborgs and prostheses.

In the ordinary medical perspective, prostheses are technical devices aimed at curing or repairing. Such medical interventions have produced several types of hybrids, mixing machines and organizations: diabetic individuals with insulin pumps; or heart patients with a pacemaker. Mechanical devices are grafted on the organic body, performing a kind of wiring of the inorganic with the organic; but these examples are only palliatives: mechanical implants function in order to compensate for the failing of organic organs, and to allow the body a "normal" (by which is meant "habitual") vital activity. In this sense, the medical operation does not aim at overcoming the ordinary human condition. It only contributes to restore what was lost (an organ, a function). The medical approach of prostheses is socially acceptable due to its proclaimed humility. By advocating the goal of merely repairing, restoring and preserving the humanity of humans, medical grafts can be viewed as in the service of human's rehumanization. Far from creating postmodern "medical cyborgs," they simply produce organorgs, tool-using ordinary humans, with restored functions.

In the new perspective of enhancement, prostheses are part of a "modeling of the human," or of human attempts to "self-transformation." (Goffette 2006: 8) If any prosthesis is an artificial addition replacing an organ or a substitute for it, we can theoretically distinguish between medical implants and enhancing prostheses: hip implants and glasses belong to the first category, while breast implants belong to the second, as their function is neither to cure nor to restore, but to perform a "cosmetic" function, to conform the individual to standards that are not engaging the

[20] See Goffette (2006: 142–143), for a comparison of the "old" medicine and the anthropotechnical paradigm.

vital prognostic of the subject. The novelty of those enhancing prostheses can be specifically identified in the deepening of the wiring and the idea of performance improvement.

However, this distinction between old repairing medicine and new enhancement practices is not always clear. Let us consider the case of Claudia Mitchell, a US military (born in 1980), who lost her entire left arm in a motorcycle accident in 2004. Mitchell was endowed with a bionic arm provided by Todd Kuiken and his team at the Rehabilitation Institute of Chicago (Neural Engineering Centre for Bionic Medicine) (Kuiken et al. 2007). This prosthesis obeys to the will of Mitchell, responding to electrical signals sent by the brain and transmitted to the nerve endings or to the muscles of the shoulder where the prosthesis is located. The prosthesis is thus characterized by a "rewiring" between the muscles, skin and nerves of the person and the receptors of the bionic arm; the technical artifact is actually inserted into the body and connected to it. In the words of Dr. Kuiken, "We've rewired Claudia. We're rewiring a human to work with a machine." (McGrath 2007) By using the verb "rewiring," Kuiken stresses the aspect of a body-machine cooperation, turning Mitchell into a posthuman cyborg. Yet at the same time, her mechanical arm is presented not as what makes her a super- or post-human, but as what allows her not to fall from her human condition into animality. Quite revealingly, Mitchell tells a journalist that, in order to peel a banana, she had to use her feet and that she "felt like a monkey." Hence, her bionic arm has clearly restored her to her lost human dignity. By repairing and restoring, prostheses re-humanized her. Despite the proclaimed wiring of a becoming-cyborg, the case of Mitchell is still closer to organorg and the general condition of humans using tools.

The case of the South African athlete Oscar Pistorius (born 1986) is somewhat different. Pistorius' legs stopping at the knee have been extended by two carbon prostheses specially designed for running. Typically, Pistorius is a case of "handicap." Pistorius had no access to the regular Olympic competition and, as a double amputee, he was assigned to the Paralympic Games, allowed to compete with one-leg runners. For sports authorities, his prostheses restored him in his lost humanity—nothing more. But in fact, at the height of his career, Pistorius strived to escape from this definition as "handicapped," and made all possible efforts to be admitted to compete with athletes without disabilities (which happened for instance in Rome in 2007). But here the highest authorities of athletics resisted. They suggested that the prostheses, far from just supplying for the absent limbs, were in reality an advantage for the South African runner, comparatively to his "organic" competitors. This decision identified Pistorius as a cyborg or enhanced body, not an organorg. By contrast, by claiming his right to compete with athletes of the Olympics (and not the Paralympics), Pistorius suggested that he was merely a human, not enhanced in any sort of way; in other terms that his prostheses were simply "medical."

16.2.4 Resisting Appropriation and Functionalism

This tension between reparation and enhancement is ritually featured in cyborg fictions. Martin Caidin's novel, *Cyborg* (1972), adapted for television under the title "Six Million Dollar Man" portrays Steve Austin: Austin, an astronaut victim of a serious accident, was not just "repaired" but also "improved" since he was made of more resistant, stronger, more flexible and also very expensive materials. A very expressive shot in Paul Verhoeven's *RoboCop* (1987) also encapsulates the transition from reparation to enhancement. RoboCop, formed from the remains of the person of the former police officer Alex Murphy (Peter Weller) atrociously mutilated, strongly suggests that the cyborg is basically providing a mechanical substitute for a broken body. But nevertheless, there is a subjective camera shot, filmed from the point of view of an half-conscious Murphy: an electronic screen intermittently opens a window of what is actually going on in the operating room; one hears the medical team cheering for having saved one of Murphy's organic arms, but the project leader cuts short and decides to amputate the remaining flesh and replace it by a new limb made of steel. Here the imperial logic of the cyborg is made explicit: a cyborg is not only providing humans with useful high-tech crutches; it insensibly leads to the amputation, or even total removal of the flesh, and its replacement by steel. Of course, expensive steel organs have an owner, and the owner is not the individual itself but others (be they, the State, a gang, or a company).

We must therefore distinguish several stages in the logic of the cyborg. Firstly, integrated wiring provides repair and restitution of the human; but beyond simple restoration of lost functions, the cyborg targets the creation of new senses or new powers: turning the individual into a super-human. Eventually, the cyborg's final end is a substitution of a cybernetic body to the biological body, of silicon to carbon, turning medicine into enhancement. A cyborg is not only repaired, it is augmented or increased. The medical purpose hides enhancing ambitions—with effects of dependence created towards the owners of the embedded technological systems.

In today's world, cyborgs are still constrained by medical regulations, by sets of normative, ethical evaluations: for instance, doping is the object of quasi universal rejection; we tend to caricature cyborgs as individuals merely implanted with gadgets (watches, jewelry, hormonal contraceptive implants or mood regulators), with exchangeable tool-limbs (surgeons with a hand-knife, craftsmen with a drilling-arm, steeds with wheel-legs...), more efficient organs (ocular zoom) or endowed with new senses (ultrasonic, embodied computing capacity).

Thus, cyborgs are depicted as human-machines or modular humans, with the effect to highlight the limitations of a purely functionalist approach of the body. Such a vision was prefigured in the work of Herbert GeorgeWells, whose *First men on the moon* depicted an "age of specialization." (Wells 1901) Chapter XXIV of the book, "The natural history of the Selenites," describes how Selenites produce individuals by focusing on the desired feature, developing other parts only as much as they are required to maintain the target function. Thus, some Selenites are all-brain, while other are merely legs. Other individuals serve as encyclopaedias, archives and books: they developed a brain that contains a large memory and only that. In this world, the organism itself is treated and shaped as a tool, more intensely loaded with social designs: hence the world of the Selenites is a society of highly advanced social insects. At the end of our analysis of the cyborg, technology seems to pursue one ultimate goal: to become "organicized," or one with the organic body. The productions of the children of Prometheus want to pass for endowments from Epimetheus: the "external organs" pretend to be akin to internal organs. At the risk of turning humans into insects.

16.3 Concluding Remarks

1. Cyborgs represent a paradox for the perspective of organology, a biological philosophy of technology that has not completely cut from its romantic origins, and that runs down from Darwin, through Kapp, Bergson or Canguilhem. In this tradition, the organism remains the primordial vital centre, from which technology flows and ultimately depends. The organism is both the origin of technology and the principle of its use. Thus, these approaches are sticking to the "organorg" perspective on artifacts: they describe quite efficiently tools, *i.e.* implements that can be localized. The organorg's tools are always temporary additions, quickly seized and quickly deposited, never irrevocably grafted. Organology is more off the mark when it comes to taking cyborgs into account: as cyborgs involve broader technical systems and threaten the integrity of the organism, in the sense that they delegate the performance of certain functions outside of the organism. Therefore, in a cyborgian world, individuals depend not only on their own organic resources, but also on more general technological networks that are controlled and regulated.
2. From the perspective of general organology, tools are only external organs. In the case of cyborgs, it seems that those external organs strive to become internalized—*i.e.* grafted or wired to the organic body: as if tools were yearning to function like glands. Such a dream can pass for an evolutionary nonsense, if we agree for instance with Henri Bergson that, in the course of evolution, tools are produced by *intelligence* (leading to humans) while glands are the products of *instinct* (leading to social insects). Then cyborgs, who strive at integrating tools into the body or at treating machines like organs, seem to confuse the products of intelligence with those of instinct. Cyborgs seem to aim at transforming humans (the highest incarnations of intelligence) into insects (the highest incarnations of instinct)—a point well illustrated by Wells' Selenites. The insect is what puts in crisis the desirability of the cyborg condition: first, the individual organism is restrained within society to the social position for which it was

designed; secondly, technology is no more a playful or optional adjunction, but it becomes mandatory, integrated into the very structure of the individual.

3. Relations between artifacts and organisms go in two opposite directions, from the outside in and from the inside out. On the one hand, indeed, techniques tend to be "organicized," to give away their proper technical nature and to become simply organic. This culminates into insect-like Selenites. But symmetrically, the organical body tends to simply renounce to perform an increasing number of functions and to replace its native organs with artificial instruments. Thus, the use of vehicles tends to make limbs unnecessary as organs of locomotion. This paves the way for leg-less humans transported on air cushions (as shown in the animated film, *Wall-E*). Similarly, the reproductive function, a fundamental biological function, tends to be increasingly dissociated from the human body: research on Medically Assisted Procreation or artificial uterus leads, ultimately, to make animal reproduction (for humans and domestic species) a fully technical operation, that is to say: external to the (organic, or more specifically, female) body.
4. The status of tools is not fully clear in the organological perspective, as organologists do not always specify which kinds of "artifacts" they describe. Simple tools, for instance, can very well pass for external or projected organs. But then, the emergence of freestanding machines suggests that they are not merely organs, but autonomous organisms, taking part to the ecological dynamics and using resources in the milieu. This raises a concern as to which models should be used to analyze the human/technical relationship: while the organology describes this relationship as mostly instrumental (living beings producing artifacts and making use of them), other biological models such as secretion, symbiosis or parasitism seem to be more accurate or at least, more evocative.

Acknowledgments In the process of writing I benefited from the incisive comments of two brilliant young philosophers of technology, Emanuele Clarizio and Victor Petit. I hope their remarks were not in vain and that I improved the final version of the text thanks to their perceptive readings.

References

Bergson, H. (1907). *L'évolution créatrice*. Paris: Alcan. English Edition: Bergson, H. (1911). *Creative evolution* (A. Mitchell, Trans.). Lanham, MD: University Press of America.

Bergson, H. (1932). *Les deux sources de la morale et de la religion*. Paris: Presses Universitaires de France. English Edition: Bergson, H. (1935). *The Two Sources of Morality and Religion* (R. A. Andrà, C. Brereton, & W. H. Carter, Trans.). New York: Henry Holt and Co., Inc.

Bonnerave, J. (2004). Le corps des musiciens. *L'Homme, 3*, 459–461.

Butler, S. (1863a, June 13). *Darwin among the machines*. Canterbury Press. Reprinted in Festing Jones, H. (Ed.). (1923). *The Shrewsbury edition of the works of Samuel Butler. Vol. 1 Canterbury settlement and other early essays* (pp. 208–210). London: Jonathan Cape.

Butler, S. (1863b, September 15). Lunaticus. From our mad correspondent. The Press. Reprinted in Jones, J. (1960). *The cradle of Erewhon: Samuel Butler in New Zealand* (pp. 195–200). Melbourne: Melbourne University Press.

Campbell, G. (1869). *Primeval man; an examination of some recent speculations* (1873 Edn.). London: Straham.

Canguilhem G. (1952). Machine et organisme. In *La connaissance de la vie* (pp. 101–127). Paris: Hachette. English Edition: Canguilhem G. (2008). Machine and Organism. In *Knowledge of life* (pp. 75–97) (S. Geroulanos, & D. Ginsburg, Trans.). New York: Fordham University Press.

Clarizio, E. (Forthcoming). Leroi-Gourhan con Google Glass: organologia delle tecnologie indossabili.

Clark, A. (2003). *Natural-born cyborgs: Minds, technologies, and the future of human intelligence*. Oxford/New York: Oxford University Press.

Clynes, M. E., & Kline, N. S. (1960, September). Cyborgs and space. *Astronautics, 26–27*, 74–75. Republished in Gray, C. H. (Ed.), with the assistance of Figueroa-Sarriera, H. J., & Mentor, S. (1995). *The Cyborg handbook* (pp. 35–41). New York: Routledge.

Darwin, C. (1871). *The descent of man and selection in relation to sex* (Vol. I). London: Murray.

Espinas, A. (1897). *Les origines de la technologie*. Paris: Alcan.

Gelman, R., et al. (1995). Distinguishing between animates and inanimates: Not by motion alone. In D. Sperber et al. (Eds.), *Causal cognition*. Oxford: Clarendon Press.

Goffette, J. (2006). *Naissance de l'anthropotechnie. De la médecine au modelage de l'humain*. Paris: Vrin.

Hacking, I. (1998). Canguilhem amid the cyborgs. *Economy and Society, 27*(2–3), 202–216.

Haraway, D. J. (1985). A manifesto for cyborgs: Science, technology, and socialist feminism for the 1980s. *Socialist Review, 15*(2), 65–107.

Kapp, E. (1877). *Grundlinien einer Philosophie der Technik. Zur Entstehungsgesschichte des Kultur as neuen Gesichtspunkten*. Brunswick: Westermann. French Edition: Kapp, E. (2007). *Principes d'une philosophie de la technique* (G. Chamayou, Trans.). Paris: Vrin.

Kuiken, T. A., et al. (2007). Targeted reinnervation for enhanced prosthetic arm function in a woman with a proximal amputation: A case study. *The Lancet, 369*(9559), 371–380.

Leroi-Gourhan, A. (1945). *Evolution et techniques. II, Milieu et techniques*. Paris: Albin Michel.

Leroi-Gourhan, A. (1964). *Le geste et la parole. Tome I : Technique et langage*. Paris: Albin Michel. English Edition: Leroi-Gourhan, A. (1994). *Gesture and Speech* (Vols. I & II: Berger A. B., Trans.). Cambridge, MA: MIT Press.

Leroi-Gourhan, A. (1983). *Le Fil du temps : ethnologie et préhistoire*. Paris: Fayard.

Lewis, M. (2013). Of a mythical philosophical anthropology: The transcendental and the empirical in *Technics and Time*. In C. Howells & G. Moore (Eds.), *Stiegler and technics* (pp. 53–68). Edinburgh: Edinburgh University Press.

Lubbock, J. (1865). *Prehistoric times, as illustrated by ancient remains, and the manner and customs of modern savages*. London: Williams and Norgate.

Marcuzzi, M. (1996). *Les Corps artificiels : peurs et responsabilités*. Paris: Aubier.

McGrath, B. (2007). Muscle memory. *The New Yorker*, 30 July 2007.

Pickering, A. (1995). Cyborg history and the world war II regime. *Perspectives on Science, 3*, 1–48.

Ruyer, R. (1946). *Éléments de Psycho-biologie*. Paris: Presses Universitaires de France.

Scheler, M. (1947). *Die Stellung des Menschen im Kosmos*. München: Nymphenburger Verlagshandlung.

Schmidgen, H. (2006). Über Maschinen und Organismen bei Canguilhem. In H. Schmidgen (Ed.), *Georges Canguilhem, Wissenschaft, Technik, Leben : Beiträge zur historischen Epistemologie* (pp. 157–178). Berlin: Merve.

Stiegler, B. (1994). La technique et le temps, 1. *La faute d'Épiméthée*. Paris : Galilée. English Edition: Stiegler, B. (1998). *Technics and Time, 1. The Fault of Epimetheus* (G. Collins, & R. Beardsworth, Trans.). Cambridge, MA: Stanford University Press.

Stiegler, B. (2004). De la misère symbolique: Tome 1. *L'époque hyperindustrielle*. Paris: Galilée. English Edition : Stiegler, B. (2014). *Symbolic misery, vol. 1: The hyperindustrial epoch* (B. Norman, Trans.). Cambridge, MA: Polity Press.

Stiegler, B. (2005). De la misère symbolique: Tome 2. La catastrophe du sensible . Paris: Galilée. English Edition: Stiegler, B. (2014). *Symbolic misery, vol. 2: The Catastrophe of the Sensible* (B. Norman, Trans.). Cambridge, MA: Polity Press.

Warwick, K. (2002). *I, Cyborg*. London: Century.

Weatherby, L. (2016). *Transplanting the metaphysical organ: German romanticism Between Leibniz and Marx*. Fordham: Fordham University Press.

Wells, H. G. (1901). The first men in the Moon. In L. E. Stover, (Ed.). (1998). *A critical text of the 1901 London first edition. With an introduction and appendices*. Jefferson: McFarland.

Chapter 17
Of Times and Things. Technology and Durability

Bernadette Bensaude Vincent

Abstract To fully accomplish the "thing turn" in the philosophy of technology this paper invites shifting the attention from humans towards the world. The concept of world here refers to the complex made of the Earth with all things and living beings, including humans; it ignores the great divide between nature and society or culture. In this worldly perspective, the thing turn means adopting the perspective of things and raising questions such as how artifacts come into being, how they intervene within the world, how they change it. Such issues are vital to prevent the alienation of technology both from nature and from human beings.

Keywords Anthropocene · Environmental activism · Nuclear waste · Plastic waste · Thing turn · Worldly perspective · Regimes of temporality · Temporalities

Given the amount of meanings, values, norms and potentials conferred to artifacts in the recent studies of technology since the "thing turn," (Miller 1998, Preda 1999, Henare et al. 2007) it may seem preposterous to claim that something is missing. Nevertheless, it is time to go further in the "back to the things themselves" move. Even though Husserl's analysis of the relations between consciousness and reality has been greatly extended and enriched by dozens of theoretical and empirical studies, the focus remains on things *for human beings*. Up to now artifacts have been explored for what they mean for humans, how they mediate our relations to the world. Accordingly, technology studies have been focused primarily on the co-shaping of objects and users or on the ethical, social and political meanings and impacts of artifacts. A narrow understanding of this topic is instantiated in the

B. Bensaude Vincent (✉)
Centre d'Etude des Techniques, des Connaissances et des Pratiques (CETCOPRA), Université Paris 1 Panthéon-Sorbonne, Paris Cedex 05, France
e-mail: bvincent@univ-paris1.fr

S. Loeve et al. (eds.), *French Philosophy of Technology*, Philosophy of Engineering and Technology 29,
https://doi.org/10.1007/978-3-319-89518-5_17

institutionalized approach – under the acronym ELSI[1] – to monitor the development of emerging technologies such as genomics, nanotechnology and converging technologies. However, the emphasis on the social, cultural, metaphysical and political dimensions of artifacts does not really take into account their temporality. As hybrids of nature and artifice, technical objects are not only carriers of human projects and values, they are also integral parts of natural cycles. They come into being through the use and the transformation of natural resources, and they endure in the world far beyond the limits of their existence as commodities. They have a history of their own.

Even Dutch philosopher Peter Paul Verbeek, who initiated a remarkable thing turn in "*What things do*," (Verbeek 2005) had the limited ambition to do justice to the concrete presence of technological artifacts in our culture. He is concerned with what things do to *us* rather than to the *world* we inhabit with them. His attempt at thinking from the perspective of things is purposefully characterized as "post-phenomenological" (Verbeek 2006, 2008) because, in his view, things are vital for elucidating our relations to the world. In this perspective the relation between human beings and their world takes centre stage, and are viewed as mutually constituting each other – human beings are what they are thanks to the ways in which they are present to the world, and their world is what it is thanks to how it appears to them (Verbeek 2005: 235).

Things essentially play the role of mediators between human beings and their environment. The major matter of concern regarding temporality is our experience of time in relation to technological innovation. This overly anthropocentric perspective has prompted a huge amount of scholarship devoted to the mutual shaping of new technologies and the tempo of life in modern societies. In addition to the cult of speed linked to transportation and communication technologies, digitalization has created a culture of instantaneity and immediacy (Castells 1996). Social theorist Harmut Rosa finally describes a threefold process of acceleration, including technology, social change and the pace of life (Rosa 2013). He points out a paradox: Technological innovations were supposed to spare our time but we are increasingly under time pressure. Rosa concludes that technology no longer serves the "project of modernity," characterized since Habermas by the ideal of autonomy and emancipation. His analysis is congruent with the recent boom of "slow" movements – slow food, slow city, slow travel, slow design, slow money, slow school, slow books, slow science, slow trading, slow living – which express a rebellion against the disciplined clock-time imposed by capitalism and globalization. By contrast, Judy Wacjman (2014) diversifies the forms and sources of time constraints in her analysis of work patterns and domestic life. Although she acknowledges the evidence that technology did not alleviate time pressure, she argues that it is not just a matter of quantity but also of quality of time.

Whatever the high value and interest of the literature on acceleration, it remains focused on how technology has shaped our pattern and experience of time. Fully accomplishing the Copernican revolution in the philosophy of technology requires

[1] For "Ethical, Legal land Societal Implications/Impacts."

shifting the attention from humans towards the world that is to the complex made of the Earth with all things and living beings, including humans. This concept of world ignores the great divide between nature and society or culture (Latour 1991). In this worldly perspective, the thing turn means adopting the perspective of things and raising questions such as how artifacts come into being, how they intervene within the world, how they change it. Such issues are vital to prevent the alienation of technology both from nature and from human beings.

A number of philosophers have already discussed such questions. For instance, Gilbert Simondon's fine-grained analysis of the process of concretization of technical objects (1958) sheds light on their genesis and ontology apart from their utility for humans. Artifacts cease to be kept hidden and forgotten because of their handiness as soon as they are no longer described as tools for performing functions according to the plans of their designers. As they come into being through a process of concretization they acquire an ontological status in a human/nature world. Simondon went beyond a "facile humanism" by adopting an object-centered perspective infused by the concept of technicity. Gilbert Hottois, on the other hand, used the concept of "operativity" in order to emphasize the intervention of artifacts upon the world. The philosophy of technology, he argued, requires more than hermeneutics because artifacts are not merely the result of an interpretation of the world, the sign of a representation; they actively contribute to shape and transform it (Hottois 1984).

As a way to go further into the investigation of the interplay between technology and the material world we share, this chapter focuses on the durability of artifacts. Though the notion of "sustainable development" has attracted a lot of public attention over the past decades, surprisingly few philosophers of technology did relate this concern to the ontological status of artifacts. Is it still the case that "the owl of Minerva spreads its wings only with the falling of the dusk"? In fact, philosophers did address this issue albeit not at the ontological level. They mainly discuss it in terms of environmental ethics. In keeping with the prevalent meaning of "sustainability," environmental ethics is mainly concerned with the Earth capacity to support human life and thus ironically favors an anthropocentric perspective. (Larrère and Larrère 2015) By contrast, the notion of *durability*, emphasizing the permanence and the endurance of artifacts, has the potential to open up a de-centered perspective. While R&D practitioners, designers and engineers routinely perform life-cycle analyses and spread concepts such as "cradle-to-cradle" design, what do philosophers have to say about the durability of artifacts?

In this chapter, the durability of artifacts is presented as a key issue that brings about a deep revision of their ontological status. I first argue that artifacts have no ultimate essence and have instead various modes of existence inscribing them in various histories and temporalities. Then, looking more precisely at the durability of two kinds of artifacts – plastics and nuclear waste – I will discuss whether they confront us with a wide range of orders of magnitude or with more complex issues of incommensurable regimes of temporality.

17.1 Prelude: Metaphysics

In a famous article of his *Principles of Philosophy*, Descartes clearly stated that he didn't "recognize any difference between artisan-made machines and the various bodies that nature alone composes." (Descartes 1641: IV §203) More concerned with the understanding of nature than with the status of artifacts, Descartes assumed that the visible mechanisms at work in machines provide a model for understanding the invisible mechanisms at work in nature. He thus firmly rejected the ontological distinction established by scholastic philosophers who claimed that natural things unlike artifacts had an intrinsic "substantial form." For Descartes, machines are neither "against-nature" nor supernatural (magic). They are ruled by the laws of nature: "And anyway all the rules of mechanics belong to physics, so that all the things that are artificial are with that [*avec cela*] natural." My literal translation of this sentence is meant to emphasize that artifacts might not entirely be identified with natural things, they could rather have a dual status: They are artificial things besides being natural ones. There is no alternative, no either/or choice. And since technology is based on natural laws, improving technology and better understanding nature are one and the same objective. Physics will "thereby make ourselves, as it were, the lords and masters of nature." (Descartes 1637: VI, 44) The rapprochement between the artificial and the natural secures and legitimizes the domination of nature by humans.

On the other hand, Descartes (1664: 36) drew a clear boundary between objects and subjects. Nature being reduced to matter – i.e. to the *res extensa* – is ontologically distinguished from the *res cogitans*, proper to subjects. His metaphysics reinforced the distinction assumed in Roman law between persons and things and prefigured Kant's moral claim that human persons must be treated as ends and never only as means. The social contract, ethics and politics, are disconnected from nature. (Serres 1990) The "great divide" between nature and society, which prompted the institution of humanities and social sciences as an independent culture, is a chief characteristic of modernity (Latour 1999).

One major instantiation of this divide is the instauration of universal time. Although initially based on a natural cycle (the duration of the tropical year), it is a pure convention answering a joint demand from science and capitalism. This abstract meantime resulting from a process of averaging the length of solar day, ignores the variable cycles of daylight in order to create regular clock hours. The social clock time has nevertheless become so naturalized that natural rhythms dependent on local environment and biology become like disruptive anomalies. (Birth 2012) Living in a modern world thus means tacitly assuming that we are (increasingly) emancipated from nature and belong to culture and society.

17.2 Entangled Histories of Plastics

Synthetic polymers have gradually replaced glass, wood, and metal in many commodities. Plastic bags, bottles, pipes, pens, phones, computers, rackets are everywhere around us. Plastic materials are so familiar that we no longer wonder about their capability to take on multiple forms and be used in all sorts of applications. Thanks to the flexibility of carbon atoms, and their ability to form bonds with other atoms in whatever direction, especially strong covalent bonds that make up long chains or macromolecules, a great variety of materials can be synthesized. Many types of synthetic polymers, each one with a variety of different formulations have been manufactured and commercialized. One of their advantages is that they can be synthesized and shaped simultaneously. Polymerization and molding, matter and form are generated in one single operation. Moreover, during this dual process it is possible to include various additives, such as plasticizers or reinforcing fibers in order to obtain specific properties and functionalities. The design of composites materials has expanded the market of plastics and secured the success and wealth of the petrochemical industry.

Despite the commercials advertising plastic items – Tupperware for instance – as icons of modern life in the 1950s, "plastics have never been modern." In Bruno Latour's sense (1991), they intermingle what modern philosophers struggled to clearly separate. When synthetic polymers replaced or displaced conventional materials in many applications, the phrase "Plastic Age" has been coined to suggest the advent of a new cultural era, comparable to the iron-age in the past. The mass consumption of plastics did actually spread the North-American lifestyle and values all over the world (Sklar 1970; Meikle 1995). As Roland Barthes (1957) argued in his review of the mythologies of modernity, plastics connote the magic of indefinite metamorphoses. Whereas gold or diamond conveys a view of permanency and eternal faith, plastics epitomize the ephemeral, the ever changing. They are so light that they lose their substance, their materiality, to become almost virtual reality. As they are malleable and indefinitely adaptable, they connote and spread such values as impermanence, flexibility, and superficiality through culture.

Plastics do have politics (Winner 1986) because of their material properties (Gabrys et al. 2013). They are not only value carriers; they also prescribe specific behaviors. Because they are cheap, light, and easily disposable, plastic bottles, cups, and syringes are specifically designed for one single use. Their normative power turns users of artifacts into consumers of disposable objects. Such consumerist behaviors often legitimized with hygienic rationales basically favor hedonistic inclinations to enjoying the present moment. They encourage a specific experience of time, as an instant detached from the flux of time, a discrete moment disconnected from the past and the future. Unlike Virginia Woolf's "moments of being" however, such instants do not go with flashes of awareness of the world around. Rather their iteration in daily routines generates a kind of protective screen-effect.

The mass consumption of plastics went on with a protective blindness concerning their whereabouts. Where do they come from, where do they go when we put them in the rubbish bin? Such trivial questions have the power to raise discomfort about our consumerist attitudes as well as ontological issues. An ontogenetic perspective on plastic objects subverts both the modern dichotomy between nature and politics and the Cartesian view of nature. Try to follow the trajectory of plastic items, and suddenly food, chemistry, children, geopolitics, become entangled with the history of technology, of capitalism, and of the Earth (Mariott and Minio-Palluello 2013).[2] The journey of a mundane artifact such as a plastic food container provides a rich narrative bringing together the depletion of resources, the conflicts between the North and the South, social unrest, the accumulation of capital in the hands of a few multinational companies, and the accumulation of plastic garbage everywhere.

The travel through space of plastic objects is also a travel through time. More precisely, the story of their pre-life and post-life intertwines many different times. Most of the 260 millions of tons of plastics annually commercialized are made out of fossil fuels, so that they irreversibly consume the results of the spontaneous degradation of organic materials, which settled in the rocks 3.4 millions of years ago. Their existence of a few days or months as disposable commodities contrasts with the geological times involved in their previous existence. The 4% of the world's oil bound to become plastic materials start new lives under the sun. After extraction, transportation, and transformation the fossils first exist as small granules made from oil and a cocktail of additives. Those resin pellets manufactured by petrochemical companies will then be melted and molded in plastic artifacts, although some of them being lost during transport or in the manufacturing site continue their granular existence among the sand of the beaches or floating on seawater and carried by streams. The ephemeral character of plastic commodities is even more delusory if we consider the accumulation of plastic detritus everywhere in the countryside and in the ocean. The nice, bright and colorful plastic toys or gadgets that we trash after one single use end up in a grand garbage patch of thousands of square meters. The micro-balls charged with highly concentrated persistent organic pollutants accumulated in the oceans since World War II create a "plastic soup," which deeply affects the biosphere of marine creatures (Gabrys 2013).

Here is the irony of the Plastic Age. The cult of impermanence and change that distinguished it rests on a deliberate blindness regarding the material condition of

[2] In a paper presented during a Conference on Plastic accumulation at University of London, James Mariott offered ice cream cornets to the audience and then started telling stories about the ice cream container left empty on the stage. On Google map he followed the trajectory of this little plastic tub manufactured by a chemical corporation listed on Stock Exchange, from the oil drill in Azerbaijan through pipelines, big tankers cruising the ocean, to the European factory, the supermarket and finally the British building where the plastic container which satisfies our appetite for ice cream would be placed in the rubbish containers and picked up by a garbage truck to a recycling station.

plastic items. These attractive objects, apparently liberated from the constraints of materiality, from gravitation and duration, are enduring. Their ephemeral existence is just the tip of an iceberg of memory. It is the upper layer of many layers of plankton buried in the seafloor or in the rocks for millions of years and it is only a small fraction of the complex biochemical and metabolic processes that constitute the carbon cycle. The brief existence of plastic artifacts as commodities is nothing but an instant abstracted from the long duration of material processes on Earth.

More importantly, the life story of any plastic object clearly demonstrates that materials have a say in what they become. On the one hand, carbon macromolecules with their bonding capacities afforded the Plastic Age, i.e. the substitution of wood, metals for plastics in most commodities. Because synthetic polymers are lighter and more malleable than the materials they displaced they have encouraged the dream of dematerialized technology (Bensaude Vincent 2013). But the plastic items that served our desire and economy do not magically disappear from the planet when get rid of them. They just disappear from our sight being shipped far away to Southern countries for complex operations of recycling with the help of toxic chemicals and cheap manpower or simply incinerated in a nearby urban power station to produce heat. Plastic toys and attractive gadgets are congealed in the eternal present of our desires, detached from their own histories, from the material world with its complex cycles of transformation as well as from the contingencies of market and fashion. Aggressive advertising campaigns presenting plastics as a cornucopia coupled with cynic strategies of programmed obsolescence have encouraged the accumulation of capital in chemical companies together with an active ignorance among consumers (Proctor and Schiebinger 2008). Such market strategies conceal that commodities are things with a life of their own, which interact with many lives around. Like all creatures in the world they are ageing, they need care and maintenance. The cult of innovation and invention leads to disregard the significance of repair and maintenance (Edgerton 2007). Technical objects are not reducible to the functions and services that they perform for us. As soon as they come into being they interact with a wide range of beings that populate the world and not just with us.

De-commodifying plastic items to objectify them is a necessary first step as long as the process of commodification of objects goes with a "genesis amnesia" cutting its products from their own past and future (West-Pavlov 2013: 128). Yet, as we will see, this is only a preliminary step.

17.3 Interlude: Ontography

The life of plastic objects confirms that the rigid boundary between nature and society is no more robust than the boundary between nature and artifact. Such ontological divides proceed from a *sub specie aeternitatis* metaphysics looking for the essence of things instead of considering their genesis and becoming. Even though plastics epitomize the artificial and have the dubious reputation of being unnatural qua synthetic they participate in the whole process of the *natura naturans*. Therefore an *ontographical* approach to artifacts may be appropriate. Why substitute the suffix "logy" for "graphy?" Just as ethnography refers to a simple descriptive report of field inquiry without any attempt at systematizing or theorizing, an ontography follows its objects and describes their trajectories. Ontography has first been promoted by Michael Lynch (2008) in keeping with the social studies of science purpose of dismantling the grand metaphysical schemes in favour of empirical case studies. "Deflating ontology" was just one side of the coin; the other one was to turn "matters of fact" into "matters of concern" by looking at how things are framed as nodes in a network of material, political, economic and legal actors. Ontography in this sense emphasizes the heterogeneous ingredients that converge in the coming into being of technical objects.[3]

In this essay the ontographical approach is meant to support two major claims.[4] First, the question about beings does not concern what lies beneath them (substance) or behind the appearance. As an attempt to identify the modes of existence of individual entities, ontography does not assume that they are deducible from other entities through a causal chain. Nor does it presuppose any hierarchy of levels of beings. Ontography pays attention to the multiple ways of being, to the variety of "modes of existence" and thus moves beyond the dilemma between realism and constructivism. Etienne Souriau (1943) promoted a "multirealism" and Latour (2013) talks of the "pluriverse." The concept of "modes of existence" not only gets rid of the either/or categories and emphasizes the plurality of ways of being but it also points to the connections between various events or actions like the grammatical modes expressed by inflections of verbs.

Second, while the logos (in onto*logy*) clearly refers to analytical and argumentative discourse, the reference to *graphein* (in onto*graphy*) means that narration is a relevant and penetrating form of discourse about things. Telling stories is a way of emphasizing the temporal dimensions of things. The plot displays the complex threads that have to be intertwined for things to come into being and then to endure into existence. Storytelling – whether it be in the form of a chronicle, a biography, a fiction, a drama, a fable or mythological tale – helps replacing the Kantian concern with the "constitution" of things by Souriau's notion of "instauration." (Souriau 1939) Instead of asking about the conditions of possibility of artifacts, let us look at how they come into being through contingent events and in the course of actions. The resilience and obduracy of artifacts, their inscription in biological, geological

[3] Graham Harman's ontography is a more ambitious attempt at promoting an alternative ontology centred on objects, and displaying the relationships between objects. The suffix "graphy" refers to a graphical representation of all the possible relations between what he considers as the four basic poles of reality (real objects, real qualities, sensual objects, sensual qualities). "Rather than a geography dealing with stock natural characters such as forests and lakes, ontography maps the basic landmarks and fault lines in the universe of objects." (Harman 2010: 125) Far from favoring a strictly descriptive approach to objects Graham promotes a "speculative realism" claiming that objects are not reducible to relations and that individual entities of various different scales are the ultimate stuff of the cosmos.

[4] This ontographical approach is instantiated in Sacha Loeve and Bernadette Bensaude Vincent (2017).

cycles and their interference with far wider cosmic process is certainly something that a conventional philosophical argumentation can express. Yet narratives provide a much thicker description of how particular things stand in the world, how they hold the place, resist and oblige humans to care for them.

17.4 The Perdurance of Nuclear Waste

In the post-war era, France has massively invested in nuclear power through a close alliance of military and civil goals with a view to assert its autarchy thanks to its own technology (gas-graphite reactors) and the uranium resources from its colonies (Hecht 1998). In the 1970s, following the first oil crisis and environmental alarms France decided to rely on nuclear technology for its supply of electricity. It resulted in the construction of 58 nuclear reactors, which today produce 73% of the electrical power consumed in France. This national choice meant to reduce the dependence from petroleum exporting countries has the additional advantage of reducing the emissions of greenhouse gas. To fuel the reactors the mining business group Areva is in charge of the exploration, extraction, and refining process of the uranium ores, with long-term contracts signed with mining companies especially in Africa. The French nuclear order thus maintains a colonial geopolitical regime (Hecht 2012). We will see that the materiality of nuclear combustibles and reactors also generates a clash of temporalities.

Nuclear plants are rather short-lived while the residues of their productive activities are extremely long-lived. The fission products of the uranium dioxide (UOX) loaded in the reactors are so radioactive that they can cause damage to all living things and will remain dangerous for a long time. According to the law of radioactive decay, the half-life of nuclides of actinides (hundred thousands years) is so out of proportion with the timescale of human life and of democratic regimes (5–10 years) that it far exceeds our capacities for anticipation and imagination. And the question of nuclear waste is a hot topic raising embarrassment and public protests.

What can be the life of the tons of dangerous waste produced year after year by nuclear reactors? It is difficult to figure out. Let us follow the trajectory of these undesirable residues thanks to an inquiry conducted by Laurence Raineau (2012) and Sophie Poirot-Delpech (2017). The assembly metal rods of used fuel extracted from the reactor are first let to cool-down near the nuclear plant for a few years. They are then packaged in huge containers of 110 tons each and trucked to La Hague, in Normandy, for further cooling underwater. Since the used fuels still contain a good proportion of uranium and plutonium and highly radioactive nuclides of actinides, technological solutions have been implemented for recycling them.[5]

[5] First, a chemical treatment of the fission products separates uranium and plutonium from non-recyclable fission products. Second, the depleted uranium is enriched. Finally it is trucked again for being reprocessed in Southern nuclear facilities, which produce a new fuel (MOX) made of a mixture of depleted uranium and plutonium that will be used to fuel advanced nuclear reactors.

However, reprocessing means that hazardous radioactive materials are handled and trucked through the country, in secret convoys to avoid publicity since plutonium is an ideal material for nuclear weapon. In addition, the attempt to close the fuel cycle cannot be successful because the fuel can be recycled only once before final disposal.

Ultimate waste, referred to as radioactive substances "for which no further utilization is predicted or envisaged" are more than toxic than the earlier ones. They are bound to spend at least the next 100 years in small packages buried in a 500 meter-deep cavern of 30 km^2. The project of containment of final waste in an Industrial Centre for Geological Storage (CIGEO) located in the North-East part of France conducted by the National Agency for the Management of Nuclear Waste (ANDRA) is well under way. It has been launched in 2000 following a decree (the 1991 Bataille Law) stipulating that the method of storage should allow the recovery of packages within 100 years. Although this measure testifies to a political awareness about the uncertainty of the future (Barthe 2006), it did not settle the question. The CIGEO project still raises public protests, and following a nationwide debate held in 2013, the opening of the industrial storage site has been simply postponed and research continues. (ANDRA 2014) To settle the noise and fury, the clock of political measures has been stopped, time frozen. Meanwhile the recalcitrant radionuclides carry on radiating and accumulating. They quietly "perdure." (Ingold 2013)

In many nuclear countries, people are desperately looking for reliable means for communicating with the future generations (400 generations for every 10.000 years) about the danger of nuclear waste sites (Galison 2010). The plans to turn storage sites into sanctuaries earnestly try to bridge the huge gulf between the lifetime of humans and that of decaying radionuclei. However, assuming that the root of the problem lies in the gap between the timescales of human history and radioactive decay, that it is just a question of orders of magnitude is still an over-simplistic view. Not only the sites of storage have local impacts on the biosphere by creating zones of exclusion for all forms of life, but also the accumulation of toxic waste is not sustainable. As long as the recycling of used fuel (MOX) and nuclear reactors based on fusion (still investigated in the megaproject of International Thermonuclear Experimental Reactor – ITER) are not working successfully, nuclear energy cannot be a renewable source of energy. Clearly, reversibility and renewability are keys to sustainability but radioactive decay is a spontaneous and logarithmic process, which resists all attempts to close the cycle.

Here is a major ontological difference between plastic waste and nuclear waste. Most synthetic polymers do not biodegrade but they are photodegradable, so that plastic debris are extremely small, mobile and pervasive. Tiny fragments of plastics are to be found everywhere on the planet. They float randomly in the oceans, thus making a "plastic soup." They occasionally melt with volcanic rocks to form 'plastiglomerates' such as the ones found in 2013 on Kamilo Beach in Hawaïi. (Corocan et al. 2014) This agglomerate of natural rock and artefacts epitomize the anthropogenic imprint on geology. Plastics have been seriously considered as serious candidate markers of the beginning of the Anthropocene by geologists (Zalasiewicz et al. 2016). Yet ultimately they decided that radionuclides would make better markers. Indeed radioactive materials, unlike plastic materials, have a statistically predictable behavior: as soon as their isotopic composition is determined their half-life is known. It does not mean that nuclear waste better fit in nature, for that matter. Even more than plastic debris they are endurable, recalcitrant and irreversible. Some plastic bottles can be recycled even though the process consumes energy and produces toxic waste, in turn. In addition, plastic debris floating in the ocean end up eaten by marine creatures. They become food for bacteria, plankton, and fish. Synthetic polymers with their toxic additives finally enter the food chain of marine creatures and the skeleton of the macromolecules starts a new existence within the biological carbon cycle thanks to microorganisms acting as silent and invisible "carbon workers." (Gabrys 2013) Whereas plastic waste create a major environmental problem because they spread toxic materials in living beings, nuclear waste are not edible, not compatible with living creatures. And even if extremophile synthetic bacteria were convinced to "eat" them, they would not dispose of their radioactive properties. Radioactive materials hold the place for ages and will still keep it after the extinction of mankind.

Disposable plastics and irreversible nuclear waste help identify the prerequisites for sustainable technologies. Technical objects have to be both durable and degradable. Hannah Arendt grasped the fine-tuned temporal condition of technical objects in her analysis of the *vita activa* with its three categories of labor, work and action. She emphasized that artifacts endure in existence after the production process and long after their uses in human societies, thus securing the "reality and reliability of the human world." (Arendt 1958: 95) The durability of artifacts plays a key role to overcome the limited lifespan of humans; it stabilizes society and the public sphere of action. However, Arendt never assumed the irreversibility of technological processes:

> This great reliability of work is reflected in that the fabrication process, unlike action, is not irreversible: every thing produced by human hands can be destroyed by them, and no use object is so urgently needed in the life process that its maker cannot survive and afford its destruction. *Homo faber* is indeed a lord and master, not only because he is the master or has set himself up as the master of all nature but because he is master of himself and his doings. (Arendt 1958: 144)

The notion of mastery inherent in the Cartesian view of technology collapses under the overwhelming evidence of the irreversibility of the products of our most brilliant technological achievements. "How can we dominate our domination; how can we master our own mastery?" For Michel Serres this pressing question results from the completion of the Cartesian project (Serres 1992: 251). Since we have become masters of space, of matter and of life, there is nothing left on Earth that does not depend on us. It is no longer possible to draw a boundary between the things that do not depend on us – natural phenomena – and those that depend on us (local and political matters). The comfortable divide underlying modernity between the "inexorable necessity" of nature and the realm of action (ethics and politics) seems to fade away.

> The lives and actions of our children soon will be conditioned in fact by an Earth that we will have programmed, decided upon, produced and modeled. [...] In the future we will live only under the conditions that we will have produced. (Serres and Latour 1995: 174)

Increased technologization and socialization of nature seems to be our destiny. The arrow of time reinforced by the common assumption of an exponential growth of information technologies as expressed by the famous Moore's law is supposed to dictate our future (Loeve 2015).

However, the arrow of time is a socio-historical construction, which has accompanied the modern ideal of mastery of nature and may consequently be questioned together with this ideal. To what extent is it possible to take into account the times of nature, to favor a model of softer technology working *with* nature instead of *upon* her or *against* her? (Larrère and Larrère 2015) And what metaphysics of time would replace the standard arrow of progress?

17.5 Conflicting Regimes of Temporalities

In 1971, US biologist and pioneer of the environmental movement Barry Commoner published an ambitious essay entitled *Closing the Circle. Man, Technology and Nature*, which turned out to be a bestseller. Unlike Rachel Carlson's *Silent Spring* (1962) Commoner did not use a poetic fable to raise public awareness about the environmental damages caused by chemical technology. He promoted ecology as the "science of planetary housekeeping." With this reference to the Greek term *oikos* (house, family) Commoner clearly regarded nature as home for humans. While maintaining a human-centered perspective, he nevertheless clearly pointed out a discrepancy between the cycles of nature and the linear time of technology:

> Stich ecological cycles are hard to fit into the human experience in the age of technology, where machine A always yield product B, and product B once used, is cast away, having no further *meaning* for the machine, the product, or the user. (Commoner 1971: 4)

For Commoner the divorce between our experience of time and nature is not due to the proliferation of innovations and the subsequent acceleration of life. It is not just a question of pace, of tempo; it is a question of *meaning*. In the linear model of technological production, the by-products of manufacturing process are meaningless. Like the *caput mortuum* of alchemists, they are worthless remains to be disposed of. In natural processes, by contrast, there is no *caput mortuum*, no waste. Everything is degraded and the building blocks entering into new combinations go through new adventures. The linear time of technology and the arrow of time epitomizing technological progress do not fit well in the cycles of nature. Concerned environmental activists have thus highlighted a fundamental discrepancy of regimes of temporality, and subsequently prompted efforts to introduce "cycle thinking" (from cradle-to-cradle) in technological innovation, through bio-inspiration. Over the past decades such efforts and campaigns have yielded a number of innovating biomimetic processes and products for promoting green chemistry as well as ecomimetic agriculture reconnecting farmers and nature. Bioeconomy – an economy based on renewable biological processes and products – is celebrated as the avenue to sustainable development (Passet 1992). The key is to insert our technological interventions within the dynamics of nature, to take advantage of spontaneous phenomena of self-organization and self-regulation. Although the bioeconomy agenda turned out to maintain the Cartesian ideal of mastering nature when it implies *redesigning* life and manufacturing synthetic organisms, it rests on the monitoring of natural processes and has to comply with natural cycles.

By contrast, another clash of temporalities has gained traction recently through the emergence and dissemination of the notion of Anthropocene (See Stewart, Chap. 14, this volume). In 200 years since the industrial revolution humans have consumed fossil resources that had been stored for 3.800000 years in the soil, thus releasing a huge amount of carbon dioxide in the atmosphere. In other terms, it is a kind of "tempophagy" – time eating (Cohen 2012). With the massive emission of greenhouse gases mankind has bridged the gulf between many orders of magnitude. The Anthropocene blurs the boundary between social history and natural history, and as such it could mark the end of modernity (Latour 2012).

Yet underlying this notion is the modern assumption of a timeline running from the Big Bang (15 billion years) to the apparition of life on the planet (4 millions years) and the history of civilization (6000 years) – a unique global time described in powers of ten. To convey a view of the gap between the powers of ten, science writers use metaphors such as the cosmic year (with humans appearing in the last second on December 31) or the week of Genesis; all of them converge on human history as though mankind were the *telos* of the existence of Earth. Unsurprisingly the narratives of the Anthropocene provide a rationale for geo-engineering projects. As the intensification of human pressure on the planet has caused damages, it seems legitimate to extend human intervention to the world itself in order to fix it. This strategy of fighting fire with fire only makes sense in the modern worldview of humans as masters of the Earth and progress as emancipation from nature, from space and time constraints.

In the narratives of the Anthropocene, the clash of temporalities resulting from human technology is seen as a quantitative difference in orders of magnitude, a leap over timescales due to the acceleration of technological power, whereas the conflict between the vector of economic growth and the cyclic time of nature is a central pillar in environmentalist movements.

17.6 From Timescales to Timescapes

How are we to understand the increasing concern with the clash of temporalities? Is it the result of the acceleration of technological innovations or the symptom of a more radical heterogeneity of regimes of temporality that calls for a deep revision of our metaphysics of time as well as of our technological practices?

The narratives of the Anthropocene do not challenge the dominant view of a global time as a universal container, which measures the existence of all beings. They even reinforce the significance of the timeline for enhancing the wide spectrum of timescales. This linear view of a global time that we take for granted is nevertheless a social convention resulting from a mixture of various ingredients: natural phenomena (from the course of celestial bodies to a specific radiation of cesium atom), technical objects (calendars and clocks), scientific concerns (precision measurement, normalization) and economic pressure (disciplining labor in industrial capitalism) (Birth 2012). The result is an abstraction, a time detached from all localities as well as from our social time (Nowotny 1994). A time viewed from nowhere. Universal time is a powerful tool ensuring the commensurability of everything so that the cycles of nature can be dutifully located in periods of geological time (Archer 2004). The heterogeneity between the cyclic regime of temporality and the linear regime is dissolved in the powers of ten.

However, reconciling the cycles of nature with our immoderate fondness for timelines is only one aspect of the issue of durable technology. The two conflicting regimes of temporality highlighted by environmental activists give a partial view on the heterogeneity of times involved in technology. Their narrative does not question the prevailing metaphysical framework since it remains focused on the polarity between humans and nature. In order to implement more durable technologies, it is necessary to pay attention to *local timescapes* rather than relying on a global survey of timescales (Adam 1998). Shifting from a global to a local perspective, looking at the crowd of agencies that inhabit the world is key to realize that everyone has its immanent temporal regime. We have to combine Michel Serres's philosophy of things with Gilles Deleuze's neo-Bergsonian metaphysics of time.

Deleuze (1980) developed a critical analysis of the notion of a metric time, of an objective linear time divided up into identical instants within which events would take place without affecting their container. He suggested multiple overlapping flows of time without any transcendent time embracing the multiplicity of times and things. If we consider time as process rather than as a universal container, things are adequately described as a constant flux time, a process of becoming. Things – whether natural or artificial – tell their times, thus delineating contingent histories. And there is no synchrony between the individual times displayed in the world of things. Together they form a strange patchwork with entangled loops and twists, a "meshwork" with knots and loose ends rather than a network with nodes and connections (Deleuze and Guattari 1980; Ingold 2013). Their timelines are traces of movements and processes, similar to the lines left by waves on the waterfront or the traces left by moving slugs on the pavement.

Serres has paid a lot of attention to material things – to solids, fluids, gas and mixtures – and to their temporality.[6] While focusing on the material basis of society

[6] Like Arendt he emphasized the role of objects in society. "When we think society we need a good philosophy of objects." Objects provide stability to the social contract, thus regulating the pace of social history: "In fact, the object, specific to *Hominidae*, stabilizes our relationships, it slows down the time of our revolutions." (Serres 1982: 87)

and culture, Serres distinguishes things from objects, more precisely quasi-objects. Humans turn things into objects when they use them to settle social links and relations: quasi-objects, are relational, always passing from hand to hand like a rugby ball and thus creating a collective. Things by contrast belong to a more obscure face of the world. They are not reducible to their causes: "We can always go from the thing produced to its conditions but never from the latter to the former." (Serres 1977a, b: 115) In *Genesis*, Serres tries to understand the emergence of things in the world without assuming the pre-established scheme of linear causality. Things come into being out of noise and disorder like a vortex in a turbulent flow that would be stabilized by repetition and redundancy. For instance, imagine that in a few minutes you could run the film of the formation of the coast of Brittany over million of years, you would see a flame. Things are like frozen pictures taken in a long process. In this respect, they are sites of memory, while being always in becoming.

In addition, Serres questions the divide between things and signs, between the hard and the soft. All things carry the marks of time while they are ageing and worn out.

> Most objects are clocks [...] New harmful blow to human narcissism: everything in the world exchange information and store it. All bodies are engraved, like the Rosette stone or shits with graphics." (Serres 1977b/2000: 217)

Material things speak and write for those who read and listen to them as Serres does in *Biogée* (2003). While "reading" living bodies, Serres insists on the multiple times embedded in them. Living things carry the memory of thousand years of life on earth in their DNA and at the same time, each of them opens up indefinite potentials of evolution.

With such philosophical resources we can move beyond the standard view of one single timeline with a scale of different orders of magnitude. "Scalism" is not sufficient to disentangle the complexity and contingency resulting from the interference of various figures of times. Just as a landscape is a composition of heterogeneous elements, a *timescape* is a composite of heterogeneous regimes of time (Adam 2004). Aside the cycle and the timeline many other forms of time are to be found: vectors, spirals, trees, strata, rhizomes and meandering rivers.... In timescapes the boundaries between natural beings and artifacts are blurred. They are all composites of various materials – flesh and bones, matrix and fibers, etc. – each one with their individual lifetimes.

Not only time is immanent to the life of every being (West-Pavlov 2013), but each habitant of the world – from volcanoes to microbes – is a nest of different times, with internal clocks. For instance, bacteria have been favorite model organisms for laboratory research because of their reproduction rate (20mn in average). Today re-engineered bacteria are used as factories to synthesize drugs or even materials such as artificial spider silk. As we re-engineer these living beings for our manufacturing purposes we overlook that their capability to perform works of interest for us, are the outcome of a long and contingent history of several millions years. While we are messing up with their own temporalities, we neglect their collective capacity of enduring in the world through mutations and adaptations to the most extreme conditions by inventing new regimes of temporality. In instrumentalizing bacteria as laboratory models because of their fast reproduction rate, in re-engineering them as tools or factories we neglect their incredible evolutionary potentials in the mid and long term. We are so focused on our time, that we tend to believe that the process of evolution would be magically erased by our project of redesigning bacteria.

17.7 Coda: Ethics

Over the past decades ethics, has become a chaperon of all research projects with a view to promote "responsible research and innovation." Ethical committees are created in research institutions to assess the compatibility of innovations with core values. Social scientists introduced upstream have to stay on board in order to anticipate the ethical, legal and societal impacts of the future technological applications.

With the kind of ethics promoted as the mandatory partner of technological innovation there is no chance to break up the arrow of time. Quite the contrary, anticipatory exercises often turn out to colonize the future, by imposing on the future generations our present standards and values. Moreover, the calls for responsible innovation are still based on the modern ideal of man in control and mastery of nature, while in our view renouncing the central position of humans and the divide between persons and things is a precondition for trying to promote a more adequate ethics of technological innovation.

Being aware of the "polychrony" of things is a safe way for promoting "responsible" technologies. It is primarily a way of preventing risks and anticipating potential disasters without adopting a techno-optimistic attitude or a catastrophist perspective. No need to anticipate the disaster, to make the future present, in order to avoid it. Jean-Pierre Dupuy's "enlightened catastrophism" (2004) turns the arrow of time upside-down but sticks to a linear "monochronic" view.

However, polychrony does not call for an ethics of responsibility. It rather provides a sound base for an *ethics of attention* to the propensity of things. In considering them as archives of a long history, we could be more respectful of them and more concerned about their maintenance and about their future. Telling the life-story of technical objects is a way to considering them as co-actors or partners who share the world with us.

An ethics and aesthetics of immanent temporalities would acknowledge the primacy of the agency and existence of all entities as the forward-moving dynamic of time itself. Such an ethics and an aesthetics would radically displace humanity as the central actor in the natural global economy, inculcating a new respect for other beings and things as co-actants, thereby contributing to an alternative ecology and *oekonomy* (in the etymological sense of management of resources) of the global system. (West-Pavlov 2013, p.122).

As Deleuze argued in his lectures on Spinoza, ethics, in contrast to morality, is closely related to ontology. Rather than "judging" actions in the name of transcendent or universal values, ethics, just like ethology, is concerned with ways of being and behaving (Deleuze 1980). Far from assuming a qualitative difference of essence between persons, animals, and things ethics determines the capabilities of every being according to their potentials (*puissance*). Camels are capable of surviving a few days without drinking, diamond is capable of cutting glass. Ethics is concerned with what things can do rather than what we must do. Technology assessment requires that we turn our attention to the register of capabilities of all things and consequently to the various times immanent in them.

17.8 Conclusion

Based on empirical studies of a sample of recent technical objects and an acquaintance with the works of Simondon, Serres and Deleuze, this chapter advocates a deep revision of our metaphysics as a condition to promote more durable technologies. The mass production of disposable items like plastics has reduced artifacts to utilities and commodities meant for consumption in discrete instants detached from the continuity of duration. The mass production of radioactive nuclides in final nuclear waste confronts us with a long duration that far exceeds our potentials of imagination and anticipation. As an alternative to the hubris underlying the Promethean sociotechnical imaginary attached to the arrow of technological progress, the environmental paradigm invites to reconciling the arrow of time with the cycles of nature. Mimicking nature or at least designing technical objects more integrated in natural cycles of nature is indeed a necessary – and still desirable – step, but it is only a small step. In order to move forward, the very notion of one universal transcendental timeline embracing all things and events needs to be questioned. While this framework proved remarkably powerful to increase our scientific knowledge of the world, it is not adequate to technological knowledge and action. Paying attention to the multiple times embedded in things and to the interplays of regimes of temporalities in all technological project is a precondition for constructing a common and durable world, shared by all sorts of beings.

Acknowledgements A preliminary version of this text has been discussed by Astrid Schwarz and Jean-Paul Karsenty during the authors' workshop held in Paris Sorbonne on 22-23 June 2015 in view of the preparation of the collective book.

References

Adam, B. (1998). *Timescapes of modernity: The environment and invisible hazards*. London: Routledge.
Adam, B. (2004). *Time*. Cambridge, MA: Cambridge Polity Press.
ANDRA. (2014). *Le débat public Cigeo*. http://www.cigéo.com/le-debat-public. Accessed 31 July 2016.
Archer, D. (2004). *The global carbon cycle*. Princeton: Princeton University Press.
Arendt, H. (1958). *The human condition* (1998th ed.). Chicago: University of Chicago Press.
Barthe, Y. (2006). *Le pouvoir d'indécision. La mise en politique des déchets nucléaires*. Paris: Economica.
Barthes, R. (1957). *Mythologies*. Paris: Seuil. English edition: Barthes, R. (1972). *Mythologies* (A. Lavers, Trans.). New York: The Noonday Press.
Bensaude Vincent, B. (2013). Plastics, materials and dreams of dematerialization. In J. Gabrys, G. Hawkins, & M. Michael (Eds.), *Accumulation. The material politics of plastic* (pp. 17–29). London: Routledge.
Birth, K. K. (2012). *Objects of time. How things shape temporality*. New York: Palgrave McMillan.
Castells, M. (1996). *The rise of the network society*. Oxford: Routledge.
Cohen, T. (Ed.). (2012). *Telemorphosis: Theories in the era of climate change* (Vol. 1). Open Humanity Press: Ann Harbour.
Commoner, B. (1971). *The circle closing. Nature, man and technology*. New York: Alfred A. Knopf.
Corocan, P., Moore, C. J., & Jazvac, K. (2014). An anthropogenic marker horizon in the future rock record. *Geological Society of America Today, June, 2014*, 4–8.
Deleuze, G. (1980, December 9). *Cours à l'Université de Vincennes-Saint-Denis*. http://www2.univ-paris8.fr/deleuze/article.php3?id_article=137. Accessed 31 July 2016.
Deleuze, G., & Guattari, F. (1980). *Mille plateaux*. Paris: Editions de minuit. English edition: Deleuze, G., & Guattari, F. (1987). *A thousand Plateaux* (B. Massumi, Trans.). Minneapolis: Minnesota University Press.
Descartes, R. (1637). *Discours de la méthode*. In R. Descartes, *Œuvres* vol. I. Paris: Adam et Tannery. English edition: Descartes, R. (1999). *Discourse on method and mediations in first philosophy* (D. A. Cress, Trans.). Indianapolis/Cambridge: Hackett Publishing Co.
Descartes, R. (1641). *Principes de philosophie*. In R. Descartes, *Œuvres* (Vol. IX). Paris: Adam et Tannery.
Descartes, R. (1664). *Le Monde*. In R. Descartes, *Œuvres* (Vol. XI). Paris: Adam et Tannery.
Dupuy, J.-P. (2004). *Pour un catastrophisme éclairé*. Paris: Seuil.
Edgerton, D. (2007). *The shock of the old: Technology in Global History*. Oxford: Oxford University Press.
Gabrys, J. (2013). Plastics and the work of the biodegradable. In J. Gabrys, G. Hawkins, & M. Michael (Eds.), *Accumulation. The material politics of plastic* (pp. 208–227). London: Routledge.
Gabrys, J., Hawkins, G., & Michael, M. (Eds.). (2013). *Accumulation. The material politics of plastic*. London: Routledge.
Galison, P. (2010). Underground Future. In M. Mostafavi & G. Doherty (Eds.), *Ecological urbanism* (pp. 304–305). Baden: Lars Müller Publishers.
Harman, G. (2010). *Towards speculative realism: Essays and lectures*. New York: Zero Books.
Hecht, G. (1998). *The radiance of France, nuclear power and national identity after World War II* (2nd edition 2009). Cambridge, MA: MIT Press.
Hecht, G. (2012). *Being nuclear. Africans and the global uranium trade*. Cambridge, MA: MIT Press.
Henare, A., Holbraad, M., & Wastell, S. (Eds.). (2007). *Thinking through things. Theorising artifacts ethnographically*. New York: Routledge.

Hottois, G. (1984). *Le signe et la technique, La philosophie à l'épreuve de la technique*. Paris: Aubier.

Ingold, T. (2013). *Making anthropology, archeology, art and architecture*. London: Routledge.

Larrère, C., Larrère, R. (2015). *Penser et agir avec la nature*. Paris: Éditions La Découverte.

Latour, B. (1991). *Nous n'avons jamais été modernes. Essai d'anthropologie symétrique*. Paris: La Découverte. English edition: Latour, B. (1993). *We have never been modern* (C. Porter, Trans.). Cambridge, MA: Harvard University Press.

Latour, B. (1999). *Politiques de la nature; Comment faire entrer les sciences en démocratie*. Paris: La Découverte. English edition: B. Latour. (2004). *Politics of nature. How to bring the sciences into democracy* (C. Porter, Trans.). Cambridge, MA: Harvard University Press.

Latour, B. (2012). *Enquête sur les modes d'existence. Une anthropologie des Modernes*. Paris: La découverte. English edition: Latour, B. (2013). *An Inquiry into Modes of Existence. An Anthropology of the Moderns* (trans: Porter, C.). Cambridge, MA: Harvard University Press.

Latour, B. (2013). *Facing Gaia. A new inquiry in natural religion*. Gifford Lectures: Edinburgh. http://www.bruno-latour.fr/sites/default/files/downloads/GIFFORD-BROCHURE-1.pdf. Accessed 31 July 2016.

Loeve, S. (2015). La Loi de Moore. Enquête critique sur l'économie d'une promesse. In M. Audétat, G. Barazzetti, G. Dorthe, C. Joseph, A. Kaufmann, & D. Vinck (Eds.), *Sciences et technologies émergentes: Pourquoi tant de promesses ?* (pp. 91–113) Paris: Hermann.

Loeve, S., & Bensaude-Vincent, B. (2017). The multiple signatures of carbon. In B. BensaudeVincent, S. Loeve, A. Nordmann, & A. Schwarz (Eds.), *Research objects in their technological setting* (pp. 185–200). Abington: Routledge.

Lynch, M. (2008). *Ontography: Investigating the production of things, deflating ontology*. Paper presented at the *Oxford Ontologies Workshop*. Retrieved from http://www.sbs.ox.ac.uk/sites/default/files/Research_Areas/Science_And_Technology/Docs/Lynch.pdf

Mariott, J., & Minio-Palluello, M. (2013). Where does this stuff come from? Oil plastic and the distribution of violence. In J. Gabrys, G. Hawkins, & M. Michael (Eds.), *Accumulation. The material politics of plastic* (pp. 171–183). London: Routledge.

Meikle, J. L. (1995). *American plastic: a cultural history. Rutgers university press*. New Brunswick: Rutgers University Press.

Miller, D. (1998). *Material cultures. Why some things matter*. Chicago: The University of Chicago Press.

Nowotny, H. (1994). *Time. The modern and postmodern experience*. Cambridge: Polity Press.

Passet, R. (1992). *L'économique et le vivant*. Paris: Economica.

Poirot-Delpech, S. (2017). Nuclear waste: An untreatable technoscientific product. In B. Bensaude Vincent, S. Loeve, A. Nordmann, & A. Schwarz (Eds.), *Research objects in their technological setting* (pp. 232–244). Abington: Routledge.

Preda, A. (1999). The turn to things: Arguments for a sociological theory of things. *Sociological Quarterly, 40*(2), 347–366.

Proctor, N., & Schiebinger, L. (Eds.). (2008). *Agnotology. The making and unmaking of ignorance*. Stanford: Stanford University Press.

Raineau, L. (2012). Regard socio-anthropologique sur les choix énergétiques d'aujourd'hui: la question du nucléaire civil. In M.-H. Parizeau & J.-P. Pierron (Eds.), *Repenser la nature. Dialogue philosophique, Europe, Asie, Amériques* (pp. 431–449). Laval: Presses de l'Université de Laval.

Rosa, H. (2013). *Social acceleration: A new theory of modernity*. New York: Columbia University Press.

Serres, M. (1977a). *La distribution. Hermes IV*. Paris: Editions de minuit.

Serres, M. (1977b). *La naissance de la physique dans le texte de Lucrèce. Fleuves et turbulences*. Paris: Editions de minuit. English edition: Serres, M. (2000). *The birth of physics* (J. Hawkes, Trans.). New York: Clinamen Press.

Serres, M. (1982). *Genèse*. Paris: Grasset. English edition: Serres, M. (1995). *Genesis* (G. James, & J. Nielson, Trans.). Ann Arbor: The University of Michigan Press.

Serres, M. (1990). *Le contrat naturel*. Paris: Editions François Bourin. English edition: Serres, M. (2012). *The natural contract* (E. MacArthur, & W. Paulson, Trans.). Ann Arbor: The University of Michigan Press.

Serres, M. (1992). *Eclaircissements, Entretiens avec Bruno Latour*. Paris: Editions François Bourin. English edition: Serres, M., & Latour, B. (1995). *Conversations on science, culture and time* (R. Lapidus, Trans.). Ann Arbor: University of Michigan Press.

Serres, M, with Latour, B. (1995). *Conversations on science, culture, and time*. Ann Arbor: University of Muchigan Press.

Simondon, G. (1958). *Du Mode d'existence des objets techniques*. Paris: Aubier (2013 edition). English edition: Simondon, G. (2016). *On the mode of existence of technical objects* (C. Malaspina, Trans.). Minneapolis: Univocal Publishing.

Sklar, R. (1970). *The plastic age (1917–1930)*. New York: George Braziller.

Souriau, E. (1939). *L'instauration philosophique*. Paris: Félix Alcan.

Souriau, E. (1943/2009). *Les différents modes d'existence*. Paris: Presses universitaires de France.

Verbeek, P.-P. (2005). *What things do*. Philadelphia: The Pennsylvania State University Press.

Verbeek, P.-P. (2006). Materializing morality: Design ethics and technological mediation. *Science, Technology, and Human Values, 31*(3), 361–380.

Verbeek, P.-P. (2008). Design ethics and the morality of technological artifacts. In P. E. Vermaas, P. Kroes, A. Light, & S. A. Moore (Eds.), *Philosophy and design. From engineering to architecture* (pp. 91–100). Dordrecht: Springer.

Wajcmann, J. (2014). *Pressed for time: The acceleration of life in digital capitalism*. Chicago: University of Chicago Press.

West-Pavlov, R. (2013). *Temporalities*. London: Routledge.

Winner, L. (1986). Do artifacts have politics?. L Winner, *The whale and the reactor*. Chicago: The Chicago University Press. http://faculty.georgetown.edu/irvinem/theory/Winner-Artifacts-Politics.pdf. Accessed 31 July 2016.

Woolgar, S., & Cooper, G. (1999). Do artefacts have ambivalence? Moses' bridges, Winner's bridges and other urban legends in S&TS. *Social Studies of Science, 29*, 433–449.

Zalasiewicz, J., Waters, C., Ivar do Sul, J., et al. (2016, March). The geologic cycle of plastics and their use as a stratigraphic indicator of the Anthropocene. *Anthropocene Review, 13*, 4–17.

Chapter 18
Transcendental Imagination in a Thousand Points

Bernard Stiegler

Abstract Horkheimer and Adorno viewed the cultural industries of their times (cinema, television) as a technological externalization of what Kant names "schematism," the operation by which imagination unifies perceptual sensibility and conceptual understanding in the temporal flux of consciousness. For them, such an industrialization of imagination was the new barbarity. This chapter argues that the conditions of possibility of such technological exteriorization are the conditions of constitution of all consciousness, namely the existence, beyond the primary and secondary retentions analysed by Husserl (i.e. conservation and remembering), of "tertiary" retentions, that is, of a technical, prosthetic memory. Unwinding the thread of tertiary retentions, the argument flows back from cinema to Husserl's analysis of the consciousness of time, to the invention of the phonograph, and ultimately to a thorough discussion with Kant regarding the three syntheses of apperception viewed as a true "cinema of consciousness." In other words, if there is an "industrial schematism," it is because schemes, as functions of tertiary retentions, are originally industrializable.

This text was originally delivered as "L'imagination transcendantale en mille points," lecture at the high school Henri IV, Paris, 2002. The original translation by George Collins has been revised by the editors.

Keywords Critical theory · Cultural industries · Ecology of mind · Exteriorization · Heidegger · Husserl · Kant · Schematism · Temporality

B. Stiegler (✉)
Institute for Research and Innovation (IRI, Paris Center of Art and Culture Georges Pompidou), Paris, France

Ars Industralis, Paris, France

Connaissance, Organisation et Systèmes TECHniques (COSTECH), Université de technologie de Compiègne (UTC), Compiègne, France

Nanjing University, Nanjing, China
e-mail: bernard.stiegler@utc.fr

S. Loeve et al. (eds.), *French Philosophy of Technology*, Philosophy of Engineering and Technology 29,
https://doi.org/10.1007/978-3-319-89518-5_18

Horkheimer and Adorno saw in Hollywood cinema, joined to radio and magazines, the risk of an actual catastrophe for human spirit, caused by an apparatus of alienation whereby "automobiles, bombs and films hold the totality together," an esthetic barbarity "subordinating all branches of intellectual production equally to the single purpose of imposing on the senses of human beings, from the time they leave the factory in the evening to the time they clock on in the morning, the imprint of the work routine which they must sustain throughout the day." (Adorno and Horkheimer 1947: 95, 104)

How would these philosophers, who were still hardly imagining the impact of the recent invention of television, have described the life of the worker – or of the unemployed worker – who, in France spends today almost 4 h watching television? And how would they have reacted to the advent of digital networks? Given the irrefutable fact that in the not too distant future these networks will be affecting profoundly the whole spectrum of the mass media and especially television, incorportating it into a new system, they would undoubtedly have described it as a worldwide apparatus of "alienation" whereby television becomes tele-action in the wake of a tele-society marching unhindered toward the "market society" so preoccupying to European social-democrats.

*

A few years ago, I wrote the following on the irreducible materiality of the image:

> The image in general does not exist. What we call the mental image, along with what I am going to call here the image-object, something always found in a story and inscribed in a technical history, are two sides of the same phenomenon that can no more be separated from each other than the signified and the signifier, which used to define the linguistic sign. Jacques Derrida's critique of the opposition of these two concepts – the postulating of the signifier as a contingent variation of an ideal invariant – is beyond challenge. Just as there is no "transcendental signification," there is no mental image in general, no "transcendental imagery" existing prior to the image-object. There remains the question of the transcendental imagination, which I will not take up here. (Derrida and Stiegler 1996: 147)

Now is the proper time to address this question of transcendental imagination.

> The undeniable difference – which is not, however, the same thing as an antithesis – between mental image and image-object, means that they are always involved with one another, neither being able to diminish the other's inherent difference. The most immediately obvious difference is that what is objective is lasting, while what is mental is ephemeral. Likewise a memory-object is lasting… while a "mental" memory is remorselessly, rapidly, effaced: living, lived memory is fundamentally unstable and always leaves us in a lurch. Death is nothing other than the total wiping out of memory. (Derrida and Stiegler 1996: 148)

From the thesis positing retentional finitude as the principle of all philosophical analysis, I derived, in the first two volumes of *Technics and Time* (Stiegler 1994, 1996), the concepts of epiphylogenesis and tertiary memory. I set out to show that when Heidegger (1927a, b), in his critique of the Husserlian view of time, which nevertheless informs his own view, posits that "the being we ourselves are" is always an heir, is always preceded by a factical already-there, by a past it has not experienced and which therefore is not its own, but which must become its own past (for such is time – see *Being and Time*, §6), the consequence, which is not acknowledged in *Being and Time*, is that beyond the primary and secondary retentions analysed by Husserl, there have to be tertiary retentions, that is to say, technical traces able to make this factical past accessible to *Dasein* as a past which is not its own, which it has not lived, and which nevertheless must become its own, which it must inherit as its own history. Such is historiality (*Geschichtlichkeit*).

What I call tertiary memory, Heidegger always called *Weltgeschichtlichkeit*. He did not, however, allow it to belong to the originary sphere of "authentic" temporality, even though this question is at the heart of the Kantian mysteries surrounding the question of the transcendental imagination.

*

"Culture" and "spirit" only begin with the fact of technics. This point of view will have serious consequences during the critique I will undertake of Horkheimer and Adorno's concept of cultural industry.

Our two philosophers characterize this concept in reference to what Kant calls the schematism of the pure concepts of understanding. The philosophy of Kant distiguishes two sources without which there is no possible knowledge for the human subject: sensibility and understanding. Schematism is the operation of imagination that allows sensibility and understanding to unite, which is to say that, by the same token, allows for the unity of consciousness itself. Now, since cultural industries are industries of the imagination, Horkheimer and Adorno depict the industrialization of imagination as an industrial exteriorization of the power to schematize, and by the very fact, in terms of a reification, an alienating becoming-thing (*chosification*) of knowing consciousness:

> The active contribution which Kantian schematism still expected of subjects—that they should, from the first, relate sensuous multiplicity to fundamental concepts—is denied to the subject by industry. It purveys schematism as its first service to the customer. According to Kantian schematism, a secret mechanism within the psyche preformed immediate data to fit them into the system of pure reason. That secret has now been unraveled. (Adorno and Horkheimer 1947: 98)

The unifying function of the imagination would thus be, after a fashion, short-circuited, eliminated by an industrialization of culture literally numbing its customer-subjects, and alienating in the most radical ways the free subject of reason, which it would in fact enslave. Hence the general "marketing" of cultural commodities would necessarily be the unleashing of the most irrational elements of the society: the most irrational, the least cultural and the most unreasonable: the most barbarian.

Horkeimer and Adorno thus accuse cinema of paralyzing the imagination and more broadly the discernment of the spectator to such a degree that he or she is no longer capable of distinguishing perception and imagination, reality and fiction.

Their discourse at this point could readily be applied to the domain of virtual reality and electronic games:

> The more densely and completely its techniques duplicates empirical objects, the more easily it creates the illusion that the world outside is a seamless extension of the one which has been relevealed in the cinema. Since the abrupt introduction of the sound film, mechanical duplication has become entirely subservient to this objective. According to this tendency, life is to be made indistinguishable from the sound film. (Adorno and Horkheimer 1947: 99)

*

Granting this, there remains the obligation of explaining why and how consciousness can to this extent be intimately transpierced and controlled by the unwinding of a film, and what truth of consciousness and "real life" is then revealed through cinema. A film is a temporal object, in the Husserlian sense of the term. It is with a critical analysis of the Husserlian theory of time (Husserl 1928) that an account of the power of cinema over consciousness can be given.

The analysis of a melody as temporal object will enable us to understand the operation of the consciousness of this melody, to the extent that this consciousness is itself a temporal flux. Husserl discovers within this flux primary retention: in the "now" of a melody, in the present moment of a musical object that flows away, Husserl shows how the present note can be a note, and not just a sound or a noise, only inasmuch it retains in itself the preceding note which remains present in it, and which in turn has retained in itself the preceding one, and so on.

This primary retention, which belongs to the present of perception, must not be confused with secondary retention, which would be for example the melody I heard yesterday, and which I can hear again in imagination by the play of memory, and which thus constitutes the past of my consciousness. Husserl, before Adorno and Horkheimer, says perception and imagination are not to be confused.

And he is right. With this distinction between primary and secondary retention, Husserl makes a crucial discovery. But this distinction soon becomes an opposition: primary retention will have nothing to do with secondary retention. Now, it is obvious that the fact of having heard a melody, whose memory is conserved by secondary retention, modifies the conditions of the flowing away of this same melody on a second audition of the same interpretation. The primary retentions are in this case retained by consciousness following criteria of selection which obviously depend on the secondary retentions of temporal objets previously perceived by the same consciousness. Were this not the case I would always hear the same thing at each audition.

Husserl absolutely refuses to envisage a composition of primary and secondary retentions, and he will, therefore, oppose them. Secondary retentions stem from the imagination, while primary retentions are part and parcel of perception: were Husserl to admit a surdetermination of the latter by the former, he would have to admit that perception is always haunted by imagination and, in this sense, inhabited by fictional reality.

Husserl must rule out the possibility that perception be, only be cinematographic, and that the perceived is nothing other than the screen of this film.

He therefore excludes a fortiori from his analyses what I have called tertiary retention (Stiegler 1996), and in particular the phonogram.

Tertiary retention is this prosthesis of consciousness without which there would be no spirit, no haunting return, no memory of unlived or non-lived past, no culture. The phonogram is such a prosthesis, but of a remarkably singular kind, in that, as a recording of a trace in a object (here an analog recording), it in turn obviously overdetermines the articulation of primary and secondary retentions.

More generally, the technical history of tertiary memories (that is, in the final analysis, the history of the *Weltgeschichtlichkeit*) overdetermines the human history of primary and secondary memories. Thus, it is only from the technical possibility of the analog recording of a melody, dating from the invention by Thomas Edison and Charles Cros of the phonograph, that it is possible for the same consciousness (1) to listen twice to the same melody, in the same interpretation – to have twice over the experience of the same temporal object – and (2) to realize by the same token that the same temporal object repeated twice affords two different experiences and thus, by this very fact, to be able to state that the play of primary retentions, that is, the phenomenon itself, is each time a different one despite the identity of the object retained – that repetition, therefore, always yields up a difference.

The experience of such an identical repetition of a temporal object only become possible, for the first time in the history of humanity, with the invention of Cros and Edison: their phonograph profoundly transformed the interplay of memory, imagination and consciousness. This transformation continues with cinema, then with television and the *Kulturindustrie* in general – exteriorizing and reifying in the same stroke the work of a supposedly "transcendental" imagination.

How was that possible?

*

In the "Transcendental Analytic" of the *Critique of Pure Reason* (Kant 1781, 1787), Kant sorts out three different syntheses: of apprehension, reproduction, and recognition. I will now show how closely linked they are to primary, secondary and tertiary retentions, and how the role played by tertiary retentions (in the constitution of consciousness, a role unacknowledged by Kant) is responsible for the power of cultural industries to "schematize everything for their customers."

My basic thesis on the Transcendental Analytic comprises two complementary arguments:

1. The passage from the 1781 edition, version A, to that of 1787, version B, witnesses to Kant's failure to articulate the three syntheses of imagination set out in A, and repeated in B, onto the transcendental unity of apperception (with imagination slipped into a subsidiary position, and the understanding regaining its absolute authority).

2. What is neither thought out nor clearly expressed in A (no more so than in B, but in the later case the problem is solved by regressing from the level of A so as to eliminate the contradiction), is the difference between primary and secondary retentions (later delineated by Husserl). Kant systematically confuses them as syntheses of apprehension and reproduction.

The "spontaneity of the understanding" is the principle of a "triple synthesis":

> For knowledge is [essentially] a whole in which representations stand compared and connected. As sense contains a manifold in its intuition, I ascribe to it a synopsis. But to such synopsis a synthesis must always correspond; receptivity can make knowledge possible only when combined with spontaneity. Now this spontaneity is the ground of a threefold synthesis which must necessarily be found in all knowledge; namely, the *apprehension* of representations as modifications of the mind in intuition, their *reproduction* in imagination, and their *recognition* in a concept. These point to three subjective sources of knowledge which make possible the understanding itself – and consequently all experience as its empirical product. (Kant 1781: A 97)

The question of this threefold synthesis, and in the first place of the first synthesis, that of apprehension, is the question of time: if the manifold of intuition is to be ordered, it is because our representations "belong to inner sense," "all our knowledge is thus finally subject to time, the formal condition of inner sense. In it they must all be ordered, connected, and brought into relation." This is why, within the manifold of intuition "the mind distinguishes the time in the sequence of one impression upon another." (Kant 1781: A 99) This distinction of temporal succession at the heart of the intuition of any phenomena whatsoever is the work and the accomplishment of the synthesis of apprehension.

Then Kant specifies the definition of the synthesis of reproduction:

> Representations which have often followed or accompanied one another finally become associated, and so are set in a relation whereby, even in the absence of the object, one of these representations can, in accordance with a fixed rule, bring about a transition of the mind to the other. (Kant 1781: A 100)

What Kant is here describing is what Husserl analyzes as secondary retention. Now, it is at the end of the next paragraph that the problem arises: Kant confounds this capacity of reproduction with that of primary retention. As a result, he must posit that the synthesis of reproduction is retention in apprehension itself:

> But if I were always to drop out of thought the preceding representations (the first parts of the line, the antecedent parts of the time period, or the units in the order represented), and did not reproduce them while advancing to those that follow, a complete representation would never be obtained: none of the above-mentioned thoughts, not even the purest and most elementary representations of space and time could arise. (Kant 1781: A 102)

In other words, Kant makes that very mistake Husserl will accuse Brentano of making. Kant is obviously referring to primary retentions whereas he believes he is describing the synthesis of reproduction as that which would render apprehension possible: "the synthesis of apprehension is thus inseparably bound up with the synthesis of reproduction." (Kant 1781: A 102) Kant thinks he's describing the synthesis of reproduction exactly when it is apprehension at stake, as the phenomenon of primary retentions which quite precisely must not be confused with secondary retentions constituting the essence of the synthesis of reproduction.

Kant confuses two forms of retention, that is, two forms of synthesis. Now, it is precisely this confusion, occuring in 1781, that makes the exposition so unclear, obliging him to rework the transcendental deduction. Indeed, it is not clear exactly what is meant, in 1781, by the synthesis of apprehension. What else can represent the obligation that the manifold "must first be run through" (Kant 1781: A 99) (for this is yet another definition of the synthesis of apprehension) if not the retention of that which is run through by that which is now running through the manifold? The only way Kant could clearly distinguish between this and a secondary retention (a reproduction in the "absence of the object," which he also defines as the synthesis of reproduction) would be if he could consider the first synthesis as the conservation of the already past in the still present and as the protention of that which is still to come.

Now intervenes the third synthesis, that of "recognition":

> For in so far as [our modes of knowledge] are to relate to an object, they must necessarily agree with one another, that is, must possess that unity which constitutes the concept of an object. (Kant 1781: A 105)

The synthesis of recognition constitutes and maintains the coherence of consciousness with itself given that is a flux, and a flux whose unity is to be guaranteed: this flux cannot contradict itself. This unification of the flux overdetermines, as synthesis of recognition, the unification of the syntheses of apprehension and reproduction through which an object may present itself unified to consciousness by the simple fact of the recognitional unification of the flux of consciousnesss:

> It is clear that (…) the unity which the object makes necessary can be nothing else than the formal unity of consciousness in the synthesis of the manifold of representations. It is only when we have thus produced synthetic unity in the manifold of intuition that we are in a position to say that we know the object. (Kant 1781: A 105)

This unification of consciousness with itself, through its objects, is the projection of apperception that Kant calls transcendental in so far as it corresponds to the *a priori* necessity expressed by a rule (a concept):

> But this unity is impossible if the intuition cannot be generated in accordance with a rule by means of such a function of synthesis as makes the reproduction of the manifold *a priori* necessary, and renders possible a concept in which it is united (…). This *unity of rule* determines all the manifold, and limits it to conditions which make unity of apperception possible (…). But [a concept] can be a rule for intuitions only in so far as it represents in any given appearances the necessary reproduction of their manifold, and thereby the synthetic unity in our consciousness of them. (Kant 1781: A 105–106)

In short, the transcendental unity of consciousness is also that of its objects, and therefore, that of the unity of the world in general. The concept unifies the diversity of what is re-produced in the empirical realm as its essence and its necessity, but only in so far as the latter are pro-duced by the concept. From out of the re-production of the past manifold, the concept abstracts its still-to-come unity: re-production is, at a deeper level, pro-duction, because the concept implements the *a priori* legality of the temporal flux constituted by the categories. This is the way in which this recogitional unification, which is also that of the very flux of consciousness, prepares, on the basis of a reproducible past, the unitary future of the flux and the objects constituted therein.

But why affirm at this point the necessity of what I have called tertiary memory? Are Kant's three syntheses indeed the translation, in the total phenomenon of consciousness, of the three retentional forms, and of that which necessarily bonds them into one? I will now address these questions.

*

The first two syntheses owe their unity to the third synthesis of "recognition," which inserts the first two forms of synthesis (i.e. the first two retentional forms), into the unified flux of consciousness, a unity that Kant calls the unity of apperception. In other words, the role of the third synthesis is to ensure the internal compatibility of all the primary and secondary retentions woven into the fabric of a consciousness that is always the same consciousness, whatever the diversity of the primary and secondary streaming through it might be, giving it form and capacity of becoming. The third synthesis is what assembles and edits the first two (which in some respects are equivalent to cinamatic "rushes" and "inserts") into a single unified temporal flux. This process is constitutive of what may be called the cinema of consciousness, projecting – being already pro-tended – toward the future.

Now can anyone avoid noticing that Kant's own flux of consciousness – which of course he uses as the object and model of the activity of all consciousness – unfolds and constitutes itself and its unity in the course of his writing of those works that make up his *oeuvre*?

How then not to notice (1) that this unity is not given, but promised, (2) that the force of Kant's work lies in the unification of materialized elements of consciousness constituting the work's literate tertiary retentions, and (3) that "Kant" is only the name of the work's author, and thus only interests us in this respect? Such a situation, from which Kant's authority proceeds, is possible only because the imagination's primary and secondary syntheses are essentially synthesizable by this synthetic flux (of consciousness) that make up an "objective" memory, such as a book – or a film.

There are two versions of the *Critique of Pure Reason* – two write-ups, that is, two archivals/syntheses of the history of the consciousness of Kant himself, and through it, of the history of philosophical consciousness – the first of which is published in 1781, and the second in 1787, each with a distinct preface, and significant modifications of the "Transcendental Analytic" taking place in the second edition, especially in the section "Transcendental Deduction of the Concepts of the Understanding." Now, what does the second preface tell us about what to make of the first write-up modified by this second edition/write-up? Roughly summerized, it tells us that the second edition changes nothing in the first, except that the former is more explicit, and that consequently, the first edition remains completely valid despite the differences in write-ups. More precisely, the preface tells us that this second edition has tried out some "corrections":

> These improvements involve, however, a small loss, not to be prevented save my making the book too voluminous, namely, that I have has to omit or abridge certain passages, which, though not indeed essential to the completeness of the whole, may yet be missed by many readers as otherwise helpful. (Kant 1787: B xlii)

These remarks are quite surprising considering that the second edition appears on some points in deep contradiction with the first. Now, these contradictions, concerning the role of the third synthesis and that of the imagination, are precisely the sign of a difficulty Kant has with solving the problem of contradiction – the contradiction of the self with itself, which is the very temporality of the self, what Deleuze called its "chasm." Yet the preface to the second edition tranquilly rolls on, explaining that

> [The] now more intelligible exposition, which, though altering absolutely nothing in the fundamentals of the propositions put forward or even in their proofs, yet here and there departs so far from the previous method of treatment, that mere interpolations could not be made to suffice. (Kant 1787: B xlii)

The two editions thus do differ considerably "here and there," but this is a simply formal discrepancy: nothing fundamental is affected. This seems so true that Kant's editors published both editions in one, as is still done today in most modern translations, and Kant suggests as much:

> This loss, which is small and can be remedied by consulting the first edition, will, I hope, be compensated by the greater clearness of the new text. (Kant 1787: B xlii)

In short, there are deep, serious contradictions between the 1781 and 1787 editions, but Kant insists on maintaining at all costs the unity of the flux of his own consciousness over this period of time, during which he have aged 6 years. What else happened than the simple passage of time? It happened that during this passing of time, events took place, and especially a public critique of the *Critique*, which would oblige Kant to rework it, that is, to rewrite the history of his own flow of consciousness "before the reading public." (Kant 1784)

Consciousness can only become self-consciousness providing it is able to exteriorize itself, to become objective in the form of traces whereby it becomes accessible to other consciousnesses. Although Kant, no more so than Husserl, resorts to some sort of tertiary retention, it is obvious that the literal recording of Kant's flux of consciousness, in so far as it leads to the writing of the *Critique of Pure Reason*, is the essential condition of the analysis of the activity of any consciousness this work aims to be. The thought of Kant can only present itself to us in the form of a book – and to himself as well, except that in his case the presentation developed in the very course of its writing, i.e. its editing, and before him: on the screen of the sheet of paper qua support of his thought, a true crutch of the understanding.

This is why, in 1996, I placed the following sentence from Kant (1784) as the epigraph to my work *Technics and Time, 2. Disorientation* (Stiegler 1996): "By the public use of one's reason I understand the use which a person makes of it as a scholar before the reading public," that is, obviously, to the extent that one writes oneself. We know that Kant never wrote a line by chance: he cannot set and identify the unity of apperception of the consciousness he is otherwise than through the possibility of inscribing, conserving and ordering the primary and secondary retentions (that is, the syntheses of apprehension and reproduction) effected by the imagination of his forgetful consciousness (whose memory is finite), in the form of tertiary retentions (the written sentences whereby the *Critique of Pure Reason* takes shape). Inscribed, set down, conserved, these sentences can be reread, criticized, analyzed, objectified, selected, and put together in new combinations. Conserving, discerning, comparing and finally editing, in the unity of a book which is also the unity of his thought, here is what Kant can do with these sentences as they are the objective materializations of his primary and secondary retentions, which can thus become manipulable.

From 1781 to 1787, Kant has all the time he needs to re-examine the past flux of his own consciousness, and to seek the persevering unity of the flux of his consciousness to come, to the extent that he was able to set down, identify and unite the manifold of his thoughts by materializing them. He thus becomes the object of himself, and can thus become the object of a re-flexive critique in which he self-affects: he can thus and only thus proceed to the examination of the conditions of his own possibility wich are also the conditions of possibility of all his objects. These are the very conditions whose "most extreme possibility" Heidegger (1927b) would accuse Kant of neglecting.

Critique analyses and synthesizes. But this is only possible because critique can manipulate, and in this case, manipulate time, that is, the play of primary and secondary retentions through their tertiary materializations.

Now these critical materializations of inner sense are just a susceptible of being manipulated by the industry of culture given that for it the consciousness of people is its raw material, objectifiable and reifiable because originarily outside of itself. And this is indeed why a critique is again necessary.

In other words, Kant can and must write that all phenomena are in me, that is, "are determinations of my identical self, (...) only another way of saying that there must be a complete unity of them in one and the same apperception." (Kant 1781: A 129) The self is in the middle of "itself," that is, in the middle of its objects and prostheses, a milieu which, by the same token, is not only itself, but also its other. And this other preceeds it, it is an already-there, a past the self has not experienced, and which can become its past only by becoming its future. This sructure of prosthetic precedence, grounding the possibility of tertiary memory and memories, is the projective support of consciousness. It is what allows a consciousnesses to inherit the past of all those who have preceded it – just like ourselves here as the reading public of Kant's works –, it is also what allows for the projection of a future. This is what we shall explore now by studying the question of schematism, which will bring up the version B of the "Transcendental Deduction."

*

In order for intuitions to be subsumed under concepts,

> There must be some third thing, which is homogenous on the one hand with the category, and on the other hand with the appearance (...). Such a representation is the *transcendental schema*. (...) Pure *a priori* concepts (...) must contain *a priori* certain formal conditions of

> sensibility (…) which constitute the universal condition under which alone the category can be applied to any object. (Kant 1787: B 177–179)

during the synthesis of recognition.

> This condition (…) is the *schema* of the concept (…) Since, however, the synthesis of imagination aims at no special intuition, but only at unity in the determination of sensibility, the schema has to be distinguished from the image. If five points be set alongside one another, thus, … , I have an image of the number five. (Kant 1787: B 179)

Just like 5 or V or 101 in the binary system used by computers, this image is empirical and therefore contingent (since the number can undifferentially be represented by all of these diverse images) and is thus entirely different from the fact of thinking of the same number. Such a thought is

> the representation of a method whereby a multiplicity for instance a thousand, may be represented in an image in conformity with a certain concept, rather than the image itself. For with such a number as a thousand the image can hardly be surveyed and compared with the concept. (Kant 1787: B 179)

Indeed, it is not blatantly obvious that in the following figure there are a thousand (1000) points:

Although the Greek for this figure is *skhema*, it is not a schema in the Kantian sense, but only an image. But then why would the geometric figure, originally image and trace (*graphein*) be called, in Greek, *skhema*? And to what extent is a number like a thousand possible, as a method "in conformity with a certain concept" for the consciousness of which it is the object, without an image?

The answer is straightforwardly, to NO extent whatsoever. A number always in some way presupposes a capability for tertiary retention (whether *via* the fingers of a counting child, the body of a magician, an abacus, or an alphanumerical system of writing) which alone can allow numbering and objectifying. This capability has a history, in the course of which it one day becomes possible to conceive of (and to conceive as a process) the number 1000. There was a time, quite recent compared to the long history of humanity, when the number 1000 was literally inconceivable to the consciousness of mankind, not yet equipped to think it.

Kant can only speak of the number 1000 (or of the figure/image above) because he has at his disposition technical and materialized systems of notation which allow for the manipulation of symbols and to set down by means of this image (from which results this word: "thousand" which is itself an – acoustic – image) the result of an operation of the understanding that passes through a joint operation of internal and external sense.

The syncronization of internal and external sense here conditions the activity of the understanding, which by the same token is submitted to the passive synthesis of its "tools." In fact, number in general can only be conceived providing it is figured in a system of traces called a system of numeration, which always refers to a gesture consisting itself in a manipulation of symbols whose nature is external – and there can be no mental arithmetic that does not result from the secondary interiorization of a calculation by symbolic manipulations, that is, by manual behavior.

French mathematician (1895–1982) Geneviève Guitel writes:

> One never grows tired of evoking the first humans as they began to count, as they awkwardly used stick to draw figures in the sand of beaches and deserts. (…) We can also practice the art of putting regular notches in a piece of wood (…) to help preserve the memory of a number.
>
> All of these material translations of numbers work through the correspondence principle, (…) but (…) things look very different if for each sheep in the herd we put a clay ball in some receptacle (…) or if we make some expressive gesture, using our body as a machine to indicate the number of fish we have caught.
>
> In the first case we have an abstract image of the herd of sheep: one clay ball per sheep. There is no need to know how to name those first numbers: we can make the correspondence silently. (…) The receptacle for the clay balls can be put into storage by the supervisor, but the supervisor knew how to count, whereas the shepherd was quite incapable of it. (Guitel 1975: 19–20)

The very conception of a number results in the enacting of gestures through the correspondence principle. These gestures allow for the production of an image which is certainly abstract, but which is an abstract image-*object*, the support and condition for the projection of a *mental* abstract image. The conception of the number one thousand presupposes written enumerations, a stage of abstraction emanating from the manipulations of symboles called "written numeration of positionality," whose schema clearly presupposes the image, even when the very possibility of the image would reciprocally presuppose the possibility of the schema – that of schematization in the sense that Kant strives to give it, that is, as a process of projection of inner sense into the tertiary memories, that is, into the images accessible to external sense.

Simondon calls such reciprocity a transductive relation (Simondon 2005) in his critique of the hylomorphism in which he considers that much of Kantian thought remains entrapped. If one should distinguish the schema from the image, it remains that there can be no manifestation of the schema without image, whether mental or not. When Kant, proposing an image for "five," traces five points on a line, thus inserting the sketch "......" into a sentence, he unfortunately forgets that the word five is already an image, hailing from a long history.

In short when it comes to transcendental imagination there can be no mental image without an image-object. In an image like that of a herd of sheep, seeen as an abstract representation materially constituted in a heap of clay balls, the first numbers as abstract entities are in fact very concrete memory supports: the flux of consciousness – in which number constitutes a determination of internal sense, suceeded by unities forming a numberable and synthesizable totality in the unity of apperception – is retentionally finite. As the memory of its own unfolding, it dissipates rapidly and must rely on external supports, on prostheses of memory which are also the fetishes of imagination and the projection screens of all its phantasms. These retentional prostheses thus bring to the flux of consciousness (i.e. to consciousness itself, for it is only flux) the spatial intuitions of the unflolding of its temporal intuitions. This is why *ars memoria* are possible.

These spatial intuitions have the advantage of being retained "objectively," allowing compressing or abbreviating of the flux: a cursive reading of the "number" depicted by the dots above is possible, but such a reading would be long and always subjected to error, whereas in the written number 1000, an image has been substituted for the operation of the unfolding of time; it has been abstracted from this unfolding to become its equivalent, once consciousness has engaged in long series of exercises – to begin with the one consisting of counting with one's body (using fingers, *digits*).

It is this general equivalence whereby time gives way to spatial figure that allows for what Marx calls the "general equivalent": capital, as currency accumulating an abstract value because of its manipulability. Money is time. The tertiary memory, of which money is the most abstract form, allowing for abstraction based on the principle of correspondence, opens by the same token the systematic exploitation of temporal operations as system of spatial equivalences: from fallible unfoldings of the flux of consciousness, enumerations become images of numbers.

*

> But if I were always to drop out of thought the preceding representations (the first parts of the line, the antecedent parts of the time period, or the units in the order represented), and did not reproduce them while advancing to those that follow, a complete representation would never be obtained: none of the above-mentioned thoughts, not even the purest and most elementary representations of space and time, could arise. (Kant 1781: A 102)

This is Kant describing primary retention, but he already believes himself to be within the synthesis of reproduction; he therefore does not see the secondary retention, since it is not exactly the same as the primary. I have shown why the retentional finitude of the flux of consciousness brings about the necessity of a third form of retention, whose consequence is the following: if the "*figurative synthesis*," the "*synthesis speciosa*," (Kant 1781: B 151) becomes in the 1787 edition the true synthesis of productive (not only reproductive) imagination – that is, transcendental imagination –; if therefore this synthesis enables to mentally draw a line to construct space, then this faculty, which is also the principle of geometric construction, could never do without drawing the line in space: without the hand.

Thales, whose revelatory experience is convoked in the 1787 preface, could never reason geometrically without gestures figuring pure space, gestures which are the a priori conditions of empirical space within this space itself. If Thales constructs the figure, and does not just follow it, he constructs a figure without which there would be no concept. The construction of the concept is that of the figure and conversely. The concept is, to be sure, accompanied by a discourse, but the discourse is itself inscribed literally: it must be just as fixed as the figure and must preserve, in sensible space, the trace of a line of reasoning regarding pure space (i.e. regarding the *a priori* conditions of possibility of intuition). Here, as for numeration, there can be no possible thought without figurations that are themselves traces, gestures of thought as thought must support itself through its inscriptions in space, inscriptions which should in turn manifest in the intuition of the empirical given a pure intuition of the formal conditions of this empirical intuition—these are, as underscored above, the crutches of the understanding, and not only of belief.

Pro-duction is a figuration and the second edition defines it as a figured synthesis (*synthesis speciosa*). If the figure were not essential here, not to say *the* essential here, then why qualify this synthesis as "figured"? Why translate *speciosa* by *figürliche*? "To figure, to give figure to" is the meaning of the verb *skhematizô*. We are examining the question of the conditions of the constitution of the schema and the role played in it by the image. Kant posits that the schema precedes the image: I hold that they are co-emergent, in other words that they are in a transductive relation. Image and schema are the two facets of an identical reality constitutive of a historical process conditioned by what I have called "epiphylogenesis" (Stiegler 1994: 183): the general system of tertiary retentions forming the milieu of the consciousness, its world as the spatialization of the time of past and passing consciousnesses qua *Weltgeschichtlichkeit*.

Heidegger (1927b) tells us that in the first version of the "Transcendental Deduction" the third synthesis is that of the future, that is, of the project. But this cannot be understood unless it is granted that the facticity of the past of *Dasein* is what opens the very possibility of its future (in "the most extreme possibility") because this past is originarily tertiarized, that is, synthesizable as a prosthesis. A true critique of the *Critique of Pure Reason*, a "new critique" if you will, must confront this question of originary exteriorization. This is what makes inheritance possible. Heidegger sees in all linear conceptions of time, dominant in both versions of the *Critique of Pure Reason*, the typical expression of metaphysics in general. But he does not see that the real problem is here with the third synthesis in so far as it presupposes exteriorization (*Weltgeschichtlichkeit*). The question of primary exteriorization, and of the projectivity resulting from it at the level of the third synthesis, is originarily linked to that of the protentional incompleteness and inadequation of the flux with itself, whitin which the edge of a judgement and the risk of a decision can be found. It is because there is inadequation of consciousness to itself, "differance" in the Derridean sense, that there can be a process of individuation in the Simondonian sense. This inadequation occurs in the situation of incompleteness in "the most extreme possibility."

Given its projective character, the synthesis of recognition, holding together as it were, the past and the future of the flux, trying to make them compatible, if not sheerly adequate (barring which there would be no this opening up that is the future), encloses in itself the whole issue of the inadequation of the flux of Kant's consciousness with itself. This is why in the second edition, this projective apparatus is transformed into a simple agent of the understanding. A secret agent, hidden in the depths of the soul.

When adequation becomes effective, there is no longer incompleteness or inadequation. The individual then bequeathes this "completed inadequation," so to speak, to the posterity of his or her heirs, in the form of tertiary memories. Through this tertiary inadequation, heirs attempt to be open to a future framed by the unfulfillment of all things ceaselessly attemting to find completion… through prostheses: the question of unfulfillment is precisely the question of prostheticity – always in the promise, however, of an absolute future.

*

The digital integration of cultural industries through the convergence of information, audiovisual and telecommunication technologies is the new framework of production and broadcast of "tertiary retentions," and a new milieu for the spirit. In the course of the twentieth century, the milieu of the spirit has become that of an industrial exploitation of the times of consciousness. It is not a question of a monstrous evolution whereby "schematism" would all of a sudden jump outside of consciousness: consciousness was never self-consciousness otherwise than by projecting itself outside itself. But in this era of information industries, and the analog and digital technologies making it possible, this exteriorized and materialized consciousness becomes the material for manipulations of flux and for projections of masses in such ways that a pure and simple annulment of "self consciousness" by its exteriorization not only becomes possible but would appear highly probable: this is what is to be thought in the homogenizing synchronization of the fluxes of consciousness by audiovisual temporal objects.

This synchronization is also responsible for the manipulation of consciousnesses in the era of audiovisual, mass industrial temporal objects. The critique of this manipulation, in other words, cannot be a denunciation of a denaturing of consciousness by cinema, but on the contrary the demonstration that consciousness functions as a cinema – which allows cinema and television to maintain their control. Consequently, the critique of cinema and television as social phenomena capable of destroying consciousness itself (this is what I call the question of an ecology of mind) calls for a new critique of consciousness itself, a redeployment of the Kantian endeavor.

The "general equivalent" as the condition of the market in which, with the cultural industries, the time of consciousnesses has itself become a commodity, is conditioned by the general equivalence of primary-secondary time within its tertiary spatializations, which can be manipulated, stocked, exchanged, and commodified.

In the industrial becoming of culture, consciousness itself is up for sale. This can always be decried as barbarian degeneracy, a monstrous state of affairs: but it is only the strict consequence of the finitude of fluxes of consciousness in general, of their originary prostheticity. Any struggle against this possibility implies its acknowledgement, that is, implies the conclusion of the preceding analyses: THERE IS NO SPIRIT WITHOUT AN OBJECTIVE RETENTIONAL MILIEU; THE HISTORY OF THIS MILIEU IS ALSO A HISTORY OF TECHNICS, THAT IS, TODAY, A HISTORY OF INDUSTRY.

The future of the spirit can only consist in a geopolitics of cultural technologies that would also be an ecology of the mind. A politics of consciousness (but what else is politics if not, preeminently, a politics of consciousness?) is necessarily a politics of technics.

References

Adorno, T., & Horkheimer, M. (1947). *Dialektic der Aufklärung*. Amsterdam: Querido Verlag. English edition: Adorno, T., & Horkheimer, M. (2002). *Dialectic of enlightment* (E. Jephcott, Trans.). Stanford: Stanford University Press.

Derrida, J., & Stiegler, B. (1996). *Echographies de la télévision: entretiens filmés*. Paris: Galilée. English edition: Derrida, J., & Stiegler, B. (2002). *Echographies of television: Filmed interviews* (J. Bajorek, Trans.). Cambridge, MA: Polity Press.

Guitel, G. (1975). *Histoire comparée des numérations écrites*. Paris: Flammarion.

Heidegger, M. (1927a). *Sein und Zeit*. Tübingen: Max Niemeyer Verlag. English edition: Heidegger, M. (2010). *Being and time* (J. Stambaugh, Trans.). Albany: Suny Press.

Heidegger, M. (1927b). *Kant und das Problem der Metaphysik*. Frankfurt am Main: F.-W. von Hermann. English edition: Heidegger, M. (1997). *Kant and the problem of metaphysics* (R. Taft, Trans.). Indianapolis/Bloomington: Indiana University Press.

Husserl, E. (1928). *Zur Phänomenologie des inneren Zeitbewußtseins (1893–1917)*. R. Boehm (Ed.). The Hague: Martinus Nijhoff. English edition: Husserl, E. (1991). *On the phenomenology of the consciousness of internal time (1893–1917)* (J. Barnett Brough, Trans.). Dordrecht: Kluwer Academic Publishers.

Kant, I. (1781 und 1787). *Critik der Reinen Vernunft*. Riga: Hartknoch. English edition: Kant, I. (1929). *Critique of pure reason* (2nd ed., N. Kemp Smith, Trans.). London: Macmillan and Co.

Kant, I. (1784). Beantwortung der Frage: Was ist Aufklärung? *Berlinische Monatsschrift, 12*(1), 481–494. English edition: An answer to the question: What is enlightment? (H. B. Nisbet, Trans.). In H. Reiss (1970). *Kant's political writings*. Cambridge, MA: Cambridge University Press.

Simondon, G. (2005). *L'Individuation à la lumière des notions de forme et d'information*. Grenoble: Jérôme Million.

Stiegler, B. (1994). *La technique et le temps, 1. La faute d'Épiméthée*. Paris: Galilée. English edition: Stiegler, B. (1998). *Technics and time, 1. The fault of epimetheus* (G. Collins, & R. Beardsworth, Trans.). Cambridge, MA: Stanford University Press.

Stiegler, B. (1996). *La technique et le temps, 2. La désorientation*. Paris: Galilée. English edition: Stiegler, B. (2009). *Technics and time, 2. Disorientation* (S. Barker, Trans.). Stanford: Stanford University Press.

Stiegler, B. (2001). *La technique et le temps, 3. Le temps du cinéma et la question du mal-être*. Paris: Galilée. English edition: Stiegler, B. (2010). *Technics and time, 3. Cinematic time and the question of malaise* (S. Barker, Trans.). Stanford: Stanford University Press.

Part IV
Innovating in Ethics, Design and Aesthetics

Chapter 19
Ethics of Technology in France

Jean-Yves Goffi

Abstract French ethics of technology has followed a most peculiar path. Generally speaking, French philosophers of technology have dismissed the methodology of Applied Ethics as widely practiced. As a result, the most notorious thinkers have been those who have expressed the most considerable reservations about the "technical phenomenon." I here attempt to establish that the most promising attempts to overcome this situation do not rise from social philosophy, but from a long-standing French tradition in the field of philosophy of life, supported by an overriding project of integration of technology into culture.

Keywords Ethics · Ethics of technology · Technology out of control · Social philosophy · Philosophy of life · Applied ethics · Technoscience(s)

In order to clarify what is meant by "ethics" the following example will be helpful (Mitcham and Duval 2000: 24). The Ford Pinto, an inexpensive US subcompact car, hastily designed in the late 1960s complied with the National Highway Travel Safety Administration (NHTSA) standards of the period but was plagued by bad design: in case of a rear-end collision, the gas tank was prone to explode, killing or severely burning the driver and the passengers. The number of fatalities due to the faulty design is disputed, but the fact remains that several law suits were filed against Ford Motor Company. In the course of one of these law suits (Grimshaw *vs.* Ford Motor Co.) a so-called "Pinto Memo" was circulated among Ford's senior management and a stolen copy was disclosed to the public. The document indicated (on the basis of a cost/benefits analysis) that it would cost more to recall the defective cars and reengineer the gas tank than to pay damages. Ford replied that the "Pinto Memo" did not reflect the company's policy and was only prepared for the NHTSA in the context of an overall study of collision effects. Nevertheless, the popular

J.-Y. Goffi (✉)
Université Grenoble Alpes, PPL, Grenoble, France

S. Loeve et al. (eds.), *French Philosophy of Technology*, Philosophy of Engineering and Technology 29,
https://doi.org/10.1007/978-3-319-89518-5_19

understanding at the time was that the document betrayed a deep contempt for ethical considerations: it was aimed at promoting what was to the best advantage of the company rather than what was right. In view of this, I will define morality as a normative rather than a merely expedient set of constraints on conduct, in both public policy and private matters, and ethics as a principled justification of these constraints.

According to Carl Mitcham, two traditions in the philosophy of technology can be distinguished: "Engineering philosophy of technology which emphasizes analyzing the internal structure or nature of technology, and humanities philosophy of technology, which is more concerned with external relations and the meaning of technology." (Mitcham 1994: IX) Philip Brey ventured a different dichotomy, based on the expressions "classical philosophy of technology" and "contemporary philosophy of technology." (Brey 2010) I have distinguished three major trends which overlap partially with those of Mitcham and Brey: (i) the attempt to build an ontology or a phenomenology of technology; (ii) the attempt to articulate the study of technology with anthropology; (iii) the attempt to evaluate technologies (or technology as a whole) (Goffi 1988). I will not try to determine which of the three formulations is to be preferred; each one has its own strengths and weaknesses. In this chapter I want to show how these different approaches to the philosophy of technology have played out in France, by stressing two major features:

(a) The ethics of technology in the French context has been shaped by a strong tension between two styles of philosophy: the humanities/classical/anthropological-evaluative style on the one hand and the engineering/contemporary/ontological-phenomenological philosophy of technology on the other hand.
(b) These tensions have been eased in an original way by philosophers who, following Bergson, have articulated philosophy of technology with philosophy of life. Whether technology is threatening life (understood as biological, spiritual, social, psychological or whatsoever) or is an expression of life is perhaps the great unifying question in the debate.[1]

19.1 Historical Background

Although it is often assumed that French philosophy of technology begun with the *Encyclopédie* of Diderot and d'Alembert, I take Claude Henri de Rouvroy comte de Saint-Simon (1760–1825) and Auguste Comte (1798–1857) as a starting point for discussion, my reason being that they focused on the social question and on ethical issues in early industrial society. Their interest in the social question is no exception: both Andrew Ure in England and, from a widely different perspective, Karl Marx in Germany (and later in England) also focused on social issues. But while Ure was inclined to give lessons in morality and Marx to read the future of society

[1] A point very nicely captured by Dominique Lecourt (2003: 6–7).

through technological development and class struggle, early nineteenth-century French thinkers used to link the social question to another one: how to terminate the French Revolution? (Saint-Simon 1820; Comte 1855) Two major features characterise their endeavor: Concluding the French Revolution was, for Saint-Simon and Comte, a political endeavor, albeit not directly. In their view, looking into the past and in the *longue durée* was essential to understand technological development. The "philosophical observation of the past" showed that the empowerment of municipalities and the development of the natural sciences introduced in Europe by the Arabs gradually gave greater impetus to industry. As a result, intellectual and material power no longer belonged primarily to the clergy, but to scientists and to the new industrial class. Terminating the French Revolution involved organizing society in order to cope with this new balance of interests first in France and in Europe, but also, in the long term, in the entire world.

Saint-Simon and Comte drew a line between theory and practice, between pure science and applications (Saint-Simon 1817; Comte 1822). They did not consider them as two opposites, but they meant that if one can define the proper aim of application, one has at the same time discovered what provides technology its moral worth as a human activity. This aim is defined as "the welfare of human beings, as individuals as well as a species" (Saint-Simon 1820) and it can be brought about by the "peaceful development of our common exploitation of the earth (*la planète humaine*)."[2] (Comte 1852).

In view of this, it is clear that early French philosophers of technology more or less dispensed with a fine-grained analysis of the internal structure of technology; they rather aimed to make sense of technology against the background of their grand philosophy of history.

Alfred Espinas (1844–1922) and Henri Bergson (1859–1941) took a different direction. This is not to say that they disregarded social and political issues but their analyses belong to a comprehensive study of the nature of action within biological evolution. While they did not develop an ethics *of* technology strictly speaking, they nevertheless developed interesting views on the relations between ethics *and* technology.

In *Les origines de la technologie* (1897), Espinas aims to establish a general theory of action, in his own words a "praxeology." Praxeology is "the science of the most universal forms and of the highest principles of action in the whole class of living beings able to move." (Espinas 1897: 8) This praxeology would comprehend a "general technology," (*technologie générale*) in the old Continental meaning of a theory of practical arts. General technology would thus be slightly less ambitious than praxeology, for it would deal only with the "sets of practical rules, of crafts or technologies [*techniques*] one can discover in mature human societies, having reached some level of civilization." (Espinas 1897: 8–9) Espinas assumed that this general theory of *technique* (general technology) should depend in turn upon life itself in its most universal principles and thus on a science of action applicable to all

[2]This view is similar to Bacon's notion of "relief of man's estate," through with a social concern perhaps missing from the Chancellor's writings.

living creatures (praxeology). From a methodological point of view, Espinas was inspired by Ernst Kapp's (1808–1896) notion of "organic projection": he considered technology as a set of organs of the social will and emphasized its unconscious origins. In his book *Les origines de la technologie*, he did not attempt to write a history of technology as such, he rather focused on the historical representations of technology even including Ancient Greek authors from the sixth and fifth centuries BC. In a broad Comtean vein, Espinas distinguished historical stages in the evolution of technology, with a physico-theological stage followed by an metaphysical-artificialist stage. This later stage was itself in turn subdivided into a human fabrication stage, followed by a divine fabrication stage. In the former physico-theological stage, the skills of crafts were viewed as expressing friendly or hostile feelings of idealized people: Gods. Action in general is seen as governed by a set of relations with the gods; it takes the form of trading with them, receiving from them a gift or a teaching, or obeying their orders. Omnipotent gods exercise their will in accordance with the nature of things: the rules and norms governing action are divine volitions; they have their own immanent morality. In the artificialist stage, nature is no longer viewed as an expression of the divine will; it is seen as the mere form and order of things such that human crafts become, so to say, self-conscious (Espinas 1897: 161). The transition between the two stages can be understood as a shift from tools (*ustensiles*) to instruments. Instruments act on matter is a more sophisticated way than tools. They gradually prompt the abstract notion of action upon an homogeneous and neutral environment. Above all, instruments are partially autonomous: intricate mechanisms may even seem to produce counter-natural movements. They tend to emancipate technology from tradition in a similar way that the Ancient Sophists attempted to get rid of tradition and conventions in the field of politics and morality. As such emancipation from tradition generates fear and distress among the most conservative circles, they develop in reaction a theory of "*fabrication divine* "(divine manufacturing) as opposed to the "*fabrication humaine.*" (human workmanship) Divine action becomes a model for any kind of action whatsoever. As far as philosophy of technology is concerned, the main interest of Espinas's analysis does not lie in his conclusions, which must probably be qualified or even rejected, but in his methodology. He did not only pay much attention to a corpus of poets, philosophers, physicians, tragic authors, and so on. He sought to provide a better understanding of technological devices in relation to the worldview of people who made and used them. He considered morality and technology as two manifestations of social will, mutually influencing each other. He did not try to make recommendations nor develop norms for a responsible use of technology; he rather insisted that technology is not a set of means for clearly defined ends: it is a social endeavor that cannot be separated from its biological roots.

Bergson also related technology to life. But as he understood life in a very different way, his evaluation of technology radically changed.[3] Bergson assumed that all forms of life are the continuation of one and the same impetus, divided into different lines of evolution (Bergson 1907). But this impetus is finite and limited for it per-

[3]I gave a short survey of his philosophy of technology in Goffi (1988: 81–86).

petually goes backward, like whirling water in a stream. This rearward movement is nothing other than the manifestation of matter. Life can thus be characterized as a "tendency to act on inert matter." (Bergson 1907: 96) In the animal kingdom, the striving to get something from inert matter takes two forms: instinct and intellect. Both are capable of inventing, but (human) intellect goes much further; a famous page of *L'Évolution créatrice* carefully outlines the stages of a progression ranging from constructed objects to manufactured and artificial instruments, such as tools and steam-engine: instinct is left back as soon as the stage of fabricated instruments is reached. But there is a price to pay for the impressive success of the intellect. By its very nature intellect operates on unorganized solids; it cannot form a clear idea of anything except of the discontinuous and the immobile (Bergson 1907: 165). This can be interpreted in a pessimistic way: ultimately, intellect has no grip on reality and technology; even though it is a product of life, the intellect remains blind to life itself. But, in a more optimistic and perhaps prophetic way, this can also mean that even if the evolution of life has stopped with the human body, with the craftsman's intelligence surrounded by some measure of intuition, the last word is yet to be said. Technology will perhaps prove able to take over the task of evolution: "Tomorrow the way will be clear, in the very direction of the breath which had carried life to the point where it had to stop." (Bergson 1932: 270)

Espinas and Bergson, despite very different assumptions, have pointed out that technology is deeply rooted in biology or perhaps rather in a "vital impulse." To borrow an expression from another philosophical idiom, they viewed technology as a "form of life." Much of contemporary French discussion of ethics and technology revolves around this issue: Can this form of life stay under control? or, in other words: Is it really a form of *life* or the *form of an uncontrollable power* hostile to life?

19.2 Technology Without Ethics

The most strident negative answer to the question of whether technology can be controlled has been articulated by Jacques Ellul (1912–1994) whose main work has had a tremendous influence on "techno-critical" thought in France and elsewhere, very often outside academic circles (Ellul 1954). To measure the significance of this book, one has to realize that one of Ellul's main target was French economist Jean Fourastié (1907–1990): Fourastié's analyses are discussed in Ellul's *La technique, ou l'enjeu du siècle* though also, more critically and perhaps even satirically in further writings (Ellul 1966). Fourastié claimed that technological progress, defined as increased work productivity, has become the fundamental and most important determinant of economic development. It leads to the development of a service based economy; the (relative) diminution of industrial jobs will lead to the development of a more humane work place environment. The service sector also creates many stable professions that give a better quality of life and many more opportunities for self-realizations than the jobs of the bygone industrial era (Fourastié 1949). What is more, this improvement will lead, according to Fourastié, to a more humane society in general: liberated from servile forms of work, people will have more time to devote to cultural aims; they will develop less competitive habits and thus become better persons from a moral point of view. This optimistic and humanistic view of technology is rejected by Ellul. He claimed that technology has become the most important factor in contemporary societies, but his interpretation of the situation is very different from Fourastié's opinion. Ellul begins by refuting the common mistakes of confusing technology and engineering, technology and applied science, technology ("*la technique*") and technical operations ("*les technique*") (Ellul 1954). He wants to characterize the technical phenomenon: the main difference between technical operations and the technical phenomenon is that the former boils down to work carried out methodically with a view to achieving a desired end. It is nothing else than the time-immemorial means-end relation. By contrast the latter is the search for "the one best way," the search for absolute efficiency, in any field of human activity whatsoever. The distinctive features of technical phenomenon are: rationality, artificiality, the automatic character of technical choice, self-augmentation, unity, universality, and autonomy. A fascinating question that arises from this view is the relation between the Ellulian notion of autonomous technology and the so-called technological determinism of Marx. Some interesting considerations raised by Ellul himself suggest that, in spite of obvious differences, the ties between the two ideas may be closer than usually believed (Ellul, 1966). In any case Ellul's characterization of the technical phenomenon has a strong impact on the ethics of technology: "Technology does not progress in terms of a moral ideal … Technology does not endure any moral judgment. The technician does not tolerate any insertion of morality in his work … It [technology] does not tolerate being halted for a moral reason." (Ellul 1977: 145, 147) At first glance, this sounds like nihilism. However since Ellul was a Christian, his position may be better understood as an attack against a particular form of idol worship or idolatry (Cf. Cérézuelle, Chap. 4 in this volume).

Ellul's radical criticism of technology or, to be more exact, of the technical phenomenon has been echoed by two French thinkers: Jean Brun (1919–1994) and Michel Henry (1922–2002), although they did not insist like Ellul on the autonomy of technology. To be sure they are not very comfortable with modernity and can be labeled as conservatives. But there is more in their works than a complaint about progress. One can perhaps define what they have in common by saying that for them technology was a symptom. Technology, in other words, cannot be understood by itself but only as an indication of some disorder or diseased condition (we are, of course, in the realm of metaphor).

To introduce Brun a quotation from Bergson may be helpful: "Sex-appeal is the keynote of our whole civilization."[4] (1907: 261) For Brun, there is more in civilization – and also in technology – than a set of well-defined means to meet vital needs and interests. Brun stated his point by using the distinction between tactics and

[4]It must be noted that this once commonly accepted translation of: "Toute notre civilisation est aphrodisiaque" does not give full justice to the original. Likewise, there is more than sexual lust in Brun's concept of desire.

strategy: "Technology does not express a vital tactics; it is a genuine existential strategy." (Brun 1981: 13) The desire expressed by technology is not a mere sexual lust: it is an aspiration to abolish any kind of individual and separate existence; by developing technology, men are seeking to reach a space and time which could be the theater of a self-creation. They want to tear themselves away from their own condition and finiteness and become new gods. Consequently, technology cannot be understood without its intentionality (*i.e.* the intention and will it carries and manifests). This intention is best expressed in myths, and as mythologies help decipher intentions they are also appropriate to understand technology. As for the technological enterprise itself, it is doomed to fail: it is impossible to overcome finiteness and desire becomes a tormentor creating only illusions and disappointments.

Another reference to Bergson may help illustrate Michel Henry's position. It is well known that, in the last lines of *l'Évolution créatrice* Bergson praises philosophy for being able to enter into contact with the creative impulse which is the basis of becoming, provided it pays attention to science understood as "a set of truths, either experienced or demonstrated," in opposition to the "barren scholasticism" that emerged by the end of the nineteenth century, around Galileo's physics (Bergson 1907: 370). In a similar vein, Henry developed a critique of contemporary science and technology in the name of his own philosophy of life rather than of evolution. So, what is life according to Henry? Life is occasionally assimilated to the Schopenhauerian will (Henry 1987: 167), while the world of technology is sometimes assimilated to the world of science (Henry 1987: 187). But on both occasions, Henry chose the easy way out: in fact, he defines life as the ability to feel and to experience itself, such that all living beings are able to speak in their own voice, to express what they are and feel like (Henry 1987: 36). To sum up, a living thing expresses its subjectivity. But with Galileo, science has excluded from its field of interest the so-called secondary qualities (even if Henry did not use this technical term): in its quest for objectivity, it has focused on things that have no inside, which are mere surfaces offered and available. What do these considerations have to do with technology? Henry does not believe that technology expresses a will to rule and dominate a valueless and purposeless world. Deliberately alluding to Descartes, he claims that we humans, are inhabitants as well as the landlords of the earth, for we are embodied living beings. Henry's point is rather that time-immemorial *tekhnê* as a manifestation of living bodies, has been replaced by a technology which only makes use of rational categories such as cause/effect and means/end. Those categories being quite distant from the norms of life technology implements its own "dead" rationality. While Henry may be reminiscent of what Jürgen Habermas says about the colonization of the lifeworld by instrumental rationality, his position is much less balanced than Habermas's, since he concluded: barbarism is not just a perversion of this technology: it is its very essence.

Neither Ellul, nor Brun nor Henry, in spite of their eloquence (or maybe because of this very eloquence) did develop an ethics of technology *per se*. They adopted such radical critical stances that they could only rely on a logic of salvation, secured either by the few (Henry), by God (Brun) or by a progressive conversion to a no-power ethics (Ellul).

19.3 An Empirical Turn *à la Française*

I will now deal with ethicists of technology who pay more attention to actual practices and actual technical devices. This seems such an obvious requirement for any self-respecting ethics of technology that some words of explanation may be useful: Ellul, Brun and Henry do mention technological practices and devices; but they only provided a sweeping evaluation of technology as a whole (as a phenomenon or a system, as an endeavor to go beyond the limits inherent into the human condition, as the expression of a world view indifferent to any form of subjectivity). They were not interested in actual practices and devices, except as illustrations to support their claims – for instance see Henry (1987: 2).

A number of philosophers of technology who are not interested in the issue of salvation through or without technology have developed alternative strategies based on empirical data (this seems a minimal requirement for a rational debate). Others, such as Gilbert Simondon (1924–1989) perhaps more ambitiously, set up to scrutinize technical objects; his approach will be discussed below.

Let us survey three examples of more empirical philosophical approaches to technologies. They deal with what would be called elsewhere professional ethics, military ethics and bioethics. It goes without saying that each field depends critically on the technologies involved.

Christelle Didier, as a lecturer at the catholic university in Lille, an institution with a long tradition of interest in social issues, has introduced engineering-ethics, a quite novel subject in France (Didier 2008). She does not adopt the usual way, beginning with mainstream normative ethical theories before going on to write a practical guide of ethical decision-making for engineers. Her approach is reflexive: she sets herself to the task of analyzing the *ethos* of engineers, which is both similar to and significantly different from the traditional *ethos* of scientists as defined in Robert K. Merton's famous papers: as engineering refers to a set of varied activities, such an approach would be doomed to fail. Didier adopts a looser notion of ethos inspired by French sociologist Pierre Bourdieu: "Ethos" designates practical principles reflecting a set of dispositions with an ethical dimension; although these principles are not intentionally structured, they are objectively systematic (Bourdieu 1980). Didier analyzes the interactions between professional codes, teaching in engineering schools and the emergence of academic publication dedicated to engineering ethics (the study focuses on France, Québec, the U.S.A. and the German Federal Republic). In short, her aim is to analyze the production procedures of ethical discourse in the field of engineering ethics and to consider more theoretical issues raised by engineering ethics. These issues include the notion of profession, the relation between professional codes and applied ethics, the problems related to risk-assessment. It is clear that Didier's ambition is not to provide engineers with a tool-kit enabling them to solve practical cases. She is more concerned with careful analyzing the controversies about the ethical issues in actual engineering practice. She considers the concept of profession to be largely irrelevant to these issues and is somewhat skeptical about the very concept of applied ethics: when wondering about the specificity of engineering practices of engineering, she points out that the concept of hybridization (between science and technology; between capital and labor) is perhaps more appropriate. She wonders how one can exercise social control over technology. She also develop critical remarks about Technology Assessment which strike miles away from Saint-Simon and Comte's naïve views of technology. Not unlike Kristin Shrader-Frechette in another context, she argues that social values other than the welfare of human beings underlie the Technology Assessment procedures and choices. In this respect, a critical examination of the actual production of values is perhaps the main issue for an Engineering Ethics worth its salt.

In a quite different style, Grégoire Chamayou takes sides in the Unmanned Combat Air Vehicle (UCAV) debate (Chamayou 2013). Without concealing his militant attitude against this military technology, he develops an acute critical analysis pointing out the simplifications, confusions (and sometimes also lies) involved in this new technological system. At first glance, unmanned combat air vehicles may look like just another weapon system (like the half-track during WWII or the combat helicopter during the Vietnam War), and this description may have been accurate in the 1940's and 1950's when only aerial target drones were considered, But the current situation is quite different: UCAVs are lethal weapons operated by a "pilot" (sometimes, a civilian) who is far away from the battlefield and lives a normal life in all respects: he goes to his combat outpost as others go to their office and may be back home in the evening. The ethical question is: "What does this mean and what does it entail when such a weapon system is adopted on a massive scale?" True, the soldiers are absent from the battlefield. But the battlefield is actually a disputed territory, and the war is actually a counter-insurgency campaign: war is replaced by the hunting of men. The hunters' lives are considered too precious to be risked in the hunt: risks are almost totally transferred to the drones. But it is not enough to have machines able to "kill from above": one must also identify targets. Another ethical issue is raised by observation drones who rely on a quantitative interpretation of suspicious behaviors in the targeted population, which dramatically increases the risk of mistakes and blunders. In addition this "brave new war" completely overturns the conventional military *ethos*: traditional military values (fortitude, self-sacrifice, honor) are based on the possibility of losing one's life in the battlefield but this is no longer the case anymore with UCAV "pilots." Waging such "wars" also results in a complete subversion of the law of armed conflicts categories: it tends to blur the distinction between armed conflicts and police operations and, as a consequence, between combatants and civilians. Last but not least, the use of UCAV increases the autonomy of state power: as the citizens are not involved anymore in armed conflicts, the State may go to war without being held accountable (and maybe surreptitiously transfer the drone technologies to the field of internal security issues: border control, illegal immigration control, demonstrations and rallies control). Chamayou does not assume that UCAV technology is autonomous in Ellul's sense (his theoretical reference are Michel Foucault and Theodor Adorno) but he rightly claims, just as Ellul would have done, that he is not just another specialist in applied ethics (Chamayou 2013: 272).

While Didier deals with the *ethos* of a profession (even if she is not satisfied with this notion), and Chamayou with a weapon system (although there is more than a weapon system to discuss), Gilbert Hottois, a bioethicist at the Université Libre de Bruxelles, developed a more ambitious project: to elucidate the situation of contemporary philosophy in connection with technology (or, better, technoscience). As he argued in his doctorate Thesis (Hottois 1979), contemporary philosophy is characterized by "secondarity (*secondarité*):" today philosophers no longer claim to have access to reality. They essentially refer to language or keep on the sidelines of texts or discourses. In a sense, philosophers have abandoned the theoretical project of making the world and history intelligible. Hottois argued that this was a consequence of the prominence of technoscience. Philosophy traditionally conceived implies that theoretical and technical frameworks be essentially different, though they do support each other. The theoretical approach inherits the task of identifying truth and meaning; the technical approach sets to operate in space and time, thus transforming them into a world with a contingent history. According to Hottois technosciences disrupt this division of labor: in the language of technosciences, signs are used to denote a potential operation or transformation. The language of technosciences is not fit to express the essence of mankind in a symbolic way: in a technoscientific environment, philosophers can still claim to speak in the name of mankind, but to no avail. This intuition is further developed in *Le signe et la technique* (Hottois 1984a). Ellul wrote a foreword to the book and some of the chapters titles are reminiscent of Ellul's terminology ("A blind autonomous growth," "Closure of the system by informatics," "The an-ethics imperative of the technical reign," etc.) Moreover, in this book, Hottois borrowed some ideas from Brun and he has been quoted with some approbation by Henry (1987: 99). Accordingly, he was sometimes considered as one of those imprecatory philosophers who curse science and technology. However such an interpretation has three weaknesses: (i) It does not take into account the fact that Hottois has a genuine expertise in the field of the issues raised by new biotechnologies. (ii) It turns a blind eye to the fact that the last section of the book was titled "A humanism without any illusions." (iii) It ignores the fact that Hottois sketches out an ethics that could be applied in the context of a world marked by technologies (Hottois 1984b). His book *Pour une éthique dans un univers technicien* is divided into a *pars destruens* and a *pars construens*. The former is a criticism of what Hottois calls the "anthropologistic assessment of technology." (Hottois 1984b 14–22) In this perspective, the assessment of science and technology is dependent on pre-assumptions about human nature. This approach does not take into account the fact that technosciences are able to modify human nature itself. Because technosciences perpetually change the reference point against which they are supposed to be evaluated they make the future opaque. Therefore Hottois calls for ethical prudence, in a way reminiscent of Hans Jonas. Mankind must be cared for and protected because it is the source of value, not because it is the supreme value. This idea has been detailed and deepened ever since. The tragic tone perceptible in Hottois's earlier writings has been somewhat mitigated, but he retains his deep conviction: seeking full control of technology is a pipedream; we can only achieve prudence through accompanying measures.

This survey of three French-speaking philosophers concerned with the ethics of technology already suggests their distance from mainstream ethics: None of them claims to work in the field of Applied Ethics; quite the contrary, they try to develop a broader approach embracing much more than moral theories.

The distance from applied ethics is reinforced by the French approach to the specific field of nanoethics. A blueprint for this approach has been provided in Bensaude Vincent and Nurock (2010). They consider that most debates have been unduly focused on the effects of nanotechnologies. In addition, the few studies who paid more attention to nanotechnologies themselves have also, Nurock and Bensaude Vincent claim, missed the point. For them, what is characteristic of nanotechnologies is less the order of magnitude – designing at the nanoscale – than the convergence of practices and knowledge involved. They consequently advocate a fieldwork ethics. While developed in an independent way, this "*éthique de terrain*" which pays more attention to the genesis of nanotechnological objects and to the values embedded in them, has been reflected in the so-called ethics "on the laboratory floor." (van der Burg and Swierstra 2013) They also insist on the need for democratic public debate about collective choices in the field of nanotechnology. This is certainly another proof of distrust in technocratic positivism, which has come under regular attacks as a result of the way various momentous post-war decisions (eg. nuclear power development) have been made in France. However the major distinctive feature in French philosophy of technology is the omnipresent reference to Simondon, be it implicit or explicit. Georges Canguilhem, whose *Machine et organisme* (Canguilhem 1952) has been seminal in the field of the philosophy of technology, directed Simondon's complementary thesis dedicated to technical objects (Simondon 1958). His key idea is that if technical objects are conceived as objects of mere utility their technical nature or technicity (*technicité*) is missed. As a consequence, technology is seen as indifferent and even hostile to culture (according to another formulation, technology belongs to civilization only (Simondon 1960–1961). In order to understand the technicity of technical objects, one should instead pay attention to the details of their genesis (a kind of *épochè* is then necessary). That is why one can find in Simondon's works careful studies of the development of external cooling fins, light-emitting diodes and other devices not usually mentioned in philosophy texts. Without going into further detail, it is clear that Simondon has built a compelling case for the reintegration of technology into culture. But this is not yet an ethics of technology. Two options are identified by Simondon. On the one hand, in a paper explicitly dealing with the ethics of technology, he suggested a threefold relation between people and technical objects which would distinguish:

- A present-oriented ethics aimed at turning destructive technologies into constructive ones, through a process of deepening technology itself.
- A future-oriented ethics in charge of the organization of a post-industrial society taking into account the claims of environmental movements.

– A past-oriented ethics which has to manage obsolete technological objects (collecting, recovering, recycling are ethical options, or at least involve ethical choices) (Simondon 1983).

However brilliant, this analysis only points to an array of issues and provides no principled answers: technology does not in itself entail an ethics, it only carries a certain internal normativity (For a different understanding, cf. Guchet, Chap. 15 in this volume).

On the other hand, as a philosopher of genesis and becoming, Simondon suggested that normativity is always a process, coming into existence from the clash between different orders and realities: an ethics of technology is never given in advance; it has to be built through the recognition of different legitimate claims (Simondon 1960–1961: 124–125).

This is a lesson which has not been forgotten by Bensaude Vincent when she criticizes so harshly the Ethical, Legal and Social Implications (ELSI) approach to the ethics of nanotechnologies, the same criticisms applying to such approaches in general (Bensaude Vincent 2012). She considers that such programs are nothing but the old technocratic governance in new clothes and she instead recommends a co-construction of technosciences based on deliberative citizen consultations (Bensaude Vincent 2009); see also Guchet (2014) for a similar argument.

19.4 Conclusion

A different story could of course, have been told. The importance of Simondon as a philosopher of invention (Simondon 2005, 2008), of Bergson as a philosopher of life and of the reflection on nanotechnologies in the recent development of the French philosophy of technology has been underscored by the mainstream of Technology Assessment. These trends converge towards a critical analysis of the way technologies are socially implemented, on the basis of a costs-benefits analysis, or of an ELSI approach. But the deficiencies and weaknesses of such methods have also been independently identified by philosophers of technology whose approach is very different such as Jean-Jacques Salomon (1929–2008), a senior official of the OECD who held a chair at the prestigious Conservatoire national des arts et métiers (CNAM) and Jean-Pierre Dupuy, who lectured on political and social philosophy and on the ethics of science and technology at the equally prestigious École Polytechnique.

The former argued that resistance to technological change cannot be dealt with by simply calculating acceptable levels of risks. Such an approach would be simplistic, as it would not take into account the values or the interests at stake (Salomon 1981: 5). The latter claims that a merely utilitarian management outlook towards the "new risks" would remain blind to the possibility of mankind causing its own destruction, and he thus seeks to build a metaphysics of the future equal to this challenge (Dupuy 2002).[5] It thus seems fair to say that Salomon and Dupuy owe more to social and political philosophy than to philosophy of technology, an observation that is consistent with the fact that it was Raymond Aron (1905–1983), a famous political scientist, who cosupervised Salomon's PhD (Canguilhem being the other supervisor) and with the debt that Dupuy acknowledges to Ivan Illich (1926–2002) when he admits that the framework of his current scholarship is a comprehensive criticism of industrial societies. In view of this, if one admits with Andrew Feenberg that "the environmental movement is the single most important domain of democratic intervention into technology" (Feenberg 1999: 93) one should not forget to mention those French philosophers who write in the field of environmental ethics (Cf C. Larrère and R. Larrère, Chap. 12 in this volume).

While French philosophers of technology would almost unanimously deny being specialists in Applied Ethics, the current trend in French philosophy of technology is towards a more cautious and empirical approach than was the case some decades ago.

Acknowledgements A preliminary version of this text has been discussed by Catherine Larrère and Bernadette Bensaude Vincent during the authors' workshop held in Paris Sorbonne on 22–23 June 2015 for the preparation of the collective book.

References

Bensaude Vincent, B. (2009). *Les Vertiges de la technoscience. Façonner le monde atome par atome*. Paris: La Découverte.

Bensaude Vincent, B. (2012). Quelle éthique pour les nanotechnologies? Bilan des programmes en cours. In C. Kermisch & M.-G. Pinsart (Eds.), *Les nanotechnologies: vers un changement d'échelle éthique?* (pp. 357–368). Bruxelles: MEM.

Bensaude Vincent, B., & Nurock, V. (2010). Éthique des nanotechnologies. In E. Hirsch (Ed.), *Traité de Bioéthique I. Fondements, principes, repères* (pp. 355–369). Toulouse: Éditions érès.

Bergson, H. (1907). *L'évolution créatrice*. Paris: Alcan. English Edition: Bergson, H. (1911). *Creative evolution* (A. Mitchell, Trans.). Lanham: University Press of America.

Bergson, H. (1932). *Les deux sources de la morale et de la religion*. Paris: Presses Universitaires de France. English Edition: Bergson, H. (1935). *The two sources of morality and religion* (R. A. Andrà, C. Brereton, & W. H. Carter, Trans.). New York: Henry Holt and Co., Inc.

Bourdieu, P. (1980). *Questions de sociologie*. Paris: Les Editions de Minuit. English Edition: Bourdieu, P. (1993). *Sociology in question* (R. Nice, Trans.). London/Thousand Oaks/New Delhi: Sage Publications.

Brey, P. (2010). Philosophy of technology after the empirical turn. *Techné, 14*(1), 36–48.

Brun, J. (1981). *Les Masques du désir*. Paris: Buchet/Chastel.

Canguilhem G. (1952). Machine et organisme. In *La connaissance de la vie* (pp. 101–127). Paris, Hachette. English Edition: Canguilhem G. (2008). Machine and organism. In *Knowledge of life* (S. Geroulanos, & D. Ginsburg, Trans., pp. 75–97). New York: Fordham University Press.

Chamayou, G. (2013). *Théorie du drone*. Paris: La Fabrique.

[5] Interestingly enough, this metaphysics is Bergsonian in its inspiration.

Comte, A. (1822). Plan des travaux scientifiques nécessaires pour réorganiser la société. Reedited in P. E. De Berredo Carneiro, & P. Arnaud, P. (Eds.). (1970). *Auguste Comte: Écrits de Jeunesse 1816–1828* (pp. 241–321). Paris-La Haye: Mouton.

Comte, A. (1852). *Catéchisme positiviste* (1966th ed.). Paris: Garnier-Flammarion.

Comte, A. (1855). *Appel aux conservateurs* (2009th ed.). Paris: Éditions du Sandre.

Didier, C. (2008). *Penser l'éthique des ingénieurs*. Paris: PUF.

Dupuy, J.-P. (2002). *Pour un catastrophisme éclairé. Quand l'impossible est certain*. Paris: Editions du Seuil.

Ellul, J. (1954). *La technique ou l'enjeu du siècle*. Paris: Armand Colin. English edition: Ellul, J. (1967). *The technological society* (J. Wilkinson, Trans.). New York: Knopf/Vintage.

Ellul, J. (1966). *Exégèse des nouveaux lieux communs*. Paris: Calmann-Lévy. English edition: Ellul, J. (1968). *A critique of the new commonplaces* (H. Weaver, Trans.). New York: Knopf. Reedited 2012, Eugene: Wipf & Stock Publishers.

Ellul, J. (1977). *Le système technicien*. Paris: Calmann Levy. English edition: Ellul, J. (1980). *The technological system* (J. Neugroschel, Trans.). New York: Continuum.

Espinas, A. (1897). *Les origines de la technologie*. Paris: Alcan.

Feenberg, A. (1999). *Questioning technology*. New York: Routledge.

Fourastié, J. (1949). *Le Grand Espoir du XX e Siècle*. Paris: Presses Universitaires de France.

Goffi, J.-Y. (1988). *La Philosophie de la technique* (1996th ed.). Paris: Presses Universitaires de France.

Guchet, X. (2014). *Philosophie des nanotechonologies*. Paris: Hermann.

Henry, M. (1987). *La Barbarie* (2014th ed.). Paris: Presses Universitaires de France.

Hottois, G. (1979). *L'inflation du langage dans la philosophie contemporaine*. Bruxelles: Editions de l'Université libre de Bruxelles.

Hottois, G. (1984a). *Le Signe et la technique. La philosophie à l'épreuve de la technique*. Paris: Aubier.

Hottois, G. (1984b). *Pour une éthique dans un univers technicien*. Bruxelles: Éditions de l'Université Libre de Bruxelles.

Lecourt, D. (2003). *Humain, posthumain. La technique et la vie*. Paris: Presses Universitaires de France.

Mitcham, C. (1994). *Thinking through technology. The path between engineering and philosophy*. Chicago/New York: The University of Chicago Press.

Mitcham, C., & Duval, R. S. (2000). *Engineering ethics* (Engineer's toolkit). Upper Saddle River: Prentice Hall.

Saint-Simon, H. (1817). L'industrie. Tome Second. Article troisième. In Grange, J. et al (Ed.). (2012) *Œuvres complètes II*. (pp. 1528 sq). Paris: Presses Universitaires de France.

Saint-Simon, H. (1820). Du Système industriel. In Grange, J. et al (Ed.). (2012) *Œuvres complètes III*. (pp. 2351 sq), Paris: Presses Universitaires de France.

Salomon, J.-J. (1981). *Prométhée empêtré. La résistance au changement technique* (1984 Edition). Paris: Anthropos.

Simondon, G. (1958). *Du Mode d'existence des objets techniques* (2013 ed.). Paris: Aubier. English edition: Simondon, G. (2017). *On the mode of existence of technical objects* (C. Malaspina, Trans.). Minneapolis: Univocal Publishing.

Simondon, G. (1960–1961). Psychosociologie de la technicité. In Simondon, G. (2004). *Sur la Technique* (pp. 27–129). Paris: Presses Universitaires de France.

Simondon, G. (1983). Trois perspectives pour une réflexion sur l'éthique et la technique. In In Simondon, G. (2004). *Sur la Technique* (pp. 313–351). Paris: Presses Universitaires de France.

Simondon, G. (2005). *L'invention dans les techniques. Cours et conférences*. Paris: Editions du Seuil.

Simondon, G. (2008). *Imagination et invention 1965–1966*. Chatou: Éditions de la Transparence.

Van der Burg, S., & Swiertstra, T. (Eds.). (2013). *Ethics on the laboratory floor*. Basingstoke: Palgrave Macmillan.

Chapter 20
Wisdom in the Technosphere

Michel Puech

Abstract The opportunities for human flourishing and the threats of human dereliction involved in contemporary technology require a new "software" in philosophy. The technological revolution runs across cultural boundaries. This chapter invites a pragmatic and pluralistic turn in philosophy and ethics for addressing the new global technosphere. The ancient and non-western notion of "wisdom" gives new momentum to this research and leads to specific methods and elements in virtue ethics and care ethics that can be harnessed for delineating a constructive ethics of technology. The morally disruptive environment of the technosphere is construed as a sapiential challenge.

Keywords Technosphere · Wisdom · Virtue ethics · Ordinary · Micro-action · Flourishing · Harmony

20.1 Flourishing in the Technosphere

The common feeling that a new material civilization is emerging with digital technologies, economic globalization, and growing ecological awareness is more than a bragging claim due to a short-sighted and ideological self-intoxication with "progress." We live in a technosphere offering unique opportunities for human flourishing as well as specific threats of human dereliction. Philosophy, human and social sciences, and the humanities at large are challenged by the new technological lifeworld. My impression is that interpreting it with the intellectual tools of the industrial era leads most of the time to confrontational social critique and a technophobic angst. One of the challenges still ahead is then a realistic, pragmatic, and constructive approach to the new technosphere in philosophical terms and particularly in ethical terms. I suggest that a renewed notion of wisdom fosters this initiative.

M. Puech (✉)
Département de Philosophie, Sorbonne Université, Paris, France
e-mail: michel.puech@paris-sorbonne.fr

S. Loeve et al. (eds.), *French Philosophy of Technology*, Philosophy of Engineering and Technology 29,
https://doi.org/10.1007/978-3-319-89518-5_20

A pluralism of voices is necessary for a substantial intellectual conversation about who we are and where we are. I am confident that European and "Continental" inspirations can harmonize with both trans-Atlantic stimulation and Asian heritage in order to compose an innovative melodic line that runs across cultural and disciplinary boundaries – exactly like the technological revolution runs across every cultural boundary. How can we create harmony between *wisdom*, an old, devalued, and individualistic aspiration, and this new vibrant, polymorphic, and assertive *technosphere*?

"Technosphere" is not a restrictive concept in my approach. It means the environment of our physical life, which is for a large part technological. Talking of "spheres" gives an idea of the pervasiveness and perhaps the inevitability of this technological environment. The image essentially expresses the awareness of a new situation for humankind. We are now used to imagining our planet as a blue sphere on the background of deep space instead of a flat solid surface under blue sky. This change in perimeter combines with the awareness of the pervasive human reshaping action on the surface of this planet: we are beginning to understand the present period as the Anthropocene, a geological time where humankind is the main shaping power at work ("Anthropocene" 2015).

There is a conceptual advantage in the notion of *sphere* due to its "spatial" connotations, one of them being the image of spheres that slot one into the other with a complex configuration of inclusion, partial inclusion, and intersection. Vladimir Vernadsky, a pioneer in sustainability issues, used this logic to define a notion of "biosphere" in 1926 (Vernadsky 1998) and then to conceive of a notion of "noosphere" that resembles the more recent notion of technosphere (Vernadsky 1945) – even if it was for him a spiritual more than a material entity.

In his seminal work, first published in 1979, René Passet made the case for the right order of spheres inclusion: the economic sphere is included in the human sphere, which is included in the biosphere (Passet 1996: 4). If we call "technosphere" the whole set of artifacts, material and immaterial, then a complex system of spheres and intersections emerges: the *technosphere* is included in the biosphere or better in the *ecosphere* (with its sustainability issues); the *infosphere* is included in the technosphere (with its digital empowerment issues); and the various *human spheres* intersect all these global spheres.

My point is not to dive into 3D geometry but rather to follow up Gilbert Simondon's description of technological systems. Simondon distinguished technological components (a diode), technological devices (a TV set), and technological systems (the TV broadcasting system). In adapting Simondon's vocabulary (Simondon 1958, pt. 1) I suggest completing this list with more global spheres, the relevant next sphere being the technosphere. This enlargement is in line with Simondon's emphasis on the "milieu" as the essential factor in technology evolution (Barthélémy 2014). Jacques Ellul's notion of a "Technological System" (Ellul 1977) is an interpretation of the technosphere as a technostructure imposing a soft ideology. The technostructure has been predominantly engaged as a domination framework. Other trends of research envisage the technosphere as an existential milieu where a particular kind of human subjects try to survive and to thrive.

The human subject in the technosphere is or yearns to be *Technosapiens*: a Homo Sapiens evolving toward wisdom, which may be its only chance to survive its own destructive power and its only opportunity to flourish in the unforgiving milieu of the technosphere (Puech 2008). Networks of human and non-human entities compose the technosphere. The "moral" nature of this environment has been captured in a sociological style by Bruno Latour (Latour 1991) and recently in a technoethical style by Peter-Paul Verbeek (Verbeek 2005, 2011). The philosophical style of Verbeek in adapting Latour's work to Don Ihde's post-phenomenology tradition in the chapter "The Acts of Artifacts" (Verbeek 2005: 147–172) exemplifies how the international academic community can draw from French twentieth century philosophy in a clarifying and constructive manner. My research on wisdom humbly aspires to participate in this effort.

Technosapiens is facing totally unexpected material circumstances: relative abundance in industrialized countries. I define relative abundance in the sense of Galbraith's "affluence" (Galbraith 1958), taking into consideration that this state of relative welfare concerning food, shelter, basic access to medicine and education, applies to only 80% of the population in industrialized countries. I emphasize the *moral* novelty of this material situation: it invites answering the challenge of abundance with a sapiential approach to modernity (Irvine 2011).

Beyond the 1970s political critique and deconstructive methods that made the success of "French theory," Foucault and Ricoeur provide rich material for this sapiential approach to modernity. Ironically, this material is largely ethical and constructive. Recent research has been sensitive to it through ambitious research on subjectivation in technical mediation (Dorrestijn 2012), as well as surveys of Foucault's lasting influence (Binkley and Capetillo Ponce 2009: 46–61) and of Ricoeur's critical theory (Kaplan 2003).

In his last publications and lectures, Foucault elaborated a model for a resistive construction of the human subject: this "technique of the self" makes particular sense in the twenty-first century technosphere. An unmistakable "ethical turn" in his works leads to a "pragmatic of the self" (Foucault 2008: 7) that transfers the question of domination from the governance of others to the governance of the self (Foucault 2008, 8). This "Foucault II" offers the planetary newcomer, Technosapiens, a robust model for transforming the domination environment into self-construction resources, through some kind of brilliant aikido move. The English translation of the French scholar Pierre Hadot's investigation into the sapiential tradition of philosophy as "spiritual exercises" bears the telling title "from Socrates to Foucault." (Hadot 1995) In his narrative theory of the self and self-constitution Ricoeur argues that the self is not a fiction produced by external constraints but that it constitutes itself through layers of interpretation and decision. Ihde's link to Ricoeur is no accident: the first leading figure in American philosophy of technology was hosted by Ricoeur, literally in his Parisian home, and he was philosophically hosted in French phenomenology (Maurice Merleau-Ponty in particular). New insights into technological praxis (Ihde 1979) highlighted the "texture of life" in the technosphere with the methods of hermeneutic phenomenology (Ihde 1971).

After decades of elaboration on this philosophical basis, the good life for Technosapiens is more and more precisely conceived as *flourishing*. This term conveniently synthesizes the sapiential effort toward self-constitution (internal flourishing) and toward ecosystemic values (external flourishing). In his ambitious theory of flourishing, Norwegian philosopher Arne Naess merges the heritage of western philosophy (Spinoza), renewed Asian ethics (Gandhi), and environmental commitment (Naess 1989). The intrinsic value of flourishing is not anthropocentric in this vision. Technosapiens's agency in harmonizing all sorts of flourishing in the natural, artificial, material and immaterial environment is only a new stage in the evolving opportunities for flourishing in the universe. Taking *flourishing* as central for a comprehensive set of values for modernity is buttressed by the recent wave of "virtue ethics" (Hursthouse 1999) and also by the emerging ethics of the technosphere (Bynum 2006; Puech 2016). "Flourishing ethics" gathers elements from a large range of traditions, including "Aristotelian roots" and "ideas suggestive of Taoism and Buddhism." (Bynum 2006: 157)

Here comes wisdom with its non-western values of humility and meditative appreciation. I intend to show in the following pages of this chapter that the morally disruptive environment of the technosphere can be construed as a sapiential challenge: is Technosapiens able to focus on flourishing as a self contributing to various flourishing spheres, or will blind instrumental rationality prevail? Seeing this predicament as a conflict between "techno" and "sapiens" defines the technophobic approach, which is still largely dominant. Seeing technology as a unique opportunity for a new "sapiens" defines the opposite effort toward a non-confrontational technoethics.

Albert Borgmann's existential analysis of modernity, despite inheriting Heidegger's pessimistic mood, paved the way to a reconstruction of "focal" activities and things in the technosphere (Borgmann 1984). In this pragmatic attempt, focusing on the dimension of the *ordinary* is key. The technosphere "touches" us; it is even intersecting our bodies, starting with medication or processed food, but also dental prostheses and in the near future more and more prostheses (mechanical as well as "bionic"). It "touches" us also through screens and keyboards, all of them framing our proximal environment as clothes used to do, and being equivalent to personal clothes in the case of the smartphone ("wearable" technology is coming). Wisdom can progressively blossom in the technosphere from the seeds of humble "ordinary technoethics" (Puech 2013), that is to say from philosophically investigating and ethically investing the most mundane mediation through artifacts (cooking, driving, using household appliances and workplace devices). How can we instantiate sapiential virtues, not in the heroic wisdom of moral prodigies, but in the most ordinary mediation with the technosphere?

20.2 Reinventing Wisdom

There is no backward or reactionary intention in a sapiential option for living in the present technosphere. The search for a new form of wisdom is clearly a forward-oriented enterprise and one of the most ambitious ones. "Wisdom studies" insist on this prospective and disruptive dimension (Kane 2010, Spence 2011). At the heart of the matter, the key ambition consists in subverting *power* as the universal reference and pivotal value of modern technology. Converging philosophical contributions point to this critical analysis. Alexander examines a modern "mantra of efficiency." (Alexander 2008) Janicaud offers a critical investigation of the growing irrationality of power. (Janicaud 1985) According to the "one-dimensional or totalizing nature of the technological hermeneutic" says David Lewin, a functional logic is supposed to reign in every mediation with the technosphere, a logic of functional "interface" that allegedly suppresses the complexity of the world, the maze of circumstances that was justifying Aristotelian *phronēsis* in ancient times (Ricoeur 1990): easy techno-power supplants ambiguity and complexity. A Ricoeurian question is then posed concerning human identity in a world submitted to a one-dimensional hermeneutic of power (Mei and Lewin 2012: 65–66); a Foucauldian answer can be found in the resolution to "design [one's] own life" (Dorrestijn 2012) through a reappropriation of power.

A simple threefold distinction (see Puech 2016) illuminates what is at stake and it highlights the neglect of a third form of power, the sapiential one:

1	Power over things	Technology
2	Power over others	Domination
3	Power over oneself	Wisdom

In face of any problem, we look for a technological solution (1). When material technology is unavailable, domination offers the alternative solution; political technology is our second-order instrumental rationality (2). When something cannot be provided by technology and cannot even be issued by an institution of power, the self has to face its own resources and change itself, which means option (3). The technosphere systematically privileges option (1) and our institutional sphere is supposed to design option (2) solutions for the remaining issues. The reader can easily explore instances by himself/herself. Let me mention just one: smoking. Option (1) solutions are the electronic cigarette, plus nicotine patches; option (2) solutions are laws and regulations that ban smoking in certain places, heavily taxing tobacco products, imposing deterrent messages and images on packages; an option (3) solution is personally quitting the addiction by constructing the internal and external environment to make it successfully. It should not be the last resort option that is hardly mentioned because power over oneself is supposedly almost extinct.

Renewed wisdom is practical and contextual: it reinstates in the technosphere the specific sense of complexity that Aristotle captured in his notion of *phronēsis* (Aristotle 350AD; Aubenque 1963; Reeve 2013). More broadly, power over oneself is revived through the de-westernization process of contemporary philosophy. The obsession with options (1) and (2) forms of power retrospectively appears as a tragic western mental restriction.

Recent trends in transcultural studies provide ample conceptual resources for a de-westernized post-modern notion of wisdom. Peter Hershock's works are landmarks in this growing academic enterprise. He confers to the Buddhist and Zen doctrines that he adapts to modernity an explicit "countercultural function" (Hershock 2005: 28), which is as simple, as innovative, and uniquely brilliant as the iconic image he uses: "reinventing the wheel." (Hershock 1999) This wisdom is first and foremost a practice: demonstrating what "buddha-nature" *does* more than understanding it or contemplating it (Hershock 2005: 68). This practice breaks the tragic cycle of command-and-control that characterizes modern technology – a circle of control says Hershock: "Thus, the better we get at controlling our circumstances, the more we will find ourselves in circumstances open to and requiring control." (Hershock et al. 2003: 595) Wisdom leads to overcoming the basic meta-attitude of command-and-control and then to recovering "our attentive resources – our capacity for meaningful and shared mutual attunement and contribution." (Hershock et al. 2003: 597)

A salient practice in this new wisdom is active meditation in the midst of ordinary activities. It is usually called by the Japanese word *samu* (Ives 1992: 36–37) and referred to Musō Soseki's definition in the fourteenth century: "People meditating on the fundamental carry out their ordinary tasks and activities in the midst of meditation and carry out meditation in the midst of ordinary tasks and activities. There is no disparity between meditation and activity." (Musō Soseki 1994: 53) A monk sweeping the courtyard or cooking in the monastery can do in it as a *samu* exercise. Obviously every humble task qualifies without any need to be a monk. Driving or shopping in full consciousness can be a self-construction and world-construction exercise, seen from the perspective of a sapiential hermeneutic of the technosphere. The irony lies in the importance of the unimportant for the technosphere of relative abundance: shopping in a mall or chatting on an online forum are neither values per se, as the ideology of consumption claims, nor anti-values per se, as resistance to consumption claims. Nothing is so futile that it cannot be taken, in the right circumstances, as a meditative exercise. Realizing the internal and external circumstances of *samu* practice in the technosphere is a never-ending task that begins everywhere.

Foucault's last formulations concerning the self elaborate on the theme "techniques de soi" (*tekhnê tou biou,* techniques of life and techniques of the self) as a determining practice in the "souci de soi" (care of the self) introduced in the third volume of his *History of sexuality* (Foucault 1984). A broader vision of Foucault's point is given by a series of papers from the same period – "L'herméneutique du sujet," 1982, "L'éthique du souci de soi comme pratique de la liberté," 1984, "Les techniques de soi," 1988 (Foucault 2001: 1172–1184, 1527–1548, 1602–1632) and by the now published last lectures at the Collège de France. Verbeek presents his idea of constructively "accompanying" technology with a clear reference to Foucault's original resistive self: "One becomes a subject not by securing a place outside the reach of power but by shaping one's subjectivity in a critical relation to it." (Verbeek 2011, 73) Shaping one's life through "self practices" goes beyond the moralization of technology traditionally promoted by mainstream philosophy of technology. Foucault unexpectedly affirmed, with a book of Ancient Greek ethics in his hands, that philosophy is a work of the self on itself (Foucault 2008: 224, 236).

Who else could dare say that the main question of philosophy is not politics, not even justice and injustice in the city, but justice and injustice as they are acted by the agency of a self, a subject? "[The] question of philosophy is not the question of politics but it is the question of the subject in politics." (Foucault 2008: 295) What is here reinvented as a "pragmatic of the self" (Foucault 2008: 7) can honestly be called wisdom and it lies precisely in the form of the third kind of power defined above, transferring the question of domination from the governance of things and the governance of others to the *governance of the self* (Foucault 2008: 8).

Foucault's basis is the Hellenistic notion of *epimeleia heautou* in Greek, "le souci de soi" in French, and self-care in English. A particular notion of *care* comes from from Heidegger's "Sorge," the existential preoccupation of the self (*Dasein*) within the world. This ordinary mundane preoccupation is in the ordinary *Dasein* a derelict form of care, lacking authenticity according to Heidegger. A different interpretation is offered by Borgmann's concept of engagement with "focal" practices, by Foucault's idea of a resistive subject who transforms governmentality pressure into self-construction techniques, and by the very practice of *samu*, the meditative reappropriation of the most ordinary technology and activity. The "exercise" (*askēsis*), that Foucault proposes as practice of the self naturally leads to *self-care in ordinary life*. The technosphere is a place where one can aim at constituting oneself as a real self (Foucault 2008: 46), endowed with all the capacities of a self as they can be listed from Foucault's references to Hellenistic wisdom – Marcus Aurelius, Seneca, Epictetus, Plutarch, Epicurus. In the technosphere, this *aksēsis* does not entail rejecting technology but rather searching for practical wisdom through and even within technology. In the Zen virtuosity of "having as not having" (and the reverse) or "acting as not acting" (and the reverse) a wise Technosapiens can find the inspiration for a free relationship – in the deepest possible ethical sense of "free" – with his/her smartphone, fridge, car, and the whole cohort of artifacts that populates our proximal technosphere.

This ethically acceptable relationship to artifacts is not derived from deontological or consequentialist reasoning. It is rooted in the ethical consistence of the agent, its self-constructive capacity. Wisdom belongs then to the third major type of reasoning in metaethics, the ethics of virtue (Hursthouse 1999: 59–62; Tiberius 2008: chap. 3). Further, wisdom ethics relies on the ethics of care and importance, formulated by Harry Frankfurt (Frankfurt 1988) and rooted in nineteenth- century philosophers R. W. Emerson's and H. D. Thoreau's theory of self-reliance. *Care* is an ethical meta-attitude or second-order attitude: it bears on the first-order attitudes and actions. First order ethical life includes rare grandiose decisions and an infinite number of micro-decisions in ordinary situations. Both are dignified subjects for caring. The essential virtues of wisdom (awareness, autonomy, harmony, humility, benevolence, courage, etc.) are virtually instantiated everyday in miscellaneous interactions with the world, a large part of it happening through a mediation in the technosphere. Self-care is the adequate virtue for this constant interaction. Slote argues that care is not simply a special part of morality but it provides the basis of an entire system of values (Slote 2007): care, self-care, and one's "management of importance" (caring about what we care about) is the foundation of a pragmatic ethics in modernity, answering the universally lamented dereliction and inconsistency of the modern ego. Frankfurt insists that care does not rely on the arbitrary valuation of certain objects but on the constitutive experience of the subject. Therefore, there is no paradox in the *self* being a *project of the self*, a perpetual auto-hermeneutic and auto-pragmatic process that recalls Ricoeur's pluralistic approach. The objective is self-consistence and, in particular, the coherence between endorsed values and actual behavior. The ethical void in the modern subject was long ago detected by Thoreau, from Stoic sources of wisdom that he shares with Foucault, and also from Asian sources of wisdom that he shares with Hershock. Adapting Buddhist wisdom to modernity, Hershock emphasizes personal consistency, reliance on self-power (Japanese *jiriki*) rather than on other-power (*tariki*), which is external assistance by other people or by magic formulas (Hershock 1999, chap. 10). In the technosphere artifacts are "helpful others" and "magical formulas" at the same time.

The correlation between wisdom virtues and the specific modern environment was already strong in Thoreau: *Walden* (1854) symbolizes the spirit with which modernity is not rejected but wisely inhabited, reappropriated by the means of a step aside, not a step backward (Thoreau 1985: 321–587). In this spirit, wisdom starts in the ordinary of the self, in its proximal sphere of action. The technosphere is not shunned by Emerson and Thoreau, contrary to a technophobic reading of their works: they systematically involve technology in the process of authentic self-building – axes and trains, agriculture and the printing press, shoes for walking in the woods, and boats for navigating the Merrimack river.

20.3 Micro-acting in Harmony

Ordinary wisdom weaves harmony in the technosphere. In this option, reappropriating technology is neither technical nor political, as it was in previous attempts to dominate domination. Using the two first forms of power described above, these attempts remained in the command-and-control hermeneutic of western modernity. Relying on the third form of power, power over oneself, a sapiential approach is also disruptive in terms of scale: it directly addresses micro-behaviors and micro-decisions in ordinary life. The truly *ethical* reappropriation of technology happens on a small scale – the individual appropriation of digital technologies on the Internet and the smartphone being paradigmatic.

Microactions are our factual and moral interface with the technosphere. "I vote 'for' Continental Edison's full range of technical interconnections every time I switch on my electric typewriter," wrote Winner (Winner 1977: 234). Writing or reading on a computer connected to the Internet means now "voting" for a lot of entities that do not even have a name for the public (tracking and advertisement "bots" for instance). Microactions like driving a car or eating a steak are ethical engagements that "vote" for the economic and institutional structures supporting these activities. The corresponding non-actions or alternative actions (walking or cycling, eating something that stands "lower on the food chain") embody different votes. This vote is political, Winner was right although his claim about bridges non suitable for buses proved to be false (Woolgar and Cooper 1999). The whole 1970s' critique has to be extended because it is *more* than a political issue. A sapiential conceptualization focuses on this technoethical "more."

In the climate change predicament, the most frequently discussed issue in "technosphere ethics," Steven Gardiner supports *ethical reappropriation* as the timely post-political and post-institutional alternative. Gardiner argues that the climate issue constitutes a "perfect moral storm" that no existing institution can resist (Gardiner 2011). An ethical turn is required and it can only take place on the level of the individual, by the citizen as a self, in its own agency. The individual person has no real power over energy, oil, or nuclear industries, and his/her personal political vote does not impact these entities, sadly enough. But he/she has full power on his/her own mobility, eating, and lifestyle practices. Furthermore, he/she has a real power of influence on his/her proximal social circle. Using this power means micro-acting for a growing harmony – ecological, economic, social.

The *voluntary simplicity* movement pioneered this form of commitment. It can be characterized as "a way of life that is outwardly simple, inwardly rich." (Elgin 1993 book's title) It relies on this distinction: "Poverty is involuntary and debilitating, whereas simplicity is voluntary and enabling." (Elgin 1993: 27) Frugality belongs to a sapiential virtue ethics of the technosphere. In her project of renewing virtue, Tiberius mentions a major change: our "rational and reflective capacities" should "function together with our emotions, moods, and desires to get us somewhere we'd like to be." (Tiberius 2008: 5–7) Reappropriation is multifarious: the self and its self-construction capacity are reinstated all along while microactions patiently reinstate harmony in the technosphere. The correspondence between the internal harmony of a (stronger) self and the external harmony of a (sustainable) technosphere and ecosphere is an achievement of wisdom.

The intimacy in our interaction with artifacts exceeds existing methods of ethical assessment in engineering ethics and standard applied ethics. Wisdom pertains to this specific level of existential experience centered on the ordinary self in its agency. Renewed wisdom tries to keep at bay the infinite discursive task of assessing competing sets of values and ideological visions. It takes a post-political shift (toward ethical reappropriation of the issues) but also a non-confrontational shift, which is even more uncanny in philosophy. In the prose of ordinary life, we incessantly adopt and adapt technologies, but not passively as it seems when the details of ordinary appropriation are neglected. A radical reappropriation often takes place in the ordinary use of the artifacts that are supplied by industrial actors and by non-profit sources. The micro-choices of users have decided between the defunct elitist encyclopedias and Wikipedia; they have decided between the paternalistic centralized French Minitel and the open Internet; they may decide between proprietary software and "free" software like LibreOffice or Firefox, between state- and corporate-controlled TV channels and the collaborative bazaar on YouTube. These choices are not really confrontational – even when newcomers feed off incumbents, it is not through direct aggression. It is more like a Darwinian progressive taking over. Opting for

LibreOffice is not a militant action, simply the software is available for free and instantly (typical *market* superiority), it is more customizable, and easier to use than its proprietary competitor. Watching YouTube videos is not direct militancy against TV and advertisement corporations but simply bypassing them and ignoring them. Using white vinegar as a cleaning and anti-scale product is not directly "fighting" industrial corporations but wisely simplifying one's life. "Texting" with mobile phones is the paradigm of a creative appropriation by users of a function that was originally implemented as a purely technical by-channel. In this case, the middle way of wisdom is neither bland adoption ("buying" the phone and with it its prescribed use) nor flat rejection (bragging about having no mobile phone). The technosphere allows ordinary non-confrontational wisdom in a myriad of cases. Their cumulative effect impacts the global and out-of-reach domination structure exactly as it produces the collective intelligence of the Web.

Non-confrontation characterizes a form of moral resistivity in the technosphere. The Dao prevails without fighting affirmed Lao Zi, referring to what we would now call ecosystemic harmony, as opposed to human disruptive command-and-control. The sage can and should do the same. This notion of non-confrontation is recurrent in Chinese (both Daoist and Confucian) and Buddhist (mostly Zen) classics. Aikido suggests that confrontation can be reinterpreted from the point of view of harmony: becoming able to perceive the broader situation (oneself, the opponent, the environment) as a scalable harmony and then becoming able to actively enter into the evolution of this harmony. Non-confrontational microactions gave birth to a new form of collaborative commons in the digital technosphere. Engagement in collaborative microactions, aiming at ecological and digital harmonization in particular, is supplanting institutional (public or corporation-driven) governance in a progressively post-political technosphere. Routinely correcting a mistake in a Wikipedia article means participating in digital harmony. This whole move does not result from a "cyber-libertarian" ideology but from the coevolution of human culture and technology on the most ordinary mediation level, through the clicks and non-clicks of ordinary people on the Web: through myriads of microactions and non-actions.

Wisdom in the technosphere is a practice of harmony consisting in two stages: (1) perceiving the immanent harmony in any situation, (2) taking action or non-action to enter into this harmony and enhance it. This attitude is diametrically opposed to the western industrial mind: (1) planning a desirable state of affairs and locating the maximum available means to impose it, (2) using the most efficient and direct means to carry out the plan. The optimism that is involved is not passivity. Harmony invites engagement in transformative action. Optimism only affirms that a transformative action toward a better harmony is always possible. Martin Luther King and Gandhi were motivated by the vision of a civil harmony that did not obtain at that time. Local and global harmonies are taken into account by an active awareness of the ecological "cost" of one's lifestyle – concerning food and its quotidian micro-decisions, concerning energy use, consumption but also production. A good reason to carefully maintain and rarely renew our electronic appliances, for instance, comes from an understanding of all the ecological and economic disharmony that would ensue, from sweatshop style factories all the way down to e-waste dumping in remote poor countries. From an "external" perspective, wisdom bets on feeling, wishing, and realizing harmony as a more efficient commitment to ecological and economic justice than the institutional and political means in use since the middle of the twentieth century that have so far only marginally improved the technosphere. From an "internal" perspective, realizing harmony in one's life and circumstances is a particularly self-rewarding exercise and at the same time it is arguably a self-justifying virtue as it leads to serenity, the most desirable consequence of wisdom.

Selecting and appropriating technology in a constructive project for one's life and for the collective remains possible in the technosphere. Some would say it is still waiting for the right ideology that would offer the right norms. I argue instead that harmony in the technosphere, on every level from the most intimate to the most global, does not depend any longer from top-down normative governance; it is invented while microacting in the technosphere, through *caminar preguntando*, generating acceptable and scalable consensus while advancing.

Every step in this process is clearly a question of *judgment*, practical judgment, an exercise that seems to be neglected theoretically as well as practically. To be or not to be on Facebook? Replacing one's phone by a new one or waiting for the next generation? Buying organic food or not? Using car, train, plane, or staying at home? What philosophers call "judgment" is needed, both in the Kantian sense (decision without the support of demonstrated knowledge) and in the Aristotelian sense (practical wisdom acknowledging the diverse parameters of a complex given situation). In deprecating wisdom our culture has progressively lost any consideration for these kinds of skills and their acquisition in ordinary life. However, it is the capacity for judgment that makes the difference between being an object in the networks of the technostructure and being the subject of a human life.

A wise self-realization is inclusive of other humans and of the environment – all kinds of environments, ecological and social environments. Holism is therefore entailed in the idea of harmony. Arne Naess's modern wisdom organically connects self-realization, central to his ethics, and the capacity of wonder in nature, equally central to his ethics (Naess 1989, 2005). Ordinary technology, and particularly digital communication technologies in one's proximal technosphere, offers multiple opportunities for an empathic rebuilding of the social, self by self, through multiple connections. Digital communication has proved to be a decisive resource for people in need – all sorts of outcasts, social, economical, cultural, and political (Katz 2008) – and for people engaged in societal protest (Castells 2012). Constructive societal action in a new form is mediated by the technosphere's networks, initiating a first sapiential action: "capacitating" the self.

Perceiving harmony, valuing harmony, and realizing harmony, the essential wisdom skills to deploy in the technosphere, are not "selfish." Wisdom is not individualism in an egoistic acceptation. Instead, it is the active engagement in the very contrary of egoism: the interdependence of all beings and of all spheres. Hershock envisions a specific dynamic for flourishing out of diversity and interdependence that is based on "the primacy of relationality." (Hershock 2012: 11) Far from individualism, we must understand the human person not as a separate rational (calculating) entity but as immersed in "an ethics of interdependence" (Hershock 2012:

16) and endowed with "a new structure of feeling" (Hershock 2012: 241) through which contributing to harmony is something felt and not concluded. The whole ecosphere and technosphere are in the scope of this ethics.

Technosapiens continuously invents itself in reappropriating the affordances of its environment. The disruptive novelty of contemporary technological mediation invites focusing more on the self's capacity and less on the various determinations that were the models for human and social sciences in the industrial era. Thus, the "potential difference," in the electrical sense of the term, between these two notions, *wisdom* and *technosphere*, can be a source of energy if and only if an appropriate circuitry is available to conduct this energy in a constructive way. Otherwise the outcome will be a short circuit.

Acknowledgements A preliminary version of this text has been discussed by Jean-Yves Goffi and Daniel Cérézuelle during the authors' workshop held in Paris Sorbonne on 22–23 June 2015 for the preparation of the collective book.

References

Alexander, J. K. (2008). *The mantra of efficiency: From waterwheel to social control*. Baltimore/London: John Hopkins University Press.

Anthropocene. (2015). *Wikipedia, The free encyclopedia*. http://en.wikipedia.org/w/index.php?title=Anthropocene&oldid=656935652

Aristotle. (350AD). *The nichomachean ethics* (W. D. Ross, Trans.). http://classics.mit.edu/Aristotle/nicomachaen.html

Aubenque, P. (1963). *La prudence chez Aristote*. Paris: Presses universitaires de France.

Barthélémy, J.-H. (2014). *Simondon*. Paris: les belles lettres.

Binkley, S., & Capetillo, J. P. (Eds.). (2009). *A Foucault for the 21st century: Governmentality, biopolitics and discipline in the new millennium*. Newcastle: Cambridge Scholars.

Borgmann, A. (1984). *Technology and the character of contemporary life: A philosophical inquiry*. Chicago: University of Chicago Press.

Bynum, T. W. (2006). Flourishing ethics. *Ethics and Information Technology, 8*(4), 157–173. https://doi.org/10.1007/s10676-006-9107-1.

Castells, M. (2012). *Networks of outrage and hope: Social movements in the internet age*. Cambridge/Malden: Polity.

Dorrestijn, S. (2012). *The design of our own lives technical mediation and subjectivation after Foucault*. Enschede: University of Twente. http://members.tele2.nl/s.dorrestijn/downloads/Dorrestijn_Design_our_own_lives.pdf.

Elgin, D. (1993). *Voluntary simplicity: Toward a way of life that is outwardly simple, inwardly rich*. New York: Morrow.

Ellul, J. (1977). *Le système technicien*. Paris: Calmann-Lévy.

Foucault, M. (1984). *Le Souci de soi. Histoire de la sexualité..., 3*. Paris: Gallimard.

Foucault, M. (2001). *Dits et écrits, 1954–1988, II: 1976–1988*. Paris: Gallimard.

Foucault, M. (2008). *Le gouvernement de soi et des autres*. Paris: Gallimard.

Frankfurt, H. G. (1988). *The importance of what we care about: Philosophical essays*. Cambridge/New York: Cambridge University Press.

Galbraith, J. K. (1958). *The affluent society*. Boston: Houghton Mifflin.

Gardiner, S. M. (2011). *A perfect moral storm: The ethical tragedy of climate change*. New York: Oxford University Press.

Hadot, P. (1995). *Philosophy as a way of life: Spiritual exercises from socrates to Foucault*. Malden: Blackwell.
Hershock, P. D. (1999). *Reinventing the wheel: A Buddhist response to the information age*. Albany: State University of New York Press.
Hershock, P. D. (2005). *Chan Buddhism*. Honolulu: University of Hawaii Press.
Hershock, P. D. (2012). *Valuing diversity: Buddhist reflection on realizing a more equitable global future*. Albany: State University of New York Press. http://search.ebscohost.com/login.aspx?direct=true&scope=site&db=nlebk&db=nlabk&AN=549600.
Hershock, P. D., Stepaniants, M. T., & Ames, R. T. (Eds.). (2003). *Technology and cultural values: On the edge of the third millennium*. Honolulu: University of Hawaii Press/East-West Philosophers Conference.
Hursthouse, R. (1999). *On virtue ethics*. Oxford/New York: Oxford University Press. http://site.ebrary.com/id/10273328.
Ihde, D. (1971). *Hermeneutic Phenomenology; the Philosophy of Paul Ricoeur*. Evanston: Northwestern University Press.
Ihde, D. (1979). *Technics and praxis*. Dordrecht/Boston: D. Reidel Pub Co.
Irvine, B. (2011). Wisdom and the good life: A philosophy of conscience. *Journal of Modern Wisdom, 1*, 4–11.
Ives, C. (1992). *Zen awakening and society*. Honolulu: University of Hawaii Press.
Janicaud, D. (1985). *La puissance du rationnel*. Paris: Gallimard.
Kane, R. (2010). *Ethics and the quest for wisdom*. Cambridge/New York: Cambridge University Press. http://site.ebrary.com/id/10412985.
Kaplan, D. M. (2003). *Ricoeur's critical theory*. Albany: State University of New York Press.
Katz, J. E. (Ed.). (2008). *Handbook of mobile communication studies*. Cambridge, MA: MIT Press. http://search.ebscohost.com/login.aspx?direct=true&scope=site&db=nlebk&db=nlabk&AN=228453.
Latour, B. (1991). *Nous n'avons jamais été modernes: essai d'anthropologie symétrique*. Paris: Éditions la Découverte.
Mei, T. S., & Lewin, D. (Eds.). (2012). *From Ricoeur to Action: The socio-political significance of Ricoeur's Thinking*. Edited by T. S. Mei & D. Lewin. Continuum studies in continental philosophy. New York: Continuum.
Musō, S. (1994). *Dream conversations on Buddhism and Zen* (T. F Cleary, Ed.). Boston: Shambhala.
Naess, A. (1989). *Ecology, community, and lifestyle: Outline of an ecosophy*. Cambridge/New York: Cambridge University Press.
Naess, A. (2005). *The selected works of Arne Naess* (H. Glasser & B. Devall, Eds., Vol. X). Dordrecht: Springer.
Passet, R. (1996). *L'économique et le vivant* (2nd éd.). Paris: Économica.
Puech, M. (2008). *Homo Sapiens Technologicus. Philosophie de La Technologie Contemporaine, Philosophie de La Sagesse Contemporaine*. Paris: Editions le Pommier.
Puech, M. (2013). Ordinary Technoethics. *International Journal of Technoethics, 4*(2), 36. https://doi.org/10.4018/jte.2013070103.
Puech, M. (2016). *The ethics of ordinary technology*. New York: Routledge.
Reeve, C. D. C. (2013). *Aristotle on practical wisdom: Nicomachean ethics VI*. Cambridge, MA: Harvard University Press.
Ricoeur, P. (1990). *Soi-même comme un autre*. Paris: Ed. du Seuil.
Simondon, G. (1958). *Du mode d'existence des objets techniques*. Paris: Aubier.
Slote, M. (2007). *The ethics of care and empathy*. London/New York: Routledge.
Spence, E. H. (2011). Information, knowledge and wisdom: Groundwork for the normative evaluation of digital information and its relation to the good life. *Ethics and Information Technology, 13*(3), 261–275. https://doi.org/10.1007/s10676-011-9265-7.
Thoreau, H. D. (1985). *A week on the concord and Merrimack rivers; Walden, or life in the woods; the Maine woods; cape cod*. New York: Literary Classics of the United States.

Tiberius, V. (2008). *The reflective life: Living wisely with our limits*. Oxford/New York: Oxford University Press. http://public.eblib.com/choice/publicfullrecord.aspx?p=737540.

Verbeek, P.-P. (2005). *What things do: Philosophical reflections on technology, agency, and design*. University Park: Pennsylvania State University Press.

Verbeek, P.-P. (2011). *Moralizing technology: Understanding and designing the morality of things*. Chicago/London: The University of Chicago Press.

Vernadsky, V. I. (1945). The biosphere and the Noosphere. *American Scientist, 33*, 1–12.

Vernadsky, V. I. (1998). *The biosphere*. New York: Copernicus.

Winner, L. (1977). *Autonomous technology: Technics-out-of-control as a theme in political thought*. Cambridge, MA: MIT Press.

Woolgar, S., & Cooper, G. (1999). Do artefacts have ambivalence? Moses' bridges, Winner's bridges and other urban legends in S&TS. *Social Studies of Science, 29*(03), 433–449.

Chapter 21
Design Made in France: Perspectives on "Industrial Aesthetics" (1951–1984)

Vincent Beaubois and Victor Petit

Abstract To be sure, the very term "design" was not much in use in the French academic field until the 1980s. However, an important line of thought concerning design had already undergone considerable development in the 1950s under the name "industrial aesthetics." This chapter tries to follow this activity in France between 1951 and 1984 and the development of an original design theory rooted in French philosophies of art and technology. This movement was based on a non-Kantian aesthetics introduced by Paul and Etienne Souriau, acknowledging technology as a real issue for aesthetics. Then, it was also connected with a philosophy placing technology in relation to its living and social milieu (Mauss, Leroi-Gourhan, Friedmann, Simondon). It is along these lines that we analyse the development of the Institute of Industrial Aesthetics, created in 1951 by Jacques Viénot, relating the continuous dialog between this specific design theory and its philosophical backgrounds. Indeed, during this period, a design theory based on technology and its socio-cultural effects has been developed, leading to an ecology of technology.

Keywords Design · Ecology of technology · Friedmann · History of design · Industrial aesthetics · Leroi-Gourhan · Simondon · Souriau · Technical culture · Technical milieu · Viénot

---

V. Beaubois (✉)
Histoire des Arts et des Représentations (HAR), Université Paris Nanterre, Nanterre Cedex, France
e-mail: vincent.beaubois@u-paris10.fr

V. Petit
Connaissance, Organisation et Systèmes TECHniques (COSTECH), Université de technologie de Compiègne (UTC), Compiègne, France

S. Loeve et al. (eds.), *French Philosophy of Technology*, Philosophy of Engineering and Technology 29,
https://doi.org/10.1007/978-3-319-89518-5_21

The very term "design" became late established in French institutions. It was only in 1984 that the Institute of Industrial Aesthetics (Institut d'Esthétique Industrielle), created in 1951 by Jacques Viénot, became the French Institute of Design (Institut Français du Design), promoting and enhancing design creation in France. However, this lack of the term "design" in the institutional field during the 1950s does not mean a lack of activity in the area of French design. Instead, by the end of the Second World War, the practice of design simultaneously grows with a significant French thought of design. The object of this paper is to account for this consideration: a thought of design before the very term "design" has even become generalized in French.

This French thought of design can be defined as a *non-technical* understanding of technology, considering technology as always linked to a cultural and human environment. This philosophical and industrial problematization and organization of the practice of design in France in the years 1940–1950 have been called "industrial aesthetics." The figure of Jacques Viénot polarized this field of thoughts concerning the aesthetics of technology, the place of this aesthetics in the development of industry and its role in the dwelling environment.[1] The singularity of this movement is to appear as symptomatic of a moment of *synthesis* in the history of European design. It highlights a tension between an aesthetics invested in industrial production and consumption (as developed in the United States) and the legacy of the European artistic avant-gardes invested in the social transformation of the dwelling environment through the development of technology.[2] The very term "industrial aesthetics" expresses this tension, translating the US "industrial design" and confronting it with the European aesthetics tradition.

The originality of the French industrial aesthetics lies in the reinvestment of these issues at the institutional level. The Institute of Industrial Aesthetics, created in the early 1950s, functioned as an original apparatus for allowing academic philosophers (Étienne Souriau, Georges Friedmann) to interact with industry (Dunlop, Philips, Arthur Martin, etc.), artists and designers (Fernand Léger, Charlotte Perriand), and representatives of public authority (especially Eugène Claudius-Petit, Minister of Reconstruction between 1948 and 1953). This kind of composite dialogue continued in the following decades, accompanying the development of design in France. From the "industrial aesthetics" of the 1950s to the promotion of an "industrial creation" in the 1980s, we will sketch a path following three cardinal concepts polarising the understanding of design during this period: *aesthetics*, *milieu* and *technical culture*. The discussion will involve, firstly, showing that the theoretical constitution of design in France inherits a reconfiguration of philosophical aesthetics by integrating an aesthetics of technology (Paul and Étienne Souriau,

[1] Jacques Viénot founded in 1948 the pioneering French design agency *Technès*, and the journals *Art Présent* in 1945 and *Esthétique industrielle* in 1951. In the same year, he founded the Institute of Industrial Aesthetics, an organization dedicated to making connections between designers and manufacturers.

[2] Including, for example, John Ruskin, William Morris. Henry van de Velde, Walter Gropius or Hermann Muthesius.

Jacques Viénot). We will then highlight the cross-development of this aesthetics with a reflection on "technical culture" trying to connect technology with cultural value. The concept of "milieu" in these debates (Georges Friedman, Gilbert Simondon) appears thus as an important contribution to shifting the question of an aesthetics of technology onto the premises of a design understood as an ecology of technologies: opening technology to cultural and vital fields is thus engaging design to create objects opened to alteration.

21.1 Industrial Aesthetics

21.1.1 Paul and Étienne Souriau: Rethinking Philosophical Aesthetics Through Technology

The originality of the very term "industrial aesthetics" manifests a double ambition: to transform simultaneously industrial creation through philosophical aesthetics and the old philosophical discipline of aesthetics – founded by Baumgarten in the eighteenth century – through the context of industrial production. This very term "industrial *aesthetics*" has often been misunderstood as suggesting a determination of design as a by-product of Fine Arts, connoting simply a decorative and cosmetic attitude to technical reality. However, this confusion is not valid if we restore a part of the evolution of French philosophical aesthetics in the second half of the twentieth century. The figures of Paul and Étienne Souriau, father and son respectively, had an important role in this recasting of philosophical aesthetics through an original industrial aesthetics thinking.

Paul Souriau took interest in the issue of an aesthetics of technology, notably in his seminal work *Rational Beauty* (*La Beauté rationnelle*) (1904) which would remain a reference for the proponents of industrial aesthetics during the 1950s. If French aesthetics, in the new century, turned to an experimental approach being influenced by Claude Bernard and Hermann von Helmholtz scientific methods, Souriau sought to rethink aesthetics by restoring its philosophical principles. He operated an aesthetic re-reading of machines and other items typically excluded from the aesthetical field by questioning particularly the principle of "purpose." While Kant denied any aesthetic claims to technical objects due to their utilitarian purpose, Paul Souriau intended to reverse Kantian aesthetics by problematizing the very notion of purposiveness.

Kant expelled the technical object from the aesthetical sphere both for the user and for the producer. From the perspective of the user, the technical object is perceived only as a means to utilitarian satisfaction. To use an object is just to follow a personal and interested satisfaction, making the universality of this experience impossible. From the perspective of the producer, the technical object is the product of an unpleasant activity carried out only for a salary (Kant 1790). In both cases, the value of the technical object is characterized by a personal appeal: its utilitarian dimension, both for the user and the producer, prohibits its admission in the aesthetical field, which is disinterested by definition.

Against Kant, Souriau reinvested the aesthetics of utility by giving it an objective value: there is a *disinterested pleasure* in usefulness (Souriau 1904: 204). Technical objects can be aesthetically experienced *by the very fact* of their usefulness, not in the sense that they are useful *to me*, but insofar as they present a viable functioning. This return to the notion of "purpose" does not have to be understood as an apology for an abstract concept, but instead as a heuristic principle exciting feelings and reason in the understanding of objects in the world. As such, according to him, there is not *one* purpose but as many purposes as singular objects. Therefore, there is, in the technical object, "as much thought, intelligence, purpose or, in a word, true art as there is in a masterpiece or a statue." (Souriau 1904: 392) Thus, Paul Souriau appeared in the 1950s as the founder of a renewed French aesthetics seeking to expand its sphere to technology (Huisman and Patrix 1961: 14).

However, even if his views differed from the Kantian aesthetics in force at that time (especially through the person of Victor Basch), Paul Souriau remained ensnared by Kant's formulation of the aesthetical problem in terms of "beauty." This is the limit his son, Étienne Souriau, points out: focusing on "purpose" as a criterion, Paul Souriau was condemned to a *static aesthetics*, an aesthetics of "work made" (which is what it is in accordance with its purpose), whereas Étienne Souriau would question "work *in the making*." (Souriau 1956)[3]

Abandoning the work made for the "making," Étienne Souriau became the main theorist of an industrial aesthetics less defined by the objects it produced than the processes it instilled in the industrial world. Presiding over the Chair of Aesthetics in the Paris-Sorbonne University, he was also one of the members of the patronage committee of the Institute of Industrial Aesthetics (Souriau 1952, 1954).

Deviating, like his father, from the psychological aesthetics advocated by Basch, he focused his works on the *genesis of form*. In this regard, he defined aesthetics neither as beauty nor art, but as a cross-discipline he called "science of forms." (Souriau 1929: 9) Thus, rethinking aesthetics involves a redefinition of the concept of "form": form is not a static essence but a certain unity engaged in a becoming. Souriau intended to reconcile the bergsonian thought of becoming with the notion of form: while Bergson reduced form to be only a "snapshot view of a transition" (Bergson 1907: 328), Souriau considered forms as units of becoming. In this way, the aesthetical study of *forms* shifts to considering processes of *formation*, what Souriau calls *instauration* of forms. Instauration process considers creation as a transitive relation between the producer and the form in becoming. To understand a form is not observe it as a final result but to understand its genesis, its instauration. This attention to the creation of coming forms resonated strongly with the practice of industrial design as the Institute of Industrial Aesthetics conceived it. According to Jacques Viénot, design did not determine some plastic quality of objects, but a singular *process* involving technical and artistic operations. Viénot did not define the industrial designer as a creator of objects, but as someone who primarily led the

[3] Souriau clearly states: "I will avoid any advocacy of the idea of purpose." (Souriau 1956: 200)

development of "prototypes," first phase of industrial production (Viénot 1939: 77). In the same way, Souriau understood the practice of industrial aesthetics as an instauration of a material dwelling environment. The industrial designer is responsible for "instaurating forms of the world where today and tomorrow's men will practically live." (Souriau 1952: 19) Thus, the Institute of Industrial Aesthetics seeks to justify its existence as an organization interlinking industrial with "industrial aestheticians": to instaurate the form of an artifact cannot be improvised and must call on real professionals.

21.1.2 Towards Implied Arts

Abandoning the classic expression "applied arts," Étienne Souriau coined the term "implied arts" to demonstrate the instaurative power of industrial aesthetics (Souriau 1952: 12; Viénot 1953: 22). The notion of applied arts is a recent one, contemporary with the aesthetic upheaval produced by industrialization. This is an abbreviation of the "arts applied to industry" phrase formulated in 1863 with the founding of the Central Union of Fine Arts Applied to Industry, which would become the Central Union of Decorative Artists 20 years later. The concept of "applied arts" suggests that technology, taken by itself, would have an aesthetic *defect* that would need to be repaired by adding an aesthetic dimension. Instead, the notion of *implied* arts involves that technology is not a separate field from aesthetics. Technology – and furthermore industrial technology – is aesthetic due to its role in the transformation of conventional aesthetics standards, creating new material forms.

The main feature of an "implied" art is to be a *reticular* art diffusing simultaneously throughout the different levels of industrial production and throughout the different elements of the dwelling environment. Industry is primarily a technical whole, it is "a complex of […] different modes of work, grouped together for particular practical convenience." (Souriau 1929: 142) To penetrate the industrial environment is venturing into a systemic universe, into a network of machines, crafts, raw materials and operations. While the French modernist avant-gardes of the early twentieth century took the single machine as a model (Duchamp, Picabia), the industrial aesthetics in the 1950s was more interested in industry as a technical and economic complex. Therefore, industrial aesthetics is constituted as a proliferating aesthetics, developing step-by-step and spreading over the entire dwelling environment. Industrial aesthetics extends "from the dam to the pen" to borrow a phrase from Viénot (1948). The work of industrial aesthetics is not focused on an isolated artifact, but polarises on the resonance existing between the different artifacts populating our environment. Industrial aesthetics designs a "society" of objects before designing individual items, a dwelling environment before isolated artifacts. Forms of refrigerators, machines tools, telephones and radios, health equipment and a thousand other devices refer to each other forming the unity of our living space. Aesthetics operates more in the relations between forms than in the form itself. Art is thus *implied* in all folds of everyday life:

> Implied is a word having a rich meaning; it means that art should be everywhere, art is implied (or should be) in all works: in a camera as much as in a teapot; in a telephone switchboard as much as in a furniture; in a typewriter as much as in a watch case. (Viénot 1953: 22)

21.1.3 Reinventing a European Tradition of Industrial Design against American Competition

As reported by Jacques Viénot, industrial aesthetics is nothing if public authorities and industry do not make industrial aesthetics a priority. According to him, the context of the major projects of reconstruction has to be used to spread industrial aesthetics and to strengthen the influence of France in a European Common Market under construction. The purpose of the journals created by Viénot (*Art Présent* in 1945 and *Esthétique Industrielle* in 1951) explicitly aims to question public authorities about design. These intentions are clear from the first issue of *Art Présent*:

> *Art Présent* is for: public authorities to inform them about the real needs of our time, to help them to know the deserving creators and their works, to decide what needs to be done or encouraged for the city, for the country. (Viénot 1945: 4)

The aim of industrial aesthetics in the 1950s is not to reform public taste, as was the case for the design movements in the early twentieth century (like *Bauhaus* or *De Stijl*), but instead to target exclusively industrial producers and governments. The Institute of Industrial Aesthetics was inaugurated in January 1951 for this purpose. Viénot wanted to develop an apparatus influencing on the interface between manufacturers and industrial designers based on the Council of Industrial Design model created in London in 1944 (Le Bœuf 2006: 91). Similar organizations developed throughout Europe, especially in Germany around the Ulm Hochschule für Gestaltung directed by Max Bill, or in Italy with the Olivetti artistic advisor Egidio Bonfante and his sponsorship-based policy. Viénot brought these figures together during the International Industrial Aesthetics Congress in Paris in 1953, seeking to make the Institute the federative outpost of European industrial design. Then, Viénot worked on creating an International Liaison Committee of Industrial Aesthetics, first step leading to the founding of the International Council of Societies of Industrial Design (ICSID, still active) in 1957 (Le Bœuf 2006: 119–127).

The wish to develop an international anchoring of design while claiming a French and European singularity in this field reflects a paradoxical historical situation of industrial design. While the origin of this practice was located in Europe through movements such as the Deutscher Werkbund or Bauhaus at the beginning of the century, the concrete industrial development of design was mainly carried out within US companies after 1930 under pressure from competition following the 1929 crash. Thus, North America appeared as the origin of a *truly industrial* design. James S. Plaut, founder and first director of the Boston Institute of Contemporary

Art, noted that American designers pictured themselves as the source of the discipline:

> Some claim that industrial design is a purely American phenomenon, rooted in mass-production conditions and owing its development to American industrial and technological skills. Some of our most famous and most successful specialists in this field claim, basically, to have invented industrial design and pictured themselves as pioneers. (Plaut 1954: 133)

A European delegation composed of representatives from ten countries, future members of the Common Market, was even mandated at the end of 1955 to visit American companies concerned with industrial design (OECE 1959: 5).

In the 1950s, this "double" origin made the understanding of the historical constitution of design as a discipline opaque: it is therefore no coincidence that some art historians, in the 1940s and the 1950s, questioned the origin and foundation of this practice. The question of a "history" of design arises in this period: to write a history of design is to found a discipline by assigning it an origin, a tradition and common events.[4] The design history books that proliferate in this period (from Pevsner to Banham) participate in a "rediscovery" of the European sources of industrial design (from William Morris to Bauhaus).

The French concept of "industrial aesthetics" is part of this movement of rooting the discipline in European field, trying to stand out from the United States model after World War II. To translate "industrial design" into "industrial aesthetics" means to assert its European identity to stand the competition of the United States in a period of a global economy. Limiting the meaning of industrial "aesthetics" to a return to the "beautiful" in the field of technology can absolutely not catch the real issues of this moment of French design: it is rather to unify and situate a practice reinventing its tradition. Thus, Viénot claimed the choice of the term "aesthetics" to translate "design": "Industrial aesthetics is a 100 percent originated in Europe discipline and expressing it by a misnomer English term seems shocking to us." (Viénot 1955: 30)[5]

The French industrial aesthetics operates as a paradoxical and constitutive moment in the history of European design, simultaneously assuming the US model as an industrial and economic achievement and rejecting the same model to consolidate a European historical specificity. It is a matter of differing oneself from the United States while comparing oneself to this country. Industrial aesthetics occurred when the aspirations of European avant-garde design were converted to fulfil market demands, requiring the accomplishment of the early century functionalist ideals in industrial and commercial terms.

[4] The publication of *Story of Modern Applied Art* (1948) by Rudolph Rosenthal and Helena Ratzka, the reissue in 1949 of *Pioneers of the Modern Movement* (1936) by Nikalaus Pevsner, renamed *Pioneers of Modern Design* for the occasion, the publications of *Art and Technology* by Pierre Francastel (1956), *Origins of Functionalist Theory* (1957) by Edward Robert Zurko, *Theory and Design in the First Machine Age* (1960) by Rayner Banham, mark the construction of industrial design as a recognized discipline.

[5] The practice of American designers connotes for him a cosmetic and too commercial approach: "Industrial Aesthetics returned from US in the form of "industrial design," that is to say, as a method of sale singularly reduced to a problem of turnover." (Viénot 1955: 30)

21.2 Towards a Technical Culture

21.2.1 Techno-logy As a Social Science of the Technical Milieu

Industrial aesthetics in the 1950s echoed another developing discipline: "techno-logy" understood literally as an epistemology of technology. As we have seen, industrial aesthetics differed from engineering because promoting a not exclusively technical view of technology. In this way, it participated in the development of the concept of techno-logy as it had developed in the social sciences in France from the work of Marcel Mauss. The latter considered technology as a total social fact engaging the whole human being (Mauss 2012: 288). The "technology" section in the journal *L'Année sociologique*, in which Mauss was involved, increased at the end of World War II. Georges Friedmann, who is sociologist, philosopher and later, a member of the Institute of Industrial Aesthetics, was in charge of this section. He particularly emphasized the importance of the psychosociology of technology (Mauss 2012: 421–429). The thought of Friedmann changed in the post war period: between the 1940s and the 1960s, he moved from the Marxist critique of capitalist illusion to a moral criticism of technicist illusion. In the pre-war period, he praised technical progress, after the war he claimed the necessity of humanising it (Petit and Guillaume, Chap. 6, this volume).

This conversion in his work specifically coincides with his commitment to the Institute founded by Viénot (Friedmann 1954). This interest Friedmann shows in technology gives birth to his concept of "technical milieu." According to him, industrial aesthetics is a way to understand that any technical milieu is also a psychosocial milieu (Friedmann 1954). This notion of "technical milieu" was developed by André Leroi-Gourhan (1945) and by Georges Friedmann (1966) as well. Both considered that human beings are constituted by this technical milieu; however, they did not understand this notion in the same way. For Leroi-Gourhan, a paleoanthropologist, "milieu" designates a general category for analysing technology: the "technical milieu" means above all the product of human evolution, the result of dynamic interplay between human groups and their environment.[6] For Friedman, a sociologist of labor, "technical milieu" refers rather to the industrial stage of technology development, in which the space and time of a "natural milieu" has dissipated. Nevertheless, for both Leroi-Gourhan and Friedmann, technology does not have to be considered under the category of "means" but in terms of "milieu." Understanding the concept of "milieu" requires making the difference between "milieu" and "environment." The term "environment" presupposes human being as *external* to this environment. Technology is precisely not a simple environment because technology is not external to our being: it is constitutive of our lives. A flint, an assembly line or a computer, as technical milieus, are at the same time biological, psychological, and social facts.

[6] See Petit and Guillaume (Chap. 6, this volume), for precisions on Leroi-Gourhan's conception of the "technical milieu."

The importance of the concept of "milieu" for thinking technology resonates also with the work of Gilbert Simondon. Simondon takes a new look at the understanding of this notion: this category does not apply only to subjects but also to technical objects themselves. According to him, a technical object becomes a *technical individual* creating its own "associated milieu" when its configuration becomes more efficient (more "concrete" following the simondonian term) and self-regulating with respect to its external milieu. The technical object creates the conditions of possibility of its functioning, such as in the case of the Guimbal turbine working only if it determines a "techno-geographical milieu" that in return conditions it (Simondon 1958: 54–55).

In his "Interviews About Technology" (*Entretiens sur la technologie*) conducted in 1965, Yves Deforge, who was a student of Simondon, highlighted the significance of the notion of "milieu" to think a techno-logy: "To summarise, we can say that technology is the knowledge of the correlations that make the technical object in agreement with itself, with the milieu and with the user." (Deforge 1966: 4) This interviews led to interrogate the intellectuals of the time about the sense of techno-logy: thus, Leroi-Gourhan explained that "it is impossible to separate techno-economics, socio-economics, and techno-logy" (Deforge 1966: 9) and Simondon pointed out that "techno-logy could be presented as including a normative aspect, an aspect of integration within culture, an aspect pretty close to aesthetics and perhaps morality." (Deforge 1966: 24) In this way, a techno-logy understood as a *human science* (Haudricourt 1964) developed, resonating with questions about industrial design engaged by Viénot and Souriau.

According to Simondon, the technical nature of an artifact is defined by *relation* to its genesis and its milieu. Connected being and connecting being, the technical object is essentially a *mediation*, i.e. the opposite of a substance. Technology is never entirely within the object, but within the object and its milieu. This point directly echoes with the industrial aesthetics issue as expressed by Georges Patrix. Patrix is an industrial designer pioneering this profession in France (especially for the use of colour) and an active member of the Institute of Industrial Aesthetics. According to him, the purpose of industrial aesthetics is not to focus on a particular object but to take into account the object in relation with its milieu. Industrial aesthetics is interested in a scale bigger than the object scale, which is why we should "go on from the little problem to the big problem which is truly the environment of Human being." (Patrix 1962: 101) Industrial aesthetics must therefore lead directly to an *environmental design* as Patrix suggested it (Patrix 1973).[7] Thus, industrial design expresses a view on techno-logy attentive to both the milieu of the object and the object as a milieu. Design deals with technical objects as they are a *milieu*, in both senses of the word: as intermediaries or mediators of relations, and as ambiance or dwelling environment. The reticular dimension of implied arts defended by Étienne Souriau thereby culminates in this notion of milieu.

[7] Patrix uses the terms "environment" and "milieu" interchangeably, leaving just floating the idea of "milieu" as not being away from us as the term "environment" would seem to presuppose.

21.2.2 Technical Culture

In the early 1980s, Jocelyn de Noblet deplored the fact that the term "design" was not widespread in the French industrial field.[8] Himself a designer, de Noblet was the director of the Research Centre on Technical Culture (Centre de Recherche sur la Culture Technique – CRCT), established in 1979, and formerly known as "Ethnotechnologie," an informal group of reflection animated by Thierry Gaudin in the margins of the French Ministry of Research where Gaudin was in charge of developing funding schemes for launching a new "innovation" policy (at the time, a new word coming from the US). On 23 July 1980, the Council of Ministers places the question of design in its agenda (Gaudin and de Vendeuvre 1981). While public authorities began to examine the question of design, the CRCT was one of the main proponents of design in France. The main task of the CRCT was to defend "technical culture," that is to say to integrate technology into the general culture. While we often speak of literary and artistic culture or scientific culture, technology is generally banned from the cultural field. The CRCT, through the journal *Technical Culture*, intended to defend the cultural value of technology, promoting its historical study and its emancipating use. Leroi-Gourhan, in his preface to the *Manifesto for the Development of Technical Culture* coordinated by de Noblet, stated that "'design' (…) is one of the best ways of promoting technical culture." (Leroi-Gourhan 1981: 6) This manifesto assumed that technical culture was the only way not to be subservient to the technical milieu:

> The person who lacks technical culture lives in ignorance of his own milieu. He is, therefore, doubly subservient: firstly he does not control his own environment, and secondly his lack of control makes him permanently socially dependent to organizations and individuals who have the skills he is missing. (De Noblet 1981a: 11–12)

The lack of technical culture promotes design and use of objects as "black boxes" (characterized in particular by the famous Kodak advertising slogan: "You Press the Button, We Do the Rest"). Following this way, the technical object becomes "as foreign as a foreign language" (Simondon 2014: 66). Against this "black box" design, authors of the manifesto understood technical culture as "the thinking at work when a technology is beyond any user control" (De Noblet 1981a: 43). Technical objects have to cease to be "foreigners" for users and the development of technical culture must go through an educational reform:

> It would be essential to instaure a general common-core syllabus devoted firstly to the 'genetic lineage' of technical objects and to the knowledge of their 'milieu', and secondly to a teaching about the 'use' and knowledge of machines. (De Noblet 1981a: 55)

[8] Despite the European adoption of the term "design" (made official by ICSID in 1959) it diffused in France with difficulty. Moreover, in the 1980s, the term "industrial aesthetics" is not understood as defended by Viénot (as implied arts) but on the contrary as applied arts: "The term 'industrial aesthetics', still commonly used, is often considered as pejorative in the fields close to the practice of industrial design because it is inadequate and confusing, but also because it points out a false practice of industrial design." (De Noblet 1981b: 15)

Jocelyn de Noblet linked the development of technical culture with the development of design: "one of the consequences of the lack of technical culture in France is our weakness in the field of industrial design." (De Noblet 1981a: 88) This thought of design would engage in what Simondon was calling for: an *action-research* into the unity of culture, in other words, into the unity of human culture and technical civilization.

This defence of techno-logy as a human science inherited from Mauss, Leroi-Gourhan, Friedmann, and Simondon, found itself formulated in the founding of the Compiègne University of Technology (*Université de Technologie de Compiègne* – UTC) in 1972 (Lamard and Lequin 2006). This university became the first engineering school to receive an industrial design training (Quarante 1981). Deforge, who we mentioned above, was lecturing at Compiègne on "*culture technique*," "*génétique des objets industriels*," "*ethno-technologie*" and "*design*" as soon as the university opened (Deldicque and Petit 2017). The evocation of this university is important if we remember that we can speak of a "Compiègne school" around the idea of "technology as anthropologically constitutive." (Havelange et al. 2003: 121) This thesis is especially embodied in the work of Bernard Stiegler, founder of UTC Costech laboratory. Extending the thoughts of Leroi-Gourhan, Simondon and Derrida, Stiegler argues that human individuation is made by technical externalization, this externalization being never preceded by any psychological interiority. This thesis is opposed to the instrumental conception of technology – reducing the object to a means – and to the anthropological conception of technology – reducing object to its use by a subject.

So, the concept of "technical culture" is more than a concept: it is primarily a program of technical, historical, and cultural understanding of objects. Defending a technical culture includes particularly the opening of black boxes to place oneself within the machine, and thus valorising the technical object, freeing it from its position of a means that is bought, degraded, and discarded.

21.2.3 From Aesthetics to Ecology

The prospect of a technical culture manifests a continuity between aesthetic issues and environmental issues in the work of Gilbert Simondon. The collection of essays entitled *On Technology* (*Sur la technique*, 2014) gathering Simondon's articles from 1953 to 1983, repeatedly highlights the critique of obsolescence and promotes a new culture leading from techno-aesthetics to eco-design.[9]

[9]The question of an ecology of technical objects was already present in *On the Mode of Existence of Technical Objects* (1958), but it was really in the 1970s that Simondon's texts made explicit references to environmentalism and ecological technologies. See "Birth of Technology" (1970), "Art and Nature" (1980) and especially his last text: "Three Perspectives for a Reflection About Ethics and Technology" (1983). He emphasized in these texts the major challenges of techno-ecological transition as energy optimization, recovery of artifacts, limiting negative externalities, etc. (Simondon 2014).

When Simondon refers to industrial aesthetics, it is to claim that "an almost essential aspect of industrial aesthetics is to organise technophany." (Simondon 2014: 39) The concept of "technophany" is constructed by Simondon as a counterpart of the concept of "hierophany" forged by the Romanian philosopher Mircea Eliade. According to Eliade, hierophany expresses the manifestation of transcendence in an object or a phenomenon in the world. (Eliade 1987: 11) Symmetrically, technophany expresses in an object the manifestation of a technical system in which it is inserted. Technophany is the expression within the object of a technical nature exceeding the scale of the object. It is the expression of the object's milieu within the object. Organising technophany means for industrial design to include data from the cultural and technical milieu of the object into its conception, and then to operate "mediations between culture and technicity." (Simondon 2014: 43) As such, Simondon shares a basic idea about aesthetics with Étienne Souriau and Jacques Viénot: "aesthetics" have to be understood in its original meaning and not much as a question of art or beauty; aesthetics is a certain mode of being, a relation with a life-milieu.[10]

The link between Simondon's aesthetics and ecology of technology is the issue of the cultural obsolescence of objects. Thus, the term of "design" – used only once by Simondon, and in a negative sense – characterises the specific depreciation of objects due to the obsolescence phenomenon: "obsolescence not only happens to things, furniture, clothes, but sometimes even to household equipment or industrial equipment, due to a difference in 'design', varying from year to year." (Simondon 2014: 346) This is the same "design" condemned by László Moholy-Nagy in 1947 when he stated that "'design' today is generally a bid for quick sale, usually nothing but an exterior cloak around a product." (Moholy-Nagy 1947: 34)

The thought of Simondon thus accompanies the nascent marriage of "design" and "environment."[11] In the early 1970s, this relation between design and environment bifurcated into two possibilities. The first one would determine the environment as an objective and quantifiable entity in the manner suggested by Buckminster Fuller (1963). The second one would identify environment as a qualitative *milieu* relative to its dweller as developed in the work of Victor Papanek (1971) or Tomás Maldonado (1972). The term "eco-design" can be reserved for this second path (*milieu*) as it gives priority to the relations between society, technology and nature. The importance of an ecology of technology becomes crucial when the life of objects become shorter while the life of wastes is longer. The philosophy of Simondon seems important for thinking design because it strives to reconcile pro-

---

[10] In *On the Mode of Existence of Technical Objects* (1958), technology and aesthetics appear as two complementary modes of existence. Technology removes objects from the world while aesthetics connects objects to the world. According to Simondon, aesthetics is always a matter of inserting or including objects in a milieu. Aesthetics is not in the single object but in the relation between object and gesture, in the relation between object and its milieu. Simondon goes even further in one of his last texts speaking of "techno-aesthetics" as an "intercategorical fusion" of aesthetics and technology (Simondon 2014: 382).

[11] Georges Patrix noted the historical correlation of the two terms: "The year 1969 saw the introduction of the word 'Design', while in1970 the word 'environment' was introduced." (Patrix 1973: 30)

duction and use in the industrial context without relying on a nostalgia for craftsmanship: "It is in this very emphasis on industrial production, in the deepening of its characteristics that an overcoming of the antithesis between the artisanal modality and the industrial one can be studied with a greater likelihood of success." (Simondon 2009: 22)

One of the characteristics of this industrial reality is its reticular dimension. Simondon invites us to change the emphasis from the machine to the infra-individual technical elements and to the networks, considering notably the modularity of objects. The distinction Simondon operates between "closed object" and "open object" turns out to be illuminating. A closed technical object is sold fully constituted: its only possible becoming is then wear and degradation. On the contrary, an open technical objects is *neotenic*: to some extent, it is always a work-in-progress, made to be remade, produced to last by evolving. The technical object opened to its milieu and to the user-repairer is, according to Simondon, "the essence of what might be called the crusade for the salvation of technology." (Simondon 2014: 401) Today's *FabLabs*, in favor of open source and creative commons, seem to pursue this endeavor of technical culture. As with Simondon yesterday, these new "amateurs" of technology strive to overcome the craft/industry divide through the deepening of industry.

21.3 Conclusion

The originality of this French thought of design stems from its interest for the place of technology in our lives. This attention to technical phenomena accounts for the initial interest of this thought for an "industrial" design. However, this thought is not a *technicist* one because it always tries to open the technical materiality to vital and cultural relations. The originality of this understanding of design is due to the development of a material milieu as a life milieu, as a dwelling milieu. Human being does not exist regardless of a technically formed milieu, especially his or her own body (Mauss 1973). Opening the dimension of sense to this relations between human beings and artifacts is to get rid of the idea of technology as means to understand how technology mixes with our gestures, how technology *is our gestures*. If this French thought of design from the 1950s to the 1980s highlighted the "industrial" side of design, it nonetheless appears to be fruitful to think more broadly design as a composite creation mixing technical, plastic, practical, economical, social, moral, etc. factors. Indeed, this authors developed a transversal thought about the design of dwelling milieu always opening technical materiality to its cultural, social and vital resonances. Opening technology to this different fields is also creating objects opened to alteration. Emphasising on both cultural and technical construction of life milieus is bringing human being into line with what is external to human being: it is to claim that *the intelligence of the object* (intelligence of a durable object opened to transformation) *is the intelligence of the subject* (intelligence of a technical culture individuated in its technical milieu).

Acknowledgements A preliminary version of this text has been discussed by Catherine Larrère and Bernadette Bensaude Vincent during the authors' workshop held in Paris Sorbonne on 22–23 June 2015 in view of the preparation of the collective book.

References

Bergson, H. (1907). *L'évolution créatrice*. Paris: Alcan. English edition: Bergson, H. (1944). *Creative Evolution* (trans: Mitchell, A.). New York: Modern Library, Random House.

Buckminster Fuller, R. (1963). *Operating manual for spaceship earth*. New York: E.P. Dutton & Co.

De Noblet, J. (Ed.). (1981a). *Manifeste pour le développement de la culture technique*. Paris-Neuilly: CRCT.

De Noblet, J. (1981b). Changements de formes. *Culture Technique, 5*, 11–29.

Deforge, Y. (1966). *Dix entretiens sur la technologie*. Paris: Nouveaux Chantiers.

Deldicque, T., & Petit, V. (2017). La recherche en design avant la "recherche en design". *Cahiers Costech, 1*, 1–35 http://www.costech.utc.fr/CahiersCOSTECH/spip.php?article25 Accessed 25 March 2018.

Eliade, M. (1987 [1956]). *The sacred and the profane: The nature of religion*. New York: Harper Torchbooks.

Friedmann, G. (1954). Quelques aspects psycho-sociologiques de l'Esthétique industrielle. *Esthétique Industrielle, 10–11–12*, 16–18.

Friedmann, G. (1966). *Sept études sur l'homme et la technique*. Paris: Gonthier.

Gaudin, T., & de Vendeuvre, E. (1981). L'enjeu du design industriel. *Culture Technique, 5*, 73–79.

Haudricourt, A.-G. (1964). La technologie, science humaine. *La pensée, 115*, 28–35. Reprinted in Haudricourt, A.-G. (1987). *La technologie science humaine, Recherches d'histoire et d'ethnologie des techniques* (pp. 37–46). Paris: Éditions de la Maison des sciences de l'homme.

Havelange, V., Lenay, C., & Stewart, J. (2003). Les représentations: mémoire externe et objets techniques. *Intellectica, 35*, 115–131.

Huisman, D., & Patrix, G. (1961). *L'Esthétique industrielle*. Paris: Presses Universitaires de France.

Kant, E. (1790). *Kritik der Urteilskraft*. English edition: Kant, E. (2001). *Critique of the power of judgement* (trans: Guyer, P., & Matthews, E.). Cambridge, MA: Cambridge University Press.

Lamard, P., & Lequin, Y.-C. (2006). *La technologie entre à l'université: Compiègne, Sevenans, Belfort-Montbéliard*. Belfort: Université de Technologies de Belfort-Montbéliard.

Le Bœuf, J. (2006). *Jacques Viénot (1893–1959). Pionnier de l'esthétique industrielle en France*. Rennes: Presses Universitaires de Rennes.

Leroi-Gourhan, A. (1973 [1945]). *Milieu et techniques*. Paris: Albin Michel

Leroi-Gourhan, A. (1981). Préface. In J. De Noblet (Ed.), *Manifeste pour le développement de la culture technique* (pp. 6–7). Neuilly-Paris: CRCT.

Maldonado, T. (1972). *Design, nature, and revolution: Toward a critical ecology*. New York: Harper & Row.

Mauss, M. (1973). The techniques of the body. *Economy and Society, 2*(1), 70–88.

Mauss, M. (2012). *Techniques, technologie et civilisation*. Paris: Presses Universitaires de France.

Moholy-Nagy, L. (1947). *Vision in motion*. Chicago: Paul Theobald.

OECE. (1959). *L'esthétique industrielle aux États-Unis*. Paris: Organisation européenne de coopération économique.

Papanek, V. (1971). *Design for the real world: Human ecology and social change*. New York: Pantheon Books.

Patrix, G. (1962). Entretien avec Georges Patrix. *Les cahiers de la publicité, 4–4*, 87–102.

Patrix, G. (1973). *Design et environnement*. Paris: Casterman.

Plaut, J. S. (1954). L'esthétique industrielle au États-Unis. *Profils, 9*, 133–134.

Quarante, D. (1981). L'expérience de Compiègne. *Culture Technique, 5*, 156–161.

Simondon, G. (1958). *Du mode d'existence des objets techniques*. Paris: Aubier. Simondon, G. (2016). *On the Mode of Existence of Technical Objects* (trans: Malaspina, C.). Minneapolis: Univocal Publishing.

Simondon, G. (2009). Technical mentality. *Parrhesia, 7*, 17–27.

Simondon, G. (2014). *Sur la technique (1953–1983)*. Paris: Presses Universitaires de France.

Souriau, P. (1904). *La beauté rationnelle*. Paris: Félix Alcan.

Souriau, É. (1929). *L'Avenir de l'esthétique. Essai sur l'objet d'une science naissante*. Paris: Félix Alcan.

Souriau, É. (1952). Passé, présent, avenir du problème de l'esthétique industrielle. In É. Souriau, C. Lalo, & P. le Goupils (Eds.), *Esthétique Industrielle* (pp. 1–27). Paris: Presses Universitaires de France.

Souriau, É. (1954). Rôle des considérations de temps dans l'Esthétique Industrielle. *Esthétique Industrielle, 10–11–12*, 12–14.

Souriau, É. (1956). *Du mode d'existence de l'œuvre à faire*. Reprinted in Souriau, É. (2009). *Les différents modes d'existence* (pp. 195–217). Paris: Presses Universitaires de France.

Viénot, J. (1939). *La République des Arts*. Paris: Horizons de France.

Viénot, J. (1945). *Art Présent, 1*. Paris: Éditions de Clermont.

Viénot, J. (1948). Du barrage au stylo. *Art Présent. Beautés de la technique, 7–8*, 9–11. & 109–111.

Viénot, J. (1951). Esthétique du machinisme. *Travail et Méthodes. Revue technique de la direction et de l'organisation des entreprises, 37*, 27.

Viénot, J. (1953). Les arts impliqués. *Esthétique Industrielle, 8*, 22.

Viénot, J. (1955). Productivité de l'esthétique industrielle. *Esthétique Industrielle, 15*, 29–36.

Chapter 22
Design and Aesthetics in Nanotechnology

Sacha Loeve

Abstract What is the status of "design" in nanotechnology? On the one hand, scientists doing nanotechnology refer to their activity as "design." On the other, the intervention of design researchers and practitioners remains confined to "the future" (i.e. societal applications and uses of nanotechnology). How are we to understand such a division of labour? To be sure it is not specific to nanotechnology but concerns the status of design in contemporary technoscience at large. However, the problem is more acute in the case of this "invisible" technology. Nanotechnology is supposed to be cut off from all sensible experience whereas design traditionally focuses on the shaping of the user's experience. After articulating the diagnosis and its implications, I question the status of a third player: "nano-art." I then draw on some resources of French philosophy of technology and aesthetics to prompt a new alliance between "techno-logy" (the study of technics) and aesthetics (the study of sensation) resulting in a re-conceptualization of design as "techno-aesthetics." The chapter closes by highlighting the political significance of such techno-aesthetic design for nanotechnology and beyond, for our everyday live amidst technoscientific objects.

Keywords Aesthetics · Aesthetical apparatus · Techno-aesthetics · Technoscience · Design · Images · Nano-art · Nanotechnology · Noumenal technology

22.1 Everywhere and Nowhere in Contemporary Technoscience

"Design" is a keyword in contemporary technoscience. It pervades the mundane parlance of synthetic biology and nanotechnology. While synthetic biology focuses on "life by design," (Bensaude Vincent 2015) the so-called "nanoworld" is depicted as a "world by design." Its atomic landscapes are like an invitation to

S. Loeve (✉)
Faculté de Philosophie/Institut de Recherches Philosophiques de Lyon (IRPhiL - EA 4187), Université Lyon 3 Jean Moulin, Lyon, France
e-mail: sacha.loeve@univ-lyon3.fr

S. Loeve et al. (eds.), *French Philosophy of Technology*, Philosophy of Engineering and Technology 29,
https://doi.org/10.1007/978-3-319-89518-5_22

enter a new world that humans would have thoroughly designed (Hanson 2012). Nano-objects and nanostructured materials are processed and studied as outcomes of a design work. Sometimes, genomes and materials themselves are called "designers" (Annaluru et al. 2014; Chong and Garboczi 2002; Shwartz 2015). Even nature is depicted as a "nanodesigner" that "invented" the molecular machinery of life more than three billions years ago (Jones 2004) – hence the popular "re-designing" phrase.

Such phrases rely on the polysemy of the word "design," which refers either to the plan, form or blueprint dictating the realization of an artifact, or to the immanent and distributed organization of a form-making process (Dilnot 1984; Buchanan 1992). Both meanings resonate with the ambivalence of synthetic biologists and nanotechnologists which oscillate between an overemphasis on the activity of human intelligence shaping a passive matter (i.e. the hylemorphic attitude criticized by Simondon (2005)) and the capture of the spontaneous activity and information embedded in molecules and genomes. In other terms, they hesitate between fabrication and piloting (Larrère and Larrère, Chap. 12, this volume).

Interestingly, the co-occurrences of "nanotechnology" or "synthetic biology" with "design" on Google outnumber those with "understanding," "knowledge," or "explanation," as well as those with "fabrication," "production," or "manufacturing." As "design" refers both to an intellectual activity and to a practical one, it seems well suited to substitute for notions belonging either to "science" *or* to "technology," and even more relevant for "technoscience" as a mode of research subverting the dualisms of representation/action, structure/operation, process/product, nature/artifice, pure/applied, praxis/poiesis, etc. (Bensaude Vincent et al. 2017). What is more, the word "design" does not suffer from the bad reputation of the later term, especially in France (Bensaude Vincent and Loeve, Chap. 11, this volume).

Enough for the *word* design in nanotechnology. But what about design as *practice*? Nanoscale scientists do "research by design" in that instead of limiting their questions to a pre-existing range of available materials, they are free to raise questions that can only be addressed by synthesizing the desired research object endowed with tailored-made properties for performing specific behaviours and functions (Marcovitch and Shinn 2014). They equate design with "technological conception" as engineers do, although unlike engineers, technoscientists generally limit their endeavors to establishing proofs of concept, i.e. to the demonstration of the feasibility of a product or process, unconcerned with the practical conditions of its actual completion. They (or their nanodevices) perform electronic, optical, magnetic, chemical or biological "design" in this sense. As to the handful of books more explicitly devoted to "nanodesign" (Rieth 2003; Ashby et al. 2009; Schommers 2013), they generally start with the basic laws of physics at the nanoscale and then address principles and rules of engineering; only toward the end do they touch upon the "broader" environmental and societal issues. They address "nanodesign" as a matter of applied science.

On the other hand, few researchers and practitioners in design intervene in nanotechnologies, or only at the margins. For instance, designers of the Laboratoire Innovation et Technologies centrées UtilisateurS (LituS) based in the nanotechnology

Fig. 22.1 *Cloud Project*, Zoe Papadopoulou, Cat Cramer. (Picture assembled by the author; courtesy of Z. Papadopoulou and C. Kramer)

center Minatec in Grenoble, is dedicated to the exploration of user's experiences of nano-embarking prototypes with a view to understand their conditions of adoption, appropriation and desirability (Delhome 2011). Design researchers also work at the IDEAs Lab® to co-construct "anticipated use scenarios" of nanotech-based products with social scientists and "potential users" focus groups. The reports of the design team are sent back to the so-called "technologists," who are then supposed to better adapt the technical functionalities to users. A strict division of labor is observed: "technologists" are the engineers in charge of translating the "technological bricks" that come out of the research lab into usable functionalities, while designers are in charge of translating these functionalities into potential social uses. Their main goal is to facilitate the appropriation of nanotechnology.

This approach to nanotechnology design is somewhat contradictory: it is supposed to be user-centric although stricto sensu *there is no user of nanotechnology*. Nanocomponents are always packaged into larger modules or systems so that users have no direct relationships with the nanoscale operations embedded in the final product. Users interact with computers, networks, screens, glasses, fridges, tires, clothes, medicines, cosmetics, food, etc., that embark or embed nanomaterials and systems, and never with nanotechnology per se. In this case design in nanotechnology is therefore not about designing the technology but about *adapting* already constituted "technological bricks" to (potential) users.

Designers in nanotechnology endorse and reinforce the divide between two understandings of design: the concept familiar in engineering (technological conception) and that of the designer (the framing of user-experience). The former is dealt with as an object-centred problem of applied science and engineering, the later as a human-centred matter of meaning, and the question is settled. In this perspective, the practice and meaning of design is trapped in the same dualisms that the technoscience is challenging (object/subject, science/society, technology/culture, etc.).

Yet another trend of design aims precisely at questioning these divides. Following "critical design," (Dunne 2006) "debate-provoking design" set up situations and performances for stirring discussion and questioning our attitudes, ethical values and judgements towards nanotech. For instance, in the "Cloud Project" (Fig. 22.1), artists Zoe Papadopoulou and Cathrine Kramer park their ice cream van in the streets, inviting people to taste ice cream frozen with liquid nitrogen. The freezing

Fig. 22.2 *NANO Supermarket*, Koert van Mensvoort. (Picture assembled by the author; Courtesy of K. van Mensvoort)

occurs so quickly as to produce nano-sized ice crystals and a very smooth ice-cream, jokingly referred to as a Grey Goo Sunday. The stated goal of The Cloud Project is to make clouds snow ice cream. Ice creams are given in exchange for a conversation about nanotechnology and geo-engineering. True or not, this street event is meant to awaken people's imagination and to invite them to discuss and debate nanotechnology, geo-engineering and climate change.

Another example is the itinerant NANO Supermarket related to Dutch designer Koert van Mensvoort's project Next Nature (Fig. 22.2).[1] People are encouraged to discuss speculative uses of nanotechnology such as NanoLift, a "physical photoshop" enabled by magnetic nanoparticles injected into the skin or a wine containing on-demand programmable nanocapsules to alter taste, smell and colour. One may also speak of "speculative design." (Auger 2013)

These projects are enjoyable, fun, and sometimes provide real opportunities for exchanging views and confronting values. But they are also revealing of a paradoxical situation: a chasm between *technical objects without uses* and *uses without technical objects*.

22.2 The Technoscientific Chasm

Laboratory nano-objects like molecular machines, electronic nanodevices or sensors/actuators are genuine technical *objects* in the sense of Simondon: they perform a definite "technical individuality." (Loeve 2010)[2] Yet most of them are designed regardless of any practical usage. Some have *applications*, however "application" differs from *use*. Use contains an unavoidable and unpredictable dimension of social

[1] http://www.nanosupermarket.org/products

[2] A technical individual according to Simondon (1958) is an object that has gone through a process of "concretization" by which the object unifies itself by integrating into its functioning scheme some specific features of its environment (here the nanoscale properties of their physico-chemical environment). The technical individual is no more an "abstract" object (the mere materialization of a theory adapted to an external environment) but a concrete object in an "associated milieu."

reconstruction and sometimes hijacking of the object's functionality. The only genuinely technical uses of nano-objects are those of scientists in the lab. In most nanotech-embarking applications, the "nano" character is made to go unnoticed and unmodified by the final user. It is made to *function*, not to be used.

Conversely, proposals of critical and speculative design are genuine *uses*. They are not reducible to "science communication" as they generate meaning by exploring ambiguities, dilemmas, unknowns and potential alternatives to "conventional" designs (i.e., to *applications*). However, designers make this possible only by designing speculative objects lacking technical individuation.

Hence a paradoxical situation: on the one hand, scientists design *actual technical objects without uses* (and only in some cases with applications); on the other, designers elicit *actual uses without technical objects* (or only with speculative ones). "Object" and "use" remain mutually exclusive. Design in nanotechnology is thus cleaved between *objects with speculative uses* (futuristic scenarios and promises) and *uses with speculative objects* (fictional proposals and projects). While the work of designers becomes almost undistinguishable from that of speculative ethicists (Nordmann 2007), scientists tend to monopolize the design of the nanoworld by framing it as an applied-science kind of engineering whose social dimensions are external to technology, limited to the context of application. This chasm has important sociocultural repercussions: it maintains the technical part of nanotech in some kind of parallel world—the "nanoworld"—out of reach for social users.

By contrast, other fields of design establish a real connection with technology. Ecodesign focuses as much on humans' awareness of their interactions with the environment as on the circulation, appropriation, production and conception of the technologies that condition this awareness. Information design focuses as much on users' experiences as on the interfaces that shape users' interactions with computational processes, devices and networks (Vial, Chap. 23, this volume). When it comes to nanotechnology, there are no "nanodesigners" in charge of designing our interactions with the nanoworld. The design of the nanoworld is handled by scientists, while designers work only on the symbolic, metaphoric and societal dimensions of the "technology-in-the-future." As a result of this division of labor users have poor interactions with the present technology, either through applications where the nano-dimenson is withdrawn from use, or through speculations where the nano-dimension is projected into "the future."

Of course engineers and designers are two distinct professional groups and their separation is the heritage of a long history. But whatever its socio-cultural reasons this separation is possible because of the existence of an irreducible *margin of indetermination* between the functioning and the uses of a technical object. As Simondon emphasized (1958), the same functioning schemes can give rise to different uses while the same use can be obtained from different functionings.[3] And just as Simondon

[3] To Simondon, use-categories refer less to the objects' intrinsic functioning than to the practical functioning of humans. From the point of view of functioning, he notes, there is more analogy between the elastic motor and a bow or arbalest than between the same motor and a steam motor, although from the point of view of use, the latter two are put into the same "motor" category (Simondon 1958: 19; Loeve 2016).

insisted that technicity[4] is irreducible to use, most designers do not want to reduce the meaning of use to the bare actualization of functioning (let aside pure functionalism as a limit-case[5]). Yet the existence of a gap does not prevent bridges to be established, quite the contrary: the gap between functioning and use affords the very space of problems and possibilities in which design operates.[6] In other terms, to design is to work with and within this zone of mismatch and indeterminacy between functioning and use.[7] The practice of design is such that it constantly redefines use-value instead of regarding it as a given, be it in the functioning of technical objects (limit-case of strict functionalism) or in socioeconomic behaviors (limit-case of pure marketing).

Therefore the question remains to understand why designers do not engage in establishing bridges between the functioning and the use of nanotechnology. As I will argue, the answer lies into a specific "distribution of the sensible." (Rancière 2004) To understand this, let us first examine the contribution of a third player, who blurs the division of labor between nanotech and design: nano-art.

22.3 Nano-tech and Nano-art

The practices of nano-art have played a key role in nanotechnology's rise to prominence since *The Beginning* (Fig. 22.3), an STM[8] image spelling the letters I.B.M. with 35 xenon atoms on nickel surface (Eigler 1990). Scientists refer to it as "art" because it signs the intentional imprint of Man in a hitherto untouched medium. *The Beginning* became soon a visually compelling evidence of the human ability to manipulate the world atom-by-atom (Schummer 2006), a powerful flagship for promoting large-scale implementation of nanotech funding initiatives (Toumey 2010),

---

[4] I.e. the degree and modality of concretization.

[5] Functionalist designers or architects like Bruno Munari, Ludwig Mies van der Rohe or Le Corbusier, were allegedly observing the principle "form follows function." In practice, they were rather opposing an aesthetics of functioning to the aesthetical masking of technicity (Antonello 2009).

[6] For Simondon it was precisely the aim of "industrial aesthetics" – the 1950–1960s French phrase for 'design' (Beaubois and Petit, Chap. 21, this volume) – to mediate between technicity and use in a particular culture or, as he had it, to "organize technophany," the aesthetical manifestation of technicity (Simondon 1960–1961: 39). For instance, the casing enclosing the mechanics of a clock watch is not only an embellishment. It is a membrane that both separates *and* links the delicate movement pieces and the social use of time. It is simultaneously a symbolic system and a protective membrane—the condition to develop, inside, the mechanical functions.

[7] To venture a "physics-for-dummies" metaphor, design operates neither like an *insulator* (isolationist discourse: science on the one hand, society on the other, and a forbidden gap between the two) nor like a *conductor*, with technical functioning dictating social use (deterministic or applicationist discourse: science-hence-application-hence-use). Design operates like *semiconductor*: it modulates gently the zone of relative mismatch between technicity and use to cross the forbidden gap and constantly redraw its borders.

[8] Scanning tunneling microscope. The image has been dubbed "*The Beginning*" retrospectively for its 1994 3D colour-publication on the IBM's STM Image Gallery website.

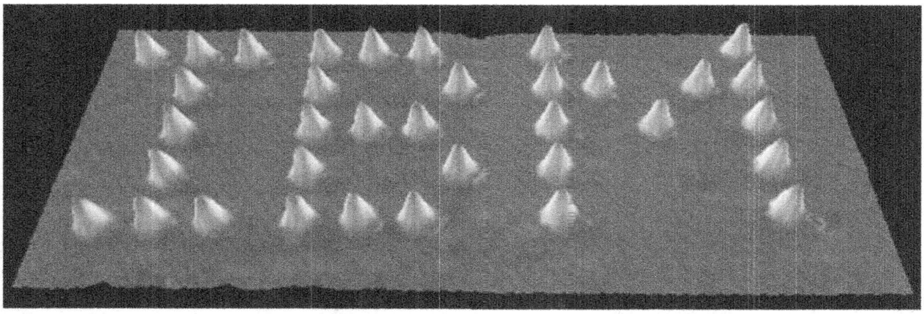

Fig. 22.3 *The Beginning*. (Don Eigler, © IBM)

an evangelistic tool for the radical proponents of a "molecular manufacture,"[9] and a worshiped icon for a growing number of scientists converting their research projects into "nano" ones.

This inaugural feat revealed the STM's potential beyond that of an observational instrument for "reading" atomic structure: as a tool for manipulating and "writing" with atoms. Since then, scanning probe microscopy manufacturers like Veeco Instruments or Omicron have encouraged (and sponsored) nano-art for promoting their instruments and exhibiting the unprecedented mastery they provide. With the rapid dissemination of scanning probe microscopy a plethora of practices consisting in tagging, drawing or signing matter at the nanoscale have been displayed (Fig. 22.4). These images mean both "I've been here!" and "I've done that!" They stage the exploration of a world of our own design, a plastic world turned synthetic, with material building-blocks as easily actionable as bits of information.

These productions were powerful marketing devices which contributed to the hype about nanotechnology and the economy of promises conditioning their funding (Audétat et al. 2015). The political function of nano-images as lures for the future is what suggests a work by visual artist Chris Robinson, entitled *Eigler's Eyes 2*. In this piece, Robinson hijacked one of Don Eigler's achievements, the "quantum corral,"[10] and replaced the ring of atoms by a circle of human figures (Fig. 22.5) as if it were our collective gaze on the nanoworld that created the phenomenon of interest. The piece thus suggests a "nano-society of spectacle" where one essentially consumes images, a "glitter science" that seeks mostly to catch attention.

The image of carbon nanotubes transistors aligned between gold electrodes used for the cover of a 2001 issue of the journal *Science* (Fig. 22.6) is not a work of nano-

[9] "Five years ago, audiences questioned whether individual atoms could be placed in precise patterns; today, I can answer that question not just with calculations, but with a slide showing the letters 'IBM' spelled using 35 xenon atoms." (Drexler 1992: 1)

[10] Where one can visualise circular standing waves of electron density, a "quantum-classical" phenomenon.

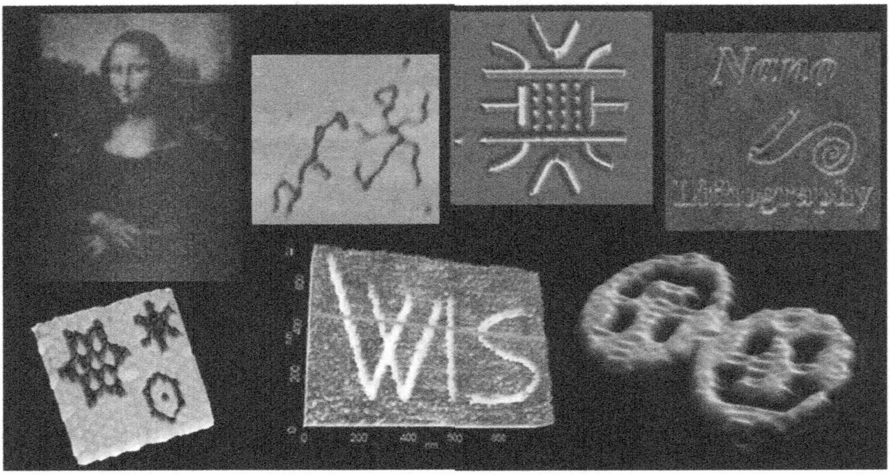

Fig. 22.4 Plastic world. (Picture processed by the author)

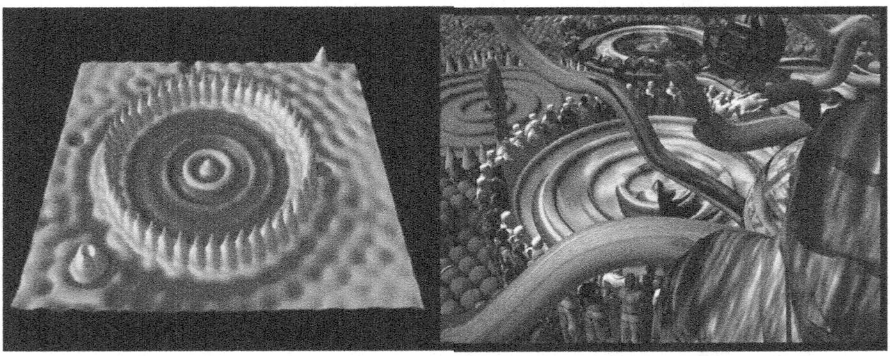

Fig. 22.5 Left: *Quantum corral* by Don Eigler (© IBM). Right: *Eigler's eyes* 2 by Chris Robinson. (Courtesy of Chris Robinson)

art but a so-called "artist's depiction," an image conveying a vision of what *could* be done in the best of possible (nano)worlds. Since the atomic structure co-exist with the representation of the macroscopic material this image had defrayed the chronicle in nano cenacles all around the world for it is a scientific nonsense. Criticisms were pointing to the trend of the most prestigious scientific journals to favour the most "sexy" images over more rigorous ones (Ottino 2003).

Another artist's depiction worth mentioning is the cover of the brochure *Nanotechnology: Shaping the World Atom by Atom* (Roco et al. 1999). This brochure[11] aimed at convincing US senators to vote the colossal budget planned for the

[11] Co-edited by the President's Committee of Advisors on Science and Technology (PCAST), the National Science and Technology Council (NSTC) and the Interagency Working Group on Nanoscience, Engineering and Technology (IWGN).

Fig. 22.6 Aberrant science

US National Nanotechnology Initiative. The cover (Fig. 22.7) displays a STM image of silicon surfacescape set against a "cosmic" background. Using central perspective, the forefront invites the viewer to "enter into a new world" decentred from the earthling referential. The Earth is seen from afar as if the viewer was landing on the bumpy surface. The background evokes the extension of the human enterprise beyond all frontiers by recycling the old heroic imagery of the conquest of space. The picture mixes the infinitely small and the infinitely vast, the outer space of the universe and the inner space of matter; it stages the nanoworld as a place waiting to be colonized.

Fig. 22.7 Cover of the US Government brochure *Shaping the world atom by atom*

Nano-art competitions are instrumentalized by research agencies to improve their public image. For instance, the nano-art imagery contest organized on a monthly basis by the Smalley Institute at Rice University and the company nanoTox® awards winning entries "based on a combinations of visual beauty and technical marvel." It is explicitly stated that "The goals of the contest are to have Fun and to promote the public's acceptance of and interest in nanotechnology: most people who are new to the topic really start to 'get it' when they see compelling pictures."[12] Nano-art also simply allows attracting kids and inspiring vocations. For instance, IBM researchers at Almaden have made a movie in stop motion with carbon monoxide molecules. A small boy named Atom falls in love with a molecule and plays with her (Fig. 22.8). The message is clear: nanotechnology is full of love and fun (Milburn 2011). One scientist comments: "If I can do this by making a movie, and I could get a thousands kids to join science rather than going to law school, I would be super happy."[13]

Many of these images have been abundantly commented (Slaattelid and Wickson 2011). In nanotech as elsewhere, science-art projects are in vogue because they are regarded as means to overcome the unfortunate divide between the "two cultures." (Snow 1959) Showcasing nanotechnology as "art" provides a cultural alibi and a

[12] http://nanoart.blogs.rice.edu/

[13] Andreas Heinrich, https://www.youtube.com/watch?v=xA4QWwaweWA (4'20).

Fig. 22.8 Playful nanotechnology. (© IBM, source: https://www.youtube.com/watch?v=oSCX78-8-q0, image capture by the author)

tool of social acceptability.[14] Symmetrically artists may use nanotech to broaden their repertoire of impressive "effects" for performing technical prowess.

22.4 Radical-Otherness-Hence-Appropriation

This detour by (the mainstream of) nano-art uncovers the shared assumption that rules the division of labor between nanotech and design.

The slogan "making the invisible visible," often used for summarising the purpose of nano-art (Raimondi 2007; Ruivenkamp and Rip 2010; Baccile and Balzerani 2013), presupposes that the nanoworld is in itself invisible, uncanny, strange, occult, counter-intuitive, etc. It is supposed to be a world apart, withdrawn in itself, radically different, occult, defying common sense and eluding any sensible intuition the kind we use in our everyday, phenomenal world. Especially because of its "quantum" nature, the nanoscale would be attainable only by an abstract science and accessible only to the initiated. As Alfred Nordmann had it, nanotech is framed as a "noumenal technology": a technology of the "things in themselves,"[15] that retreats from human access, perception, and control (Nordmann 2005). Thus nano-art, be it practised by artists or by scientists, aims at the appropriation of the nanoworld: if the nanoworld is noumenal it has to be rendered phenomenal; if it is invisible it has to be rendered visible; if it is uncanny it has to be rendered familiar; if it is scary it has to be rendered fun, etc. It aims at making the nanoscale culturally appropriable, at making it fit in our daily lives.

[14] This way of resorting to "art" to show technoscience "in culture" can also be regarded as an implicit admission of failure of scientific and technical culture (Lévy-Leblond 2010).

[15] Although qualifying atoms and molecules as *noumena* is incorrect from a Kantian point of view (since they are objects of science we constitute them as phenomena in space and time and subject them to principles of causality, conservation, etc.), Nordmann uses this phrase to emphasize the disruption of the traditional sequence "representation hence technical agency."

Yet designers seem to start from the same pre-assumption as they seek to translate the "uncanny otherness" of the noumenal into the phenomenal. The role of nanodesign would be to build up mediations in order to turn these noumenally constituted technologies into sensible, appropriable, meaningful, and debatable things (Delhome 2011).

I refer to such implicit preconceptions as *radical-otherness-hence-appropriation*. Its rhetoric is reminiscent of that of twentieth-century physics popularization, showcasing relativity and quantum physics as remote universes inaccessible to common sense, and scientists as priests accessing the hidden reality of things. It corresponds to a public image of science based on the pre-assumption of a gap that should be bridged by popularizers to educate the public.[16]

In the case of a technoscience like nanotechnology, this rhetoric is misguided. The very idea of an "epistemic rupture" between science and lay knowledge, emphasized particularly by Gaston Bachelard (1938), does stand for technoscientific objects. Nano-objects are not "theoretical entities" prone to provoke debates between realists and instrumentalists; they are conceived (and visualized) as mundane and actionable building blocks that afford functionality and performance (Bensaude Vincent et al. 2011).[17]

In brief, do we need nano-art for making the nanoworld visible and familiar when nanotechnology already does it? In taking up the rhetoric of the radical otherness of the nanoscale, the mainstream of nano-art condemns itself to practices that are mimetic to the technoscience instead of trying to make a difference. As to design, it finds itself stuck between nanotech, which monopolizes the design of the nanoworld, and nano-art, which tends to aestheticize it.

22.5 The Nanotechnological Sensible

An epistemic version of the "radical-otherness-hence-appropriation" argument is that "nanotechnology produces visible representations of the invisible." I would like to dispute this claim.

First, it relies on a misunderstanding of the novelty of nanotech. Indeed, from inertia to electromagnetic waves to atoms and molecules to quarks and black holes, it is modern science, not (nano)technoscience, that has populated the human world with visible representations of entities and processes that unfold beyond the reach of our senses—but obviously not beyond our ability to *represent* them. If there is a

[16] In doing so, science communication was maintaining and even enlarging the gap between lay commonsense and a sacralized science while lamenting about the public's "deficit of knowledge" (Bensaude Vincent 2001).

[17] To give only two examples taken from quantum physics, in the nanoworld Heisenberg's principle of indeterminacy is no longer a mysterious property as it becomes a tool for making quantum confinement to enhance the electronic behavior of nanoparticles. "Schrödinger's cat" is no longer a thought paradox when it is used as a working laboratory device consisting of two coupled ultracold atoms with one measuring the other (Raimond et al. 2001).

novelty in nano compared to quantum physics (which does "smaller than nano") or to chemistry (which has been working at the nanoscale since ever), it is precisely the possibility of *experiencing* individual molecular objects through devices enacting the sensitivity of this supposedly "noumenal" nanoworld. It is the possibility of a *molecular aesthesis*. The relevant novelty of nanotech does not lay in the "Nth industrial revolution" so fervently promised but in the new technical *milieu of experience* it affords by the instauration of unprecedented sensitive relationships with and within materiality.

Second, nano-objects are neither invisible nor visible; they would rather be "a-visible," though not "a-sensible." They are indifferent to the dichotomy of the visible and the invisible that haunts the history and epistemology of modern science. The a-visible things, forces and processes of the nanoscale can be visualized, but they can also be listened and touched (and maybe one day smelled and tasted). The nanotechnological sensible is not restricted to the visual domain. *Touch* seems to be the prevailing sensory modality. Touch presents all kinds of nuances in the nanoworld, from caress to strike, from tact to prehension. Scanning probe microscopes operate like blind persons using a cane or reading Braille. The STM deciphers the properties of a conductive material by probing its electronic surface cloud. The vibrating cantilever of the atomic force microscope (AFM) brushes past slightly the material to sketch the main lines of its surfacescape in non-contact mode; in intermittent contact mode, it gropes around and gives gentle touches to localize the disposition of objects; in contact mode, it scrapes the surface to detail or alter its corrugation. Nanobiophysicists use optical or magnetic tweezers to grab motile proteins and feel their forces in motion. Nanotribology studies textures and friction at the nanoscale. Molecules recognise and bind to each other by mutual contact and prehension, etc.

However, just like vision, touch does not suffice to define the intrinsic character of the nanotechnological sensible. Its defining character is its *transmodality*. Nanotechnological operations present many kinds of transductions between different sensory modalities: visualizing tact, listening a surfacescape, touching light, etc. Transmodality is not specific to nano (for instance we usually translate written symbols into sounds while reading or visualize sounds when writing music) but the reversal: nano-percepts are specifically transmodal, and monomodal only when translated for human access. A nano-object or process is not audible nor visible nor touchable nor audible but all this altogether: it should rather be said "transible."

Most important, nanotech-generated images are *not* representations. They belong to a different regime of imaging, which I have named elsewhere "imaginaction" (Loeve 2011) or "the regime of image-objects." (Loeve 2009)[18] As Michel Foucault argued, representation is "the dissociation of the sign and resemblance." (Foucault

[18] These concepts were inspired by Bergson's theory of perception as a process occurring into things according to action and not into their representation (Bergson 1896), and by Simondon, for whom perception is only one phase in the life of images (Simondon 2008). For both these two philosophers, images are not limited to the visible and can exist outside, before and after perception (Loeve 2011, 2015).

1966: 70) Representing requires a deliberate estrangement from sensible likeness, the construction of a distance between "object" and "image." This distancing, expressed in the "re" of re-presentation, is both a key principle of the scientific ethos (critical spirit and organized scepticism with regard to sensible data) and a concrete operation implemented in scientific instruments and settings. For instance, electron microscopes and spectroscopy techniques inscribe the trace of a distant interaction between a carefully prepared sample and a radiation emitted by the apparatus and then transmitted and/or diffracted by the sample. That the curve, spectrum or diffraction pattern re-presents the invisible properties of the sample means that its features are in principle distinguishable from those of the technical apparatus displayed to produce them. Both the realist's claim that the picture does represent some of the real properties of the sample and the positivist's claims that "this is just a representation" can only occur in the regime of representation. Whether something can be said about the real or about its representation, in both cases the real stands in the distance of "aboutness," independently from the instrumentation allowing its objectification (Nordmann 2006).

Now, in scanning probe microscopy constructing a distant picture of nature—representing—is no longer the problem. This family of instruments emblematic of nanoscale research has also been dubbed "local probe" or "near-field" microscopy to emphasize its difference versus other and subsequently dubbed "far-field" microscopes (from optical to electron) and spectroscopy techniques. In far-field microscopy, imaging and interpreting images means constructing a distance. In near-field, imaging is done in proximal rather than in distal mode, and interpreting the image means accounting for the tightest interaction between the technical conditions of imaging and the operative behaviour of the object. An STM image of an atom is no more a model or a distant trace of an interaction with the sample's atoms; it is the interactive contexture of an object in a particular milieu configured by particular imaging conditions. The experimental image is literally the surface of the object co-acting with the probe mechanism of the instrument. It is an image-object, whereby the manipulation of the object and the production of an image are one and the same process. The functioning of the instrument and the features of the object are no more separable.

A scanning probe image is both object-oriented and experience-oriented (Bueno 2008): it harbors information on the object as well as on the experimental mode of accessing it. But it is not only *human* experience-oriented. It is also about what it is like to perceive a nano-object from the perspective of another nano-object. In a number of nanotech experiments, an electron, a photon, a spin, an atom, a molecule, a surface or a nanoparticle is not only investigated as a *target* of study but also as a *vector* or *proxy* for addressing another object. It is not only an object for a subject (the human knower) but also an object for other objects. The observed object can shift to observing system and serve as a detector, sensor, probe, tool or actuator interacting with another object at its own scale and in its own mode. Nano-objects are not noumenal, they rather extend the phenomenal to the relations between objects.

Of course, modern science deals with relations between objects, but only from the foreign perspective of the subject constructing a distant representation of them (a causal model for instance), not from that of objects sensing other objects. Modern science deals with interobjectivity but not with the interobjective sensible. It produces "objectivations *of*" while nanotechnoscience generates "objectivations *with*," whereby interobjective relationships are not represented from afar but participated from within. Nanotech's image-objects and local probes provide humans with access to the ways objects access their environments. This process of participation to interobjective relationships does not imply dry and narrow realism. It involves many imaginings, narratives, and never goes without recourse to metaphor (Bogost 2012) – the act of experiencing something in the terms of another, for instance: how a copper surface looks like for a cobalt atom (Stroscio and Celotta 2004). Metaphors, here, are not only to be understood figuratively but sometimes literally, when they lead to a functioning analogy by which things actually work together – for instance, a molecular "machine." (Browne and Feringa 2006)[19]

To recap, nanotech does not make visible representations of the invisible; it rather create a trans-sensibility below representation and beyond the subject-object correlation.[20] It is not a technology beyond the sensible, but a technology that redefines the sensible. The nanotechnological sensible extends to relations between objects that can be shared with human subjects, be they technically or imaginatively instrumented. It delineates a new techno-aesthetic milieu.

22.6 Three Techno-aesthetic Experiments

Here I comment three projects that go beyond instrumental relationships between art and technology and do not consider the nanoscale as "otherness" to be appropriated into everyday experience; they rather consider the nanotechnological sensible as a new experiential space to be explored and questioned while providing an intellectual and emotional grip on its technological constitution.

1. Paul Thomas addresses the relation between humans and objects from the perspective of touch at the nanoscale (Thomas 2009; Hawkins and Straughan 2014).

[19]Although I maintain Simondon's definition of the metaphor as a "relation of identity" based on likeness versus analogy as an "identity of relation" based on operations (Simondon 2005: 108), in my perspective, the regime of imaginaction displays a continuous spectrum between the two.

[20]Following Quentin Meillassoux (2010) the proponents of "speculative realism" (Graham Harman, Levi Bryant, Ian Grant, Ray brassier) have tackled what they dub the "philosophies of access": philosophies that, after Kant, phenomenology and the linguistic turn, have renounced to state anything about the real in favor of a discourse on human access to reality. Speculative realists repeatedly argue that these philosophies commit the fault of letting the object-object relationships unthought at the benefit of subject-object or subject-subject correlations (Harman 2011). Without going too deep into this debate, let me just stress that it is possible to think human modes of access that borrow objects' modes of access to other objects. Thinking this kind operational decentring is one of the major challenges of a philosophy of technoscientific objects today.

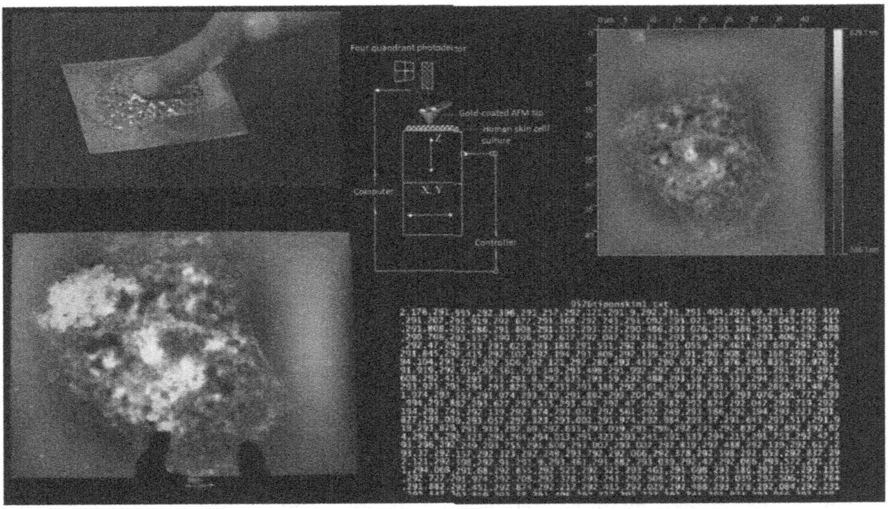

Fig. 22.9 *MIDAS project*, Paul Thomas. (Courtesy of P. Thomas)

His project *MIDAS* uses an AFM for probing the relationships between skin and gold. The AFM cantilever, coated with gold, scans a sample of human skin culture cells (Fig. 22.9). The experiment challenges three preconceptions regarding touch:

- *Subjectivity*: Relying on the metaphor of touch attached to the AFM, the artist revisits the contact between skin and gold the other way round: it is gold that explores skin and generates data on its topography. His installation displays different possible renderings of those data in our macroscale sensory modalities: Touch, with a haptic interface, sonic transductions of the AFM data, and large-format visual projections that the visitor can modify through haptic interfaces. Both sensibilities, machinic and human, are made not identical but analogue, the one serving to understand the other and reciprocally in the process that Thomas refers to as a "deterritorialization of our interface with the world."
- *Charnel character*: Touch is usually regarded as the sense of immediacy and sensuality. Instead of exploiting the seductive character of images, the artist chose to show the list of raw data (van der Waals force measures) aside the haptic, visual and sonic renderings of them, without forgetting the algorithms that allow generating them—all things that scientists tend to show less and less. By giving to see and to manipulate the various possible means of transduction of the skin-gold relationship, the project produced a distancing of touch.

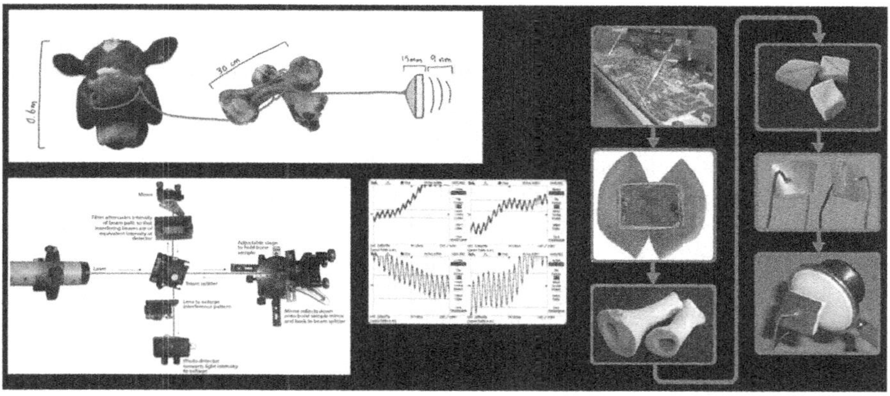

Fig. 22.10 *Bone transducers*, Boo Chapple and William Wong. (Courtesy of Boo Chapple)

– *Antitypy*[21]: Rather than a simple means of contact touch reveals to be a topologically singular space, rich of physical and chemical events (forces, particle exchanges, reactions and transitions, e.g. how gold atoms are transferred to the skin). The problem is no longer to render the nanoworld accessible to our familiar sense of space, it is to interpret space anew. In his view, the imaging methods of nanotechnology remain too dependent on visual conventions still belonging to the regime of representation (in particular perspective). He opposes an interstitial and transitional space to the occulocentric and perspectivist understanding of space, a space-between: between humans and objects, between several scales, and between various sensory modalities as well.

2. In *Can you hear the femur play?* artist Boo Chapple and biomedical engineering physicist William Wong relate their troubles with engineering "bones audio speakers at the nanoscale." (Chapple and Wong 2008) The project starts from their reading of biology papers on bone piezoelectricity arguing that the bone matrix reacts to mechanical stress by emitting electrical signals captured by the stem cells that regenerate bone tissues. In order to make bone piezoelectricity audible Chapple tries to craft an electro-acoustic transducer out of bones bought from the butcher (Fig. 22.10).

The project lasted for 3 years, with considerable technical hitches. Connected to an electrical circuit, the transducers alone did not work, so they tried using various microphones and oscilloscopes to detect and amplify the sound. Each time they thought getting something they realized it comes from interferences with background noise.[22] Doubts aroused about the reality of the phenomenon, about the scientific

[21] Resistance of matter to a penetrative force.

[22] At some point, they even built a Michelson interferometer, a sophisticated instrument insensible to electromagnetic perturbations. The apparatus detects a range of frequencies between 300 and 3000 Hz. Given the frequency range of the human ear (from 20 to 20,000 Hz), the piezoelectricity

literature, about trust in science, and about the artist's obsession to "possess" the phenomenon in a "macrosensorial" way. They had "the map, but not the territory," Chapple says. These deceptive results push them to question the role of technology: they had thought getting a tool to accomplish the prowess of "macrosensoriality." Instead of that, the functioning of the device revealed the incommensurability of the forces to which it connects the human operator. They finally managed to amplify the sound of the electrified bones with a simple physician's stethoscope, ending up with a hybrid setup of high- and low-tech, of physicist's and physician's apparatuses.

As the project diverts a biological process from its natural function and turns it into a technical device, Chapple and Wang explicitly raise the issue of the instrumentalization of life: what does it mean to impose technical norms on life? But while this issue is usually mixed to that of the commoditization of life, here it is addressed for itself (Guchet 2014: 309), as "technicization" rather than "instrumentalization," one could say. Instead of raising this issue from the viewpoint of a defined society (capitalist, utilitarian, predatory, etc.) they build an experiential and narrative apparatus that allows questioning the technicization of life as a practice, as "a means in itself" rather than as a means for something else. The project does not condemn nor promote the instrumentalization of life, it "makes it strange," and exposes this strangeness to different kinds of questioning—aesthetical: how does it feels to experiment life through technology? Anthropological: what does it tell about the relationships of human beings to other species? Ontological: what does it tell about the relationships between life and technology?

3. *Pantoffel für Pantoffeltierchen* ("Slippers for sleepy animals") by Grit Ruhland, is a few tenth microns sculpture realized in collaboration with researchers at the Max Planck Institute of Dresden. Made of 60 layers of liquid photopolymer that hardens when irradiated with laser at certain wavelength, it has required considerable work of mask modelling and laser beam programming. In the exhibition, one does not see the sculpture but the glass slide that is supposed to host it; above it, an electron microscope image represents the invisible object (Fig. 22.11).

At first sight, the view of the slipper provokes a feeling of identification and familiarity. It is an everyday-life object, which we habitually use to feel comfort and warm. But when we learn that only a unicellular organism named *Paramecium* has the good shape and size to fit in the slipper, the feeling of appropriation is impeded. The use-meaning of the object vanishes before reconstituting itself anew, but not for us, humans: for the bacteria in question, thanks to a linguistic circumstance, as it occurs indeed that in German, the vulgar name of *Paramecia* is *Pantoffeltierchen*, "animals-slippers." The sculpture, dedicated to *Paramecia*, plays with the meanings conferred by use and by naming: the defeated human use-meaning finally re-weaves itself for non-humans (*Paramecia*) by the mediation of humans who named this bacteria "animals-slippers" because of the foot-shape of its unicellular body. It is thus stimulating exercise of "disanthropocentrism" that, paradoxically, includes the human.

---

should have been audible, but it was not.

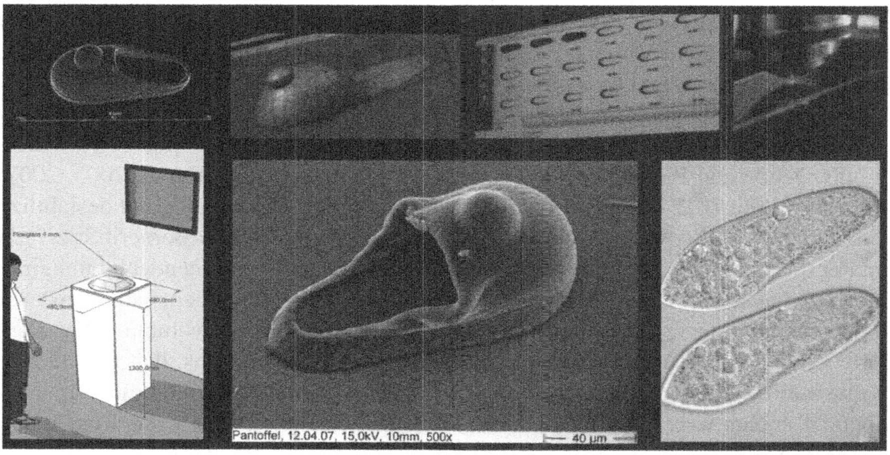

Fig. 22.11 *Slippers for sleepy animals*, Grit Ruhland. (Courtesy of G. Ruhland)

The three art/technology-works described provide interesting cues to rethink the role of design in nanotechnology in that they do not separate the technical and the aesthetic and do not reconcile them either In my view, what they do is rather to experiment the making and unmaking of possible "distributions of the sensible." (Rancière 2004)

22.7 Distributions of the Sensible

Philosopher Jacques Rancière defines the "distribution (or sharing) of the sensible" (*partage du sensible*) as "the implicit law governing the sensible order that parcels out places and forms of participation in a common world by first establishing the modes of perception within which they are inscribed." (Rancière 2004: 85) The distribution of the sensible shapes aesthetics "understood in a Kantian sense—re-examined perhaps by Foucault—as the system of *a priori* forms determining what presents itself to sense experience." (2004: 12) By producing "a system of self-evident facts of perception based on horizons and modalities of what is visible and audible as well as what can be said, thought, or done," (2004: 85) the concept refers to the recognition of a shared and common world of perception and, simultaneously, the delimitations defining the respective parts and positions of social actors. "This apportionment of parts and positions is based on a distribution of spaces, times, and forms of activity that determines the very manner in which something in common lends itself to participation and in what way various individuals have a part in this distribution." (2004: 12) Therefore the distribution of the sensible introduces politics at the core of aesthetics, "a delimitation of spaces and times, of the visible and the invisible, of speech and noise, that simultaneously determines the place and the stakes of politics as a form of experience." Politics, Rancière writes, "revolves around what is seen and what can be said about it, around who has the ability to see and the talent to speak, around the properties of spaces and the possibilities of time." (2004: 13)

It is possible to give a technological twist to Rancière's political view of aesthetics. The concept of "aesthetical apparatuses," prompted by philosophers of art Jean-Louis Déotte and Pierre-Damien Huyghe is a good candidate (Déotte 2005, 2007; Huyghe 2006, 2012). It refers to all devices that shape, reconfigure and destabilize common sensibility. Not only do Déotte and Huyghe agree with Bernard Stiegler's major thesis of the technologically constituted character of transcendental imagination – what Kant called "schematism," the way by which concepts are translated into sensible images and reciprocally (Stiegler, Chap. 18, this volume) – but they provide documented historical analyses of these reconfigurations, that they reintegrate into the history of art. From pencil to cinematograph to digital imaging, aesthetical apparatuses sketch a history of human sensibility which, far from being harmonious, progressive or linear, is rather intrinsically subject to conflict – conflicts between different ways of feeling, different world-apparatus ensembles, as much as between our imaginative and our technical apparatuses always working in a close but tensed interplay (Hui 2015). In other words there is no *a priori* logic of the sensible determined by the inner structure of the human subject nor is aesthetics *a posteriori*, determined by the technological environment. The logic of the sensible is constituted *a praesenti* in our collective interplay and struggles with and within our technical milieu.

For instance, if human access to nanoscale processes depends on choices of transmodality, who is going to chose which perceptive modality allows users access to it and on what grounds? If users' experiences of nano-enabled products depend on delimitations between what is visible and invisible, tangible or intangible, sayable and unsayable, audible and inaudible, who – artists, citizens, designers, engineers, scientists, philosophers, social scientists, industrialists, users, etc. – is granted the legitimacy to determine which features should be rendered sensible? The distribution of social roles of the nanoworld is contingent on techno-aesthetic partitions framing the relationships between the sensible and the sayable, and these partitions should be a matter of public and pluralistic deliberation.

Thus, rather than an ontological given, the division of labor between nano-engineering and design could be understood as a particular distribution of the sensible. By taking up this concept, designers could engage more actively in the design of our interactions with the nanoworld and reflect on its political dimension.

22.8 Reconsidering Design As Techno-aesthetics

The status of design in nanotechnology is paradoxical. In many laboratories around the world, thousands of researchers are designing new experiences enabled by nanotechnology: listening to materials, touching sounds, writing with atoms, communicating with molecules, etc. By contrast in our everyday world, nano-embarking applications are not especially exciting or futuristic; they are unimaginative and poor in design. Most of them invite users to behave as passive consumers of active functions.[23]

Instead of engaging with the nanoworld, designers work only on the symbolic, metaphoric and societal dimensions of future applications and potential uses of nanotechnology, disconnected from its present mode of existence, whose material and operative dimensions are kept out of reach of designers, let to scientists. The intervention of design remains external to the design of the nanoworld. As a result, users are connected to the nano-dimension only by the halo of promises and fears symbolizing "the future."

My proposition is that design should reconsider and perhaps reclaim its role as techno-aesthetics. As I have argued, there is nothing ontologically inaccessible in nano-objects and processes once we look at them through a techno-aesthetic prism. The partition of design between scientists designing an allegedly "invisible" or "noumenal" technology and designers working on its appropriation into everyday life is contingent on distributions of the sensible. The partitions between the inter-objective relationships that are rendered sensible and those that are not, between the aesthetics of functioning and the aesthetics of use, or between those who manipulate transmodality and those who are provided with readymade perceptual access, should not be considered as given but as partitions for designers to play with.

Because there is no natural matching between functioning and use, design is always based on techno-aesthetic choices or "organizations of technophany," as Simondon had it (1960–1961: 39). In the case of nano-enabled use experiences, design choices go beyond the technical object, and extend to the way we may or may not perceive that with which the object connects us: other living subjects, scales, objects, milieus. Design in nanotechnology is thus crucially concerned with the question of determining what should be rendered sensible as a political issue in our natural and artificial environment (e.g. toxicological issues, environmental impacts, etc.).

Techno-aesthetics can be defined as the study and design of apparatuses that transform the functioning of the sensible. As a field of study, it would be an alliance between "techno-logy" understood as the study of technics – another French tradition (Sigaut 1985) – and aesthetics understood as the study of sensation and feeling.[24] As a design practice, it would afford intellectual, practical and emotional grasps on nano-enabled experiences. A "good" techno-aesthetic design would not be merely an "engineered" or "technical" aesthetics: It would have to be a "techno-logical" aesthetics or, if the neologism were not too pedantic, an "aesthechnology," a technicity exploratory of itself. In other terms, techno-aesthetics would have to be reflexive in order to be politically relevant and to question the technically enabled distributions of the sensible instead of merely imposing them. Rather than "making the insensible sensible" like mainstream nano-art claims to do, designers would

[23] For a large consumer products inventory, see the project on Emerging Nanotechnologies at the Woodrow Wilson Center: http://www.nanotechproject.org/cpi/

[24] For an elaboration of techno-aesthetics from Baumgarten, Bergson, Simondon and Deleuze see Loeve (2011).

have to ask *what* should be rendered sensible and *why*, to make *intelligible* the *ways* in which the sensible is technically operated and partitioned, and to render these partitions problematic, debatable, and perhaps reversible.

The design of nanotechnology is still to be invented.

Acknowledgements A preliminary version of this text has been discussed by Victor Petit and Xavier Guchet during the authors' workshop held in Paris Sorbonne on 22–23 June 2015 for the preparation of the collective book. I want to thank them as well as Anne Guénand and Charles Lenay for their useful comments.

References

Annaluru, N., Muller, H., Mitchell, L. A., Cai, Y., et al. (2014). Total synthesis of a functional designer eukaryotic chromosome. *Science, 344*(6179), 55–58.

Antonello, P. (2009). Beyond futurism: Bruno Munari's useless machines. *Avant Garde Critical Studies, 24*(1), 315–336.

Ashby, M. F., Ferreira, P. J., & Schodek, D. L. (2009). *Nanomaterials, nanotechnologies and design. An introduction for engineers and architects*. Burlington: Elsevier.

Audétat, M., Barazzetti, G., Dorthe, G., Joseph, C., Kaufmann, A., & Vinck, D. (Eds.). (2015). *Sciences et technologies émergentes:pourquoi tant de promesses ?* Paris: Hermann.

Auger, J. (2013). Speculative design: Crafting the speculation. *Digital Creativity, 24*(1), 11–35.

Baccile, N., & Balzerani, M. (2013). Invisible and intangible, when art meets nanoscience. *Plastik: Art & Science,* 3. ISSN 2101-0323.

Bachelard, G. (1938). *La formation de l'esprit scientifique: contribution à une psychanalyse de la connaissance objective*. Paris: Vrin.

Bensaude Vincent, B. (2001). A genealogy of the increasing gap between science and the public. *Public Understanding of Science, 10*(1), 99–113.

Bensaude Vincent, B. (2015). Life by design: Philosophical perspectives on synthetic biology. In *BIO web of conferences* (4, p. 00015). EDP Sciences.

Bensaude Vincent, B., Loeve, S., Nordmann, A., & Schwarz, A. (2011). Matters of interest: the objects of research in science and technoscience. *Journal for General Philosophy of Science, 42*(2), 365–383.

Bensaude Vincent, B., Loeve, S., Nordmann, A., & Schwarz, A. (2017). *Research objects in their technological setting*. Abington: Routledge.

Bergson, H. (1896). *Matière et mémoire*. Paris: Alcan. English edition: Bergson, H. (1911). *Matter and Memory* (trans: Paul, N. M. & Palmer, W. S.). London: Swan Sonnenschein & Co.

Bogost, I. (2012). *Alien phenomenology, or, what it's like to be a thing*. Minneapolis: University of Minnesota Press.

Browne, W. R., & Feringa, B. L. (2006). Making molecular machines work. *Nature Nanotechnology, 1*(1), 25–35.

Buchanan, R. (1992). Wicked problems in design thinking. *Design Issues, 8*(2), 5–21.

Bueno, O. (2008). Visual evidence at the nanoscale. *Spontaneous Generations, 2*(1), 132–133.

Chapple, B., & Wong, W. (2008). Can you hear the femur play? Bone audio speakers at the nanoscale. *Leonardo, 41*(4), 355–359.

Chong, K. P., & Garboczi, E. J. (2002). Smart and designer structural material systems. *Progress in Structural Engineering and Materials, 4*(4), 417–430.

Delhome, T. (2011). Innovation et interfaces: de l'intelligence ambiante à l'intelligence sensorielle. In A. Sauvageot, X. Bouju, & X. Marie (Eds.), *Images & mirages @nanosciences* (pp. 247–256). Paris: Hermann.

Déotte, J.-L. (2005). *Appareils et formes de la sensibilité*. Paris: L'Harmattan.
Déotte, J.-L. (2007). *Qu'est-ce qu'un appareil? Walter Benjamin, Jean-François Lyotard, Rancière*. Paris: L'Harmattan.
Dilnot, C. (1984). The state of design history. Part II: Problems and possibilities. *Design Issues, 1*(2), 3–20.
Drexler, E. K. (1992). New technologies for a sustainable world. Hearing in front of the U.S. Senate Committee on Commerce, Science, and Transportation's Subcommittee on Science, Technology, and Space, June 26, 1992. Foresight Update, 14. http://www.islandone.org/Foresight/Updates/Update14/Update14.1.html#anchor634804, Accessed 25 March 2018
Dunne, A. (2006). *Hertzian Tales: Electronic products, aesthetic experience, and critical design*. Boston: MIT Press.
Foucault, M. (1966). *The order of things. An archaeology of the human sciences* (English trans: 2002). London: Routledge.
Guchet, X. (2014). *Philosophie des nanotechnologies*. Paris: Hermann.
Hanson, V. L. (2012). Amidst nanotechnology's molecular landscapes: The changing trope of sub-visible worlds. *Science Communication, 34*(1), 57–83.
Harman, G. (2011). *The quadruple object*. Winchester/Washington, DC: Zero Books.
Hawkins, H., & Straughan, E. R. (2014). Nano-art, dynamic matter and the sight/sound of touch. *Geoforum, 51*, 130–139.
Hui, Y. (2015). Induction, deduction and transduction: On the aesthetics and logic of digital objects. *Networking Knowledge, 8*(3), 1–19.
Huyghe, P. D. (2006). *L'art au temps des appareils*. Paris: L'Harmattan.
Huyghe, P. D. (2012). *Le cinéma avant après*. Grenoble: De l'incidence éditeur.
Jones, R. A. L. (2004). *Soft machines. Nanotechnology and life*. New York: Oxford University Press.
Lévy-Leblond, J.-M. (2010). *La science (n') e(s)t (pas) l'art. Brèves rencontres*. Paris: Hermann.
Loeve, S. (2009). *Le concept de technologie à l'échelle des molécules-machines. Philosophie des techniques à l'usage des citoyens du nanomonde*. PhD thesis, University Paris Ouest.
Loeve, S. (2010). About a definition of nano: How to articulate nano and technology. *HYLE–International Journal for Philosophy of Chemistry, 16*(1), 3–18.
Loeve, S. (2011). Sensible atoms: A techno-aesthetic approach to representation. *NanoEthics, 5*(2), 203–222.
Loeve, S. (2015). La notion d'objet relationnel dans les nanotechnologies (avec et après Simondon). *Cahiers Simondon, 6*, 47–109.
Loeve, S. (2016). L'ATP synthase: un moteur moléculaire? (petit récit technique). In V. Bontems (Ed.), *Gilbert Simondon ou l'invention du futur* (pp. 131–145). Paris: Klincksieck.
Marcovitch, A., & Shinn, T. (2014). *Towards a new dimension. Exploring the nanoscale*. New York: Oxford University Press.
Meillassoux, Q. (2010). *After finitude: An essay on the necessity of contingency*. London: Bloomsbury Publishing.
Milburn, C. (2011). Just for fun: The playful image of nanotechnology. *NanoEthics, 5*(2), 223–232.
Nordmann, A. (2005). Noumenal technology: Reflections on the incredible tininess of nano. *Techné, 8*(3), 3–23.
Nordmann, A. (2006). Collapse of distance: Epistemic strategies of science and technoscience. *Danish Yearbook of Philosophy, 41*, 7–34.
Nordmann, A. (2007). If and then: A critique of speculative nanoethics. *NanoEthics, 1*(1), 31–46.
Ottino, J. M. (2003). Is a picture worth 1,000 words? Exciting new illustration technologies should be used with care. *Nature, 421*(6922), 474–476.
Raimond, J. M., Brune, M., & Haroche, S. (2001). Manipulating quantum entanglement with atoms and photons in a cavity. *Reviews of Modern Physics, 73*(3), 565–582.
Raimondi, S. (2007). *NanoArt. Vedere l'invisibile*. Milan: Skira.
Rancière, J. (2004). *The politics of aesthetics: The distribution of the sensible*. London/New York: Continuum.

Rieth, M. (2003). *Nano-engineering in science and technology. An introduction to the world of nano-design*. Singapore: World Scientific Publishing.
Roco, M. C., et al. (1999). *Nanotechnology: Shaping the world atom by atom*. Washington, DC: National Science and Technology Council Committee on Technology, Interagency Working Group on Nanoscience, Engineering and Technology. http://www.wtec.org/loyola/nano/IWGN.Public.Brochure/IWGN.Nanotechnology.Brochure.pdf, Accessed 25 March 2018.
Ruivenkamp, M., & Rip, A. (2010). Visualizing the invisible nanoscale study. *Science Studies, 23*(1), 3–36.
Schommers, W. (2013). *Nanodesign: Some basic questions*. Singapore: World Scientific Publishing.
Schummer, J. (2006). Gestalt switch in molecular image perception: The aesthetic origin of molecular nanotechnology in supramolecular chemistry. *Foundations of Chemistry, 8*(1), 53–72.
Shwartz, M. (2015). *Designer carbon could give batteries a makeover*. http://www.futurity.org/designer-carbon-batteries-932482/. Accessed 18 June 2015.
Sigaut, F. (1985). More (and enough) on technology! *History and Technology, an International Journal, 2*(2), 115–132.
Simondon, G. (1958). *Du mode d'existence des objets techniques* (2nd ed. 1989). Paris: Aubier. English edition: Simondon, G. (2017). *On the Mode of Existence of Technical Objects* (trans: Malaspina, C.). Minneapolis: Univocal Publishing.
Simondon, G. (1960–1961). Psychosociologie de la technicité. In G. Simondon (2014), *Sur la technique* (pp. 27–129). Paris: Presses Universitaires de France.
Simondon, G. (2005). *L'individuation à la lumière des notions de forme et d'information*. Paris: Millon.
Simondon, G. (2008). *Imagination et invention*. Paris: L'Éclat.
Slaattelid, R. T., & Wickson, F. (2011). Imag(in)ing the nano-scale: Introduction. *NanoEthics, 5*(2), 159–163.
Snow, C. P. (1959). *The two cultures and the scientific revolution: The Rede Lecture*. Cambridge: Cambridge University Press.
Stroscio, J. A., & Celotta, R. J. (2004). Controlling the dynamics of a single atom in lateral atom manipulation. *Science, 306*(5694), 242–247.
Thomas, P. (2009). Midas: A nanotechnological exploration of touch. *Leonardo, 42*(3), 186–192.
Toumey, C. (2010). 35 atoms that changed the nanoworld. *Nature Nanotechnology, 54*(4), 239–241.

Chapter 23
Ontophany Theory: Historical Phenomenology of Technology and the Digital Age

Stéphane Vial

Abstract For over 20 years, the concept of the *virtual* has prevailed in French digital studies. Yet two decades of daily cultural integration with interfaces have demonstrated that virtuality is only one of many aspects of our interactive experience with digital devices. A need therefore exists for new concepts more apt to address the philosophical complexity of the digital phenomenon and the significance of our interactions with calculated matter as they are true existential experiences of phenomenological significance. In this chapter I explain why I have suggested introducing the phenomenological concept of *ontophany* (manifestation of being). In close relationship with a comprehensive and broadened understanding of Bachelard's notion of "*phenomenotechnique*," I examine the hitherto unidentified technicality of this manifestation process. Prior to their existence as tools in uses, technologies are first the perceptual structure of our existence; they are the "devices" or the invisible matrixes, produced by culture and history, into which our potential experience-of-the-world is cast. Not only do the following theoretical propositions seek to contribute, philosophically, to *Internet Studies* and to a better understanding the Digital Age, they also hope to give rise to a broader deliberation on technology and perception, as they relate to an approach I would characterize as a *historical phenomenology of technology*.

Keywords Digital media · Digital studies · Design · Environment/milieu · Internet studies · Historical phenomenology · Ontophany · Virtual

---

S. Vial (✉)
PROJEKT Design Lab (EA 7447), University of Nîmes, Nîmes, France

ACTE Institute, Sorbonne Paris 1 University, Paris, France
e-mail: stephane.vial@unimes.fr

S. Loeve et al. (eds.), *French Philosophy of Technology*, Philosophy of Engineering and Technology 29,
https://doi.org/10.1007/978-3-319-89518-5_23

We have learned to take things at interface value. –Sherry Turkle

This chapter is the outcome a dual intellectual filiation and a dual professional expertise.

Intellectually, its influence is rooted both in existential phenomenology, from which I have chosen to adapt a number of fundamental concepts (including the notion of being-in-the-world), and in the French tradition of a "historical epistemology" of science (Lecourt 1969) whose key figures include Gaston Bachelard, Jean-Claude Beaune, Bertrand Gille and Jean-Pierre Séris.

Professionally, I have drawn upon the twofold experience fostered as an interaction designer at LEKTUM, the Paris-based web design agency I founded and managed from 2009 to 2014, and as an educator and design researcher, originally as a philosophy professor at École Boulle, a Paris-based school of Art and Design, and currently as Associate Professor of Design and Digital Media at the University of Nîmes.

L'être et l'écran (Vial 2013) is both the synthesis and the fruit of these filiations and experiences. Its objective, both phenomenological and epistemological, is to provide the philosophical analysis of technology, in general, and of digital technologies, in particular, with a conceptual renewal that relies upon the observation of experience (phenomenological component) and the history of technology (epistemological component).

For over 20 years, French digital[1] studies (Quéau 1993; Lévy 1995; Missonnier and Lisandre 2003; Tisseron 2008, 2012) have seen the *virtual*, as a *philosophical* concept, prevail. My work postulates that although the above concept is *philosophical* in origin, it fails to be relevant at grasping the nature of the digital phenomenon, *philosophically*. Two decades of daily cultural integration with interfaces have demonstrated that virtuality (or simulation), is one of many aspects of our interactive experience with digital devices. A need therefore exists for new concepts; ones more apt at penetrating the philosophical complexity of the digital phenomenon, and more likely to enlighten us as to the significance of our interactions with interfaces, given that these encounters constitute a phenomenological and existential experience.

Thus, I have suggested introducing the concept of *ontophany*, whose etymology merges (without any particular hierarchical distinction) the dimensions of being (*ontos*) and of appearance (*phaïnô*). It bears witness to my profound attachment to Bachelard's notion of "*phenomenotechnique*," which I believe the term "ontophany" revives and broadens into a form of comprehensive phenomenotechnique. I wish to examine the hereto unidentified technicality of the manifestation process through the prism of the contemporary digital field. Not only do the following theoretical propositions seek to contribute, philosophically, to *Internet Studies* and to a better understanding the Digital Age, they also hope to give rise to a broader deliberation on technology and perception, as they relate to an approach I would charac-

[1] Until now, French digital studies have been developed outside of the field of philosophy, with a loose, and often awkward, appropriation of the latter's concepts.

terize as a *historical phenomenology of technology*. According to this approach, technology is no longer a body of objects isolated from their subject; technical nature becomes an intrinsic aspect of subjectivity (among others) which varies in relation to its historical context. Man is as much part of the machine as the machine is part of man. This marks a departure from post phenomenology (Ihde 1990), which stipulates that technologies mediate our relationship to things according to four broad categories of "Human-Technology Relations": Embodiment Relations (e.g. eyeglasses), Hermeneutic Relations (e.g. thermometers), Alterity Relations (e.g. robots), and Background Relations (e.g. automated heating systems) (Ihde 1990). The techno-transcendental phenomenology put forth hereafter seeks to render the overall transcendental technical nature of appearance, as historically determined by an era's technical culture, more perceptible. As a new ontophanic milieu, the *Digital* Age therefore represents an optimal field of observation.

23.1 Ontophany As an Hypothesis, or Comprehensive Phenomenotechnique

23.1.1 Another Look at "Phenomenotechnique"

Gaston Bachelard first introduced the notion of phenomenotechnique in 1931, in a short article entitled "Noumène et microphysique" (*Noumenon and microphysics*) (Bachelard 1931–1932). This entirely fabricated concept sheds light upon one of the fundamental characteristics of modern science: scientific work consists not in describing phenomena as if they preceded their dedicated theories, but in constructing them wholly, using technological devices that afford them an appearance and an existence as phenomena per se. Nuclear physics offer an excellent illustration. In 1911, Ernest Rutherford surmised that an atom's mass is concentrated in its central core, the nucleus, and that electrons determine only the atom's size. However, because an atomic nucleus is made of matter one million billion times denser than ordinary matter (an atomic nucleus is a thousand times smaller than an atom, yet contains 99.97% of its mass), a phenomenal observation of the nucleus seemed impossible. So it remained until 1932, when John Cockcroft and Ernest Walton suggested projecting particles, electrically accelerated to high speed, onto the nucleus in order to disintegrate it, and thus, observe it. The first particle accelerator was born, and it became a foundational instrument in nuclear physics.

As a scientific reality, the atomic nucleus first existed theoretically as a hypothesis; a technical instrument then brought it to exist phenomenologically. Therefore, "in modern science, an instrument is truly a reified theorem," (Bachelard 1933: 140) in the sense that, as our example demonstrates, the particle accelerator is the theory of the atom, technically embodied. Hence, "a measuring instrument always ends up as a theory: the microscope has to be understood as extending the mind rather than the eye." (Bachelard 1938a, b: 240) In other words, technical instruments developed through scientific reasoning are at the heart of the active theoretico-practical phenomena elaboration process.

> Phenomena must (…) be carefully selected, filtered and purified; they must be cast in the mould of scientific instruments and produced at the level of these instruments. Now instruments are just materialized theories. The phenomena that come out of them bear on all sides the mark of theory. Truly scientific phenomenology is therefore essentially a phenomenotechnology. Its purpose is to amplify what is revealed beyond appearance. It takes its instruction from construction. (Bachelard 1934: 13)

What we must essentially understand is that – as phenomena – scientific realities do not exist beyond of the devices capable of revealing them: in order to *appear*, they require an *appliance* (or *device*). "Therefore, here, the phenomenon is a device-dependent phenomenon." (Bachelard 1951: 5) Instead of discovering exogenous phenomena, science builds them from within, when it creates the instruments capable of materializing theories. Hence, phenomenotechnique refers to the constructivist technique of phenomena expression. The major philosophical lesson to be retained here is: technological materialization is a criterion for phenomenal existence. In modern science, a phenomenon must be technologically, or at least plausibly, fabricated in order to exist per se. In other words, technology is able to engender phenomenality, or the potential to appear.

23.1.2 Ontophany, or the Transcendental Technicity of Appearance

Although they encompass scientific phenomena, universal phenomena are not device-dependent: their appearance does not rely upon the use of devices. Quite the contrary, it seems as though the latter are already in place. Nonetheless, I wish to demonstrate that they are not quite natural and independent, appearing at their own licence. Just as knowledge forms through interactions with its object, perception is also the result of interactions with the phenomenon. Analogously, just as science's terms of application are technological, perception relies upon equivalent technical conditions of execution.

My hypothesis is the following, if technical feasibility is a criterion for phenomenal existence, then this must apply beyond the boundaries of scientific phenomenon; universal phenomena also owe their phenomenality to technical factors. Technical determination is indeed one of phenomenality's overlooked foundations; technical influence should not be interpreted as exogenous, as if phenomena were affected extrinsically. I seek to demonstrate that appearance is in itself a phenomenotechnical process, one which is technically determined and intrinsic to the phenomenality of phenomena.

The phenomenality of phenomena refers to the way being (*ontos*) appears (*phaïnomenon*), and holds the particular characteristic of feeling-of-the-world. I elect to call this *ontophany* which, according to the etymological sense initiated by Mircea

Eliade (Eliade 1965), means that something is revealing itself to us. Thus, if we suppose that all ontophany is by definition technical ontophany,[2] or possesses, at the very least, a technical dimension, then, we may postulate that perception's a priori conditions are not transcendental, as suggested by Kant, but technical, as posited by Bachelard. Consequently, technology can be defined as an ontophanic matrix, a general perceptual structure that determines how beings appear. As such, this structure is not a component of our cognitive aptitude's inner framework (it does not structure knowledge a priori); it contributes instead to the outer framework of our technical culture (which I choose to call a techno-transcendental structure).

The idea of ontophany is therefore no more than the idea of a comprehensive phenomenotechnique which encompasses all phenomena; meaning, that all phenomena, not just scientific phenomena, are phenomenotechniques. In turn, a transcendental technicity of appearance exists, implying that every phenomenal manifestation or "phany" possesses an a priori technical feature. Thus technologies, prior to their existence as the tools that form our uses, are first the perceptual structure of our existence; they are the "devices" or the invisible matrixes, produced by culture and history, into which our potential experience-of-the-world is cast. Consequently, all perception is the result of a technical scheme which inconspicuously determines the very manner we feel-in-the-world at a given time.

23.2 From Technical System to Ontophanic Milieu

23.2.1 The Phenomenological Structure of Technical Revolutions

Although my focus is primarily phenomenological, it rests upon an epistemological foundation essential to its meaning. This foundation is the historical theory of technical revolutions, which arises with the confrontation of two great works: Bertrand Gilles' *Histoire des techniques* (1978) and Thomas Kuhn's *The Structure of Scientific Revolutions* (1962).

At a time in contemporary French philosophy of technology, when most base their work upon Simondon (1958, 1960-1961), I have chosen instead to base mine upon Gille. The latter, besides pre-empting the dogmatic incantations of the former, extends beyond the boundaries of history. In my opinion, Gille's history of techniques possesses philosophical value, owing not least to his notion of a "technical system," which, in approaching the history of techniques through a problematized bias, advances a genuine philosophy of history. Thus, even its author concedes that it is more "technical epistemology" (Gille 1979) than history of techniques. Also, in order to avoid any confusion, my use of the term "technical system" bears no affiliation to Jacques Ellul's "technician system," which suggests that technology is a com-

[2] One must not mistake the notion of ontophany as we suggest it here with that of "technophany," dear to Gilbert Simondon, whose elaboration was also based upon Mircea Eliade.

prehensive system responsible for its own growth in a quest for maximum efficiency. It refers instead only to the expression as defined by Bertrand Gille; a "technical system" is the cohesive whole forged by a period's dominant techniques (e.g. the first industrial technical system was made up of a coal/steam engine/metal alliance).

Analogical reasoning then allows us to apply Thomas Kuhn's interpretation framework to Bertrand Gille's historical model in order for the latter to project itself as a philosophy of technology. The correlation can be expressed thus: just as the *scientific revolution* represents a shift in "paradigm," as imagined by Thomas Kuhn; a *technical revolution* represents a shift in "technical system," as apprehended by Bertrand Gille. In this light, the history of techniques can be viewed as a history of technical revolutions, or the part-substitutive part-cumulative succession of different technical systems.

However, the theory would be incomplete without its phenomenological layer. My position is that at each technical revolution, both a *systemic revolution* (i.e. a shift in technical systems) and an *ontophanic revolution* (i.e. a renewal of the structures of perception) occur. In the transition from one technical system to another, unfamiliar materials (wood, steel, petroleum, electricity, information) or novel inventions (the scotch yoke, the steam engine, the particle accelerator, the computer, Internet, etc.), alter not only the object of our perception, they also adjust the act of perceiving from within its core phenomenological dynamic. The technical culture redefines and renegotiates the phenomenality of beings (their ontophany) itself.

Observing the sky in the active silence of the Renaissance's wood and water mechanisms ("eotechnic" ontophany, Mumford 1934) or observing it from within the first Industrial Revolution era of steam engines and omnipresent metal (mechanized ontophany), bear little in common with the qualitative experience of the sky in a time of near constant immersion in interactive situations (digital ontophany). The being-in-the-world inherent to eotechnic ontophany, notable for its bodily proximity to nature and the silence of its instruments, is not that engendered by mechanized ontophany, marked by the violence of machines and the comprehensive mechanization of existence, nor is it the being-in-the-world particular to digital ontophany, determined by swift calculations, fluid procedures, and immersion in interfaces.

Hence, we must consider that for every technical system shift, a corresponding ontophanic shift occurs, which is then followed by the qualitative renewal of our sentiment of being-in-the-world.

23.2.2 The Ontophanic Sentiment or Technology As "Umwelt"

What emerges from the preceding analysis is that an era's technical system, behaving akin to a techno-transcendental structure, shapes the phenomenal quality of the world we experience. It is not so much the object of perception, but the act of perception, which changes; when the ontophanic quality of an experience-of-the-world is transformed, the manner of feeling-in-the-world is itself reiterated. By this, I mean the qualitatively registered, or felt, aspect of the being-in-the-world experience. Henceforth, it shall be referred to as the *ontophanic sentiment*, which is understood as the perceived feeling and the experienced sentiment of the world's presence. The ontophanic sentiment should be deemed the result of a process which possesses both a subjective psychological dimension and an objective techno-historical one.

The subjective psychological component can be theoretically linked to *qualia*. Qualia refers to the different qualitative and subjective aspects of our mental states which are ineffable, intrinsic, private and directly accessible (Leyens 2000). They are "the qualitative and phenomenal characteristics of sensitive experiences, by virtue of which these resemble and differ from each other as they do." (Dennett, 1992; as cited par Leyens, 2000: 773) In short, they are what we feel *in a unique manner* whenever we perceive something, and they constitute the first dimension of the ontophanic sentiment. The second is the techno-historical dimension, which is the result of the group's objective conditions of life at a given time.

This latter dimension is of particular interest to me; my hypothesis being that within this dimension, the technical culture inside of which we exist (i.e. the mechanical culture of the early twentieth century, or the digital culture of the early twenty-first century) possesses a phenomenological influence over the *qualia* we perceive. Why would this be? Because reality or being are always particular and fortuitous; they respond to a period's technical context. Being-in-the-world or being-here (*Dasein*) is not a disconnected metaphysical condition, alienated from a century's context and anchored into the spirit as if the spirit itself were unchanging and isolated. Being-in-the-world or being-here are fundamentally different whether we inhabit the eotechnic technical system or the digital technical system.

This justifies why *technical systems* are not only superior levels of technical coherence and consolidation but must also be understood as genuine *ontophanic milieus*. By this, I do not mean "technical milieus" in the sense that it seems to have acquired within the French "simondonian" mind-set but rather that of perceptual milieus (*Umwelt*), as introduced in Jacon von Uexküll pioneering works. The latter's milieu theory holds, in my opinion, and as I will strive to demonstrate, more phenomenological than biological significance. Uexküll's *Umwelt* is often mistakenly translated as "environment," yet from the beginning of *Mondes animaux et monde humain* (A Foray into the Worlds of Animals and Humans), after his description and analysis of the "the tick's world," the author establishes a firm distinction between "milieu" or "self-world" (*Umwelt*) and "environment" or "surroundings." (*Umgebung*) The first is an animal's species-specific perceptual world, and depends entirely on a species' sensory devices; the second is "the surroundings we observe around him" (Uexküll 1934: 29), as made up of elements drawn from our own perceptual world (Fig. 23.1).

Once the ideas of Gille and Uexküll are combined, they allow me to support the following: just as animals live out their existence in their own perceptual milieu, which is determined by their *specific* (in the sense of *species*-specific) sensory devices, humans exist within their own perceptual milieu, which is qualitatively correlated with the systemic technical devices of their times. These *ontophanic milieus*,

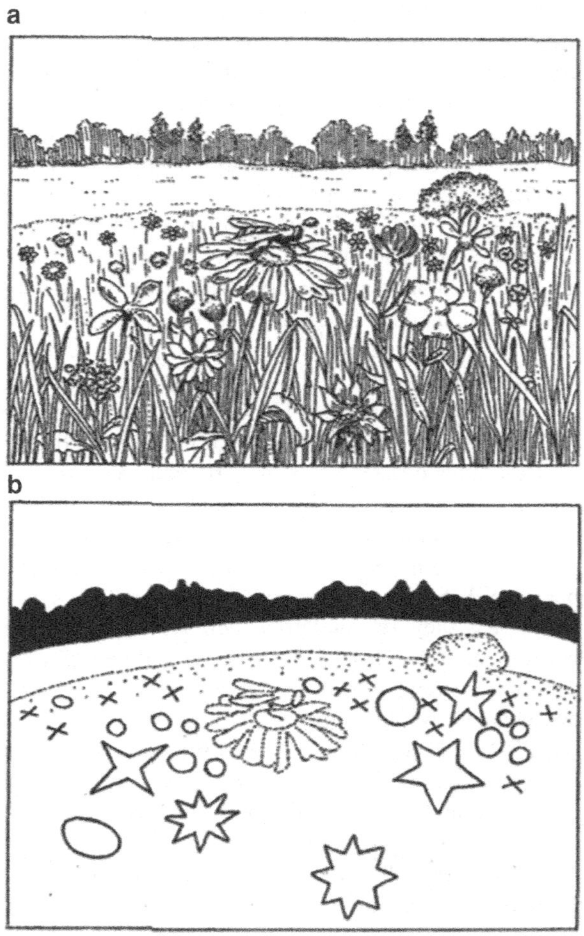

Fig. 23.1 *Umgebung (environment)* and Umwellt (milieu). (Source: J. Von Uexküll, *Théorie de la signification,* plates 5a et 5b)

are essentially technical *Umwelts*, or techno-perceptual milieus. Although "there is a link between Uexküll's *Umwelt* and the concept of a technical milieu," (Petit 2013) one can see that my understanding of milieu, as a concept, rests upon a novel phenomenotechnique significance that grants it the phenomenological weight insufficiently present in Simondon's "milieu" and Bernard Stiegler's "technical milieu."

Just as Uexküll urged us to imagine each animal as surrounded by "a sort of soap bubble that represents its milieu, and fills itself with all the characteristics the subject need access," we must imagine human beings from a given historico-technical period as occupying a sort of phenomenological soap bubble, or techno-perceptual vessel that is profoundly unique and characteristic of that period. Every given ontophanic milieu possesses its own particular ontophanic sentiment, made up of unique and singular qualia, which cannot be replicated in a different ontophanic milieu. We can no more fathom "what is it like to be a bat," (Nagel 1974) than what it would be to perceive the world in an eotechnic ontophanic context such as that which prevailed during the Renaissance. Just as, according to Uexküll, animals discern "a whole new world taking shape in every bubble," humans also experience a new world in each ontophanic milieu.

Thus, each generation learns to perceive the world from within its own phenomenological bubble, establishing its perceptual relationship to reality by means of existing technical devices. As such, we are all *Technological Natives*, that is, phenomenologically speaking, natives of the structuring and dominant technology that configures our psycho-cognitive mechanisms relative to the world: railroad and photography natives, electricity and telephone natives, computer and internet natives, etc. If the expression *Digital Natives* is to have meaning, it must be phenomenological; as being born and raised in the digital soap bubble is phenomenologically different from being born and raised in the mechanical soap bubble. Being is therefore " being tech-born:" existence is a techno-perceptual emergence into the presence of things. To capture the techno-perceptual dimension of presence is to gain access to the *ontophanic sentiment*, and discover how our technical devices fashion the marrow of our world.

23.3 Digital Ontophany and Its Categories

The digital technical system has induced an unprecedented ontophany. We are faced with new ways of being: the algorithmic and interactive procedures we navigate through in the interfaces, icons and avatars we employ in our various simulated environments; the uncountable connection, navigation and notification actions we accomplish on our networks; or, the multiple interactions used to exchange daily with connected objects and other intelligent devices. Digital artefacts modify our perceptive habits, gradually establishing a new phenomenological "soap bubble" around us.

This new *ontophanic milieu*, which we progressively culturally integrate, is that of *digital ontophany*. The virtual is an undeniable attribute of this new *Umwelt* – digital devices produce digitally simulated beings – but it is merely one among many others. In order to consider the digital's phenomenological complexity, one must transcend the virtual and invent new concepts. In this section, I suggest we analyse *digital ontophany* by means of 11 categories, more logical than phenomenological. These do not intend to objectively describe how the digital phenomenon appears, technically and scientifically (though this viewpoint deserves consideration). Instead, the categories aim to reveal what the digital phenomenon subjectively establishes from an ontophanic perspective, that is, to consider it as it is experienced by the subject, from the point of view of its unique phenomenal response.

23.3.1 Noumenality: The Digital Phenomenon Is a Noumenon

Every technical revolution is a one of matter. The digital revolution is that of calculated matter. Part-mathematical, part-electronic, calculated matter is made up of electronic signals coded as binary data, or numbers. This singular state of matter, which operates at an invisible level, matches what Kant called a noumenon: a phenomenon deprived of phenomenality, imperceptible to humans as it is situated beyond the boundaries of plausible experience. Like quantum processors, digital processors are first and foremost noumena.

23.3.2 Ideality: The Digital Phenomenon Is Programmable

Calculated matter is above all an ensemble of idealities, or reasonable beings reliant upon programming languages. The latter can be defined as formal languages made up of symbols which allow a problem to be reduced to an algorithm. Everything a computer is able to do, be it a mainframe or a pocket device, is the result of lines of code. Calculated matter is, by nature, logical; this is why, as Lev Manovich (2013) described it, the digital age is one where *Software Takes Command*. Additionally, this explains why programming, which Pierre Levy (1992) labelled a "one of the fine arts," is one the most influential activities of our time.

23.3.3 Interactivity: The Digital Phenomenon Is an Interaction

The digital noumenon is not an obscure phenomenon, accessible only to programmers. Its programmable quality renders calculated matter fundamentally interactive, it can be actioned by a user, whose move leads to a systematic reaction, or response. Interaction is just that: reacting to a reaction which provokes a new reaction one must react to. To live in the digital era is to live in the midst of interactions, to be immersed in a potentially infinite relationship with calculated matter, as if it were a partner who always bounces something back. To design a digital interface is to design these interactions (Moggridge 2007).

23.3.4 Virtuality: The Digital Phenomenon Is a Simulation

The virtual describes digital devices' (particularly those equipped with a graphic interface) ability to produce *computer-simulated* realities. From a perceptive standpoint, it marks our move from a "culture of calculation" to a "culture of simulation."

(Turkle 1995) In the digital era, we work on a virtual *desktop*, file virtual files, draw with virtual paintbrushes… all of which are *computer-simulated*. The virtuality of digital interfaces is the most obvious part of the mould into which our perception is cast; at the phenomenal level, it is the *visible* agent of the calculated matter operating invisibly at a noumenal level. This explains the persistent, and inaccurate, amalgamation between the digital and the virtual.

23.3.5 Versatility: The Digital Phenomenon Is Unstable

No programmer in the world is capable of writing a program that will function bug-free on its first try. This is why numerous tests and "debuggings" occur prior to the launch of a new software or application. The bug and calculated matter are consubstantial, it is the digital phenomenon's versatility. Living within digital ontophany means living with unstable matter that one must occasionally reboot or restart. Having grown accustomed our machines' functional contingencies, we have gradually integrated that "bugs happen" into our perceptions and adapted accordingly. Floridi (2017) recently called that "the unsustainable fragility of the digital."

23.3.6 Reticular Nature: The Digital Phenomenon Is "Other-Phanic"

Potentially available social links within a group depend upon the devices that accept their activation, and when they are activated, allows them to be phenomenalized in a way that bears the ontophanic stamp of the device. Like the telephone (which enabled us to speak without seeing each other), the Internet engenders a new ontophany of others (or "otherphany"), making it possible to communicate directly without speaking to or seeing each other (text messages, tweets, instant messaging). Digital otherphany is therefore radically new (Vial 2014a, b), and consists in a paradoxical ambivalence which blends presence with absence: the other is here without being here (Missonnier et Lisandre 2003).

23.3.7 Instant Reproducibility: The Digital Phenomenon Is Replicable.

It has become so commonplace that we have already forgotten its extraordinary quality. Calculated matter makes it technically possible to *instantly* create a potentially infinite number of copies of a single element (text, image, sound, etc.); whatever the element may be, the processor deciphers it as a deliberate sequence of 0 s and 1 s. Not only is this, in the history of matter, a previously unseen property; from a phenomenological perspective, it is a prodigious perceptual characteristic. Let us simply recall the time Renaissance printers required to complete a single copy of Homer's *Odyssey,* as compared to that which is today necessary to duplicate the same text a hundredfold and email it to a hundred recipients.

23.3.8 Reversibility: The Digital Phenomenon Is Retractable

The entire physical universe is subject to entropy, in other words, to growing disorder. On the scale of life, death is an illustration of the universe's fundamental irreversibility. However, one of the digital phenomenon's ontophanic properties is the potential to backtrack. In the land of calculated matter, it is always possible to "undo" (ctrl-z) or "redo." (shift-ctrl-z) From the perspective of phenomenological reception, this places before the user an event of near-supernatural proportion: it appears to overturn the irreversible essence of our physical world. Having grown accustomed to this ontophany of reversibility, we occasionally come to regret its absence from our non-digital experiences.

23.3.9 Destructibility: The Digital Phenomenon Can Be Annihilated

Calculated matter can vanish. It takes no more than an electrical power outage for that which has not been registered into memory to literally disappear from the realm of reality. Where has the data gone? It was no more than a sequence of 0 s and 1 s awaiting registration; it volatilized the very moment the electrical current ceased to flow through the microprocessors' millions of transistors. It has vanished. Calculated matter is decidedly odd, introducing into our experience-of-the-world an ontophany of disappearance that our faculties of perception are beginning to fathom and work with.

23.3.10 Fluidity: The Digital Phenomenon Is Thaumaturgical

Digital thaumaturgy is the near-miraculous phenomenology wherein things have lost their previous heft in order to become light and fluid, magically complying with our desires and expectations. Not only are our text messages sent and delivered more speedily than by post, but it has become easier, simpler, and more expeditious to write and transmit a message. All that can be accomplished digitally can be done with greater fluidity and ease. The digital phenomenon frees us from an important portion of reality's resistance capacity. Like a thaumaturgical king, it accomplishes miracles, or simply works wonders.

23.3.11 Playfulness: The Digital Phenomenon Is Game-Like

Things that spontaneously stimulate amusement and spark a "playsurable" experience can be deemed playful (Vial 2014a, b). Playfulness is a technological device's (the realm of objects) ability to create a playful attitude (realm of the subject), that is the capacity to stimulate play within a psyche. The adoption of a playful attitude is natural and near-immediate when opposite an interface. The digital realm is not merely subject to the gamification processes (Genvo 2011), it is intrinsically playful: based upon all the preceding characteristics (interactivity and reversibility, in particular) it spontaneously favours a playful attitude and stimulates our aptitude for play. Play is an essential component of every digital phenomenon; and for this reason, we live in an increasingly gamified world.

23.4 Conclusion

Under no pretence do these 11 categories constitute an exhaustive analysis of digital ontophany, which deserves to be further developed. These 11 categories are merely a first conceptual foray into understanding the unprecedented phenomenality of digital beings. They allow us to rise above "digital dualism," a belief according to which "the digital and the physical are separate spheres." (Jurgenson 2012: 84) Such an exceedingly metaphysical belief splits the contemporary world in half along an invisible boundary. On either side of this limit, two spheres exist: the first would be the digital/online/on-screen domain; the second would be the physical/disconnected/off-screen domain. The word *virtual* is used to qualify the former, and *real* is attributed to the latter. Ontophany theory offers an efficient conceptual filter for deconstructing the profane metaphysics of reality and virtuality. Despite a lack of heuristic properties (it has produced no knowledge, nor does it bear any scholarly weight), the latter remains widespread not only among most of the digital era's users, including the media and public authorities, but also within the greater scientific community, in the humanities, who continue to distinguish between the real and the virtual as if evidence of their distinction was scientifically available.

In addition, ontophany theory offers designers and computer scientists elements of vocabulary destined to better orient a number of fundamental choices related to the design of digital products. Thus, I have included below, with my most heartfelt thanks, a simplified version of logician and computer scientist, Jean Sallantin's, work: a logical model that offers a rational, comprehensive view of digital ontophany's

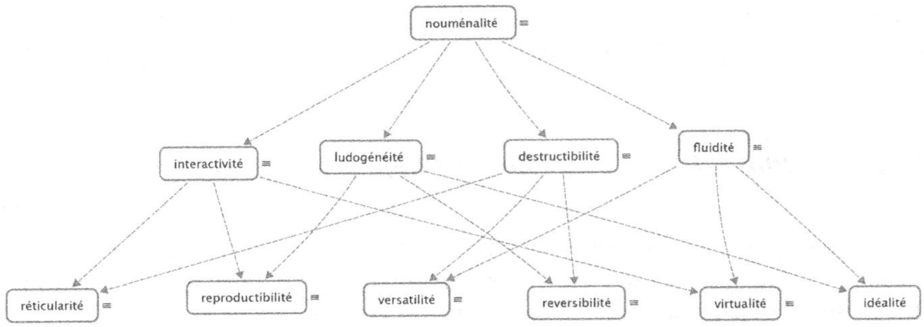

Fig. 23.2 Logical model of digital ontophany: Jean Sallantin's contribution (Jean Sallantin is a computer scientist and research director emeritus at the CNRS. He founded *Forum des Débats pour le bien commun* (Forum of debates for the common good), an association whose intent is to produce digital engineering for high level ethical debates, most notably it has developed the tools *Hypostasis* et *Dialoguea*. More information online: http://forum-debats.fr)

categories, that is, the means by which digital matter appears to humans, and their interdependencies (Fig. 23.2).

This diagram reveals the dependence structure between the categories of digital ontophany. Arrows signify that notions are dependent for either the proof or the negation. The presence of noumenality implies that of all other categories, its absence implies that of the four categories which depend upon it, and hence of all categories.

Digital design is currently the most innovative form of design; it dictates the shape of digital ontophany. Without design's creative spark, a number of calculated matter's most fundamental properties would not exist; the virtuality of graphic interfaces, is a, example. At a time where interactivity is widespread, the future of our being-in-the-world has never been so connected to design. Henceforth, its task is the creative exploitation of calculated matter's ontophanic capacities in order to produce hitherto undreamt meaningful human experiences. For those wondering whether they should participate in the digital revolution, the answer is simple: we must participate in its design. Therein lies the responsibility of digital design, it must better our experience of digital ontophany, regarded as a perceptual environment whose digitally centred phenomenality is fundamentally hybrid: digital and non-digital, online and offline, on-screen and off-screen. This is the environment our children are being raised in and, as they absorb new perceptual structures, where they acquire a sense of reality – their own.

Acknowledgements A preliminary version of this text has been discussed by Victor Petit and Ronan Le Roux during the authors' workshop held in Paris Sorbonne on 22–23 June 2015 for the preparation of the collective book.

References

Bachelard, G. (1931–1932). Noumène et microphysique. *Recherches philosophiques*, I, 1931–1932, pp. 55–65. In Bachelard, G. (1970). *Études*. Paris Vrin.

Bachelard, G. (1933). *Les Intuitions atomistiques*. Paris: Boivin.

Bachelard, G. (1934). *Le Nouvel Esprit scientifique*. Paris: Presses Universitaires de France, Quadrige. English edition: Bachelard, G. (1984). *The New Scientific Spirit* (trans: Goldhammer, A.). Boston: Beacon Press.

Bachelard, G. (1938a). *La Formation de l'esprit scientifique*. Paris: Vrin. English edition: Bachelard, G. (2002). *The Formation of the Scientific Mind: A Contribution to a Psychoanalysis of Objective Knowledge* (trans: McAllester Jones, M.). Manchester: Clinamen Press.

Bachelard, G. (1938b). *Psychanalyse du feu*. Paris: Editions Gallimard, Folio. English edition: Bachelard, G. (1987). *The Psychoanalysis of Fire* (trans: Ross, C. M.). Boston: Beacon Press.

Bachelard, G. (1951). *L'Activité rationaliste de la physique contemporaine*. Paris: Presses Universitaires de France.

Dennett, D. C. (1992). *Consciousness explained*. New York: Back Bay Books.

Eliade, M. (1965). *Le Sacré et le Profane*. Paris: Gallimard. English edition: Eliade, M. (1987). *The Sacred and the Profane: The Nature of Religion* (trans: Trask, W. R.). Harvest/HBJ Publishers.

Genvo, S. (2011). Penser les phénomènes de "ludicisation" du numérique: pour une théorie de la jouabilité. *Revue des sciences sociales*, n°45.

Floridi, L. (2017). The Unsustainable Fragility of the Digital. and What to Do About It. *Philosophy & Technology* 30(3), 259–261.

Gille, B. (1978). *Histoire des techniques*. Paris: Editions Gallimard, Bibliothèque de la Pléiade.

Gille, B. (1979). La notion de "système technique" (essai d'épistémologie technique). *Culture technique*, n°1, Centre de recherche sur la culture technique, Neuilly-sur-Seine.

Ihde, D. (1990). *Technology and the lifeworld: From garden to earth*. Bloomington/Indianapolis: Indiana University Press.

Jurgenson, N. (2012). When atoms meet bits: Social media, the mobile web and augmented revolution. *Future Internet, 4*, 83–91.

Kuhn, T. S. (1962). *The structure of scientific revolutions*. Chicago: The University of Chicago Press.

Lecourt, D. (1969). *L'épistémologie historique de Gaston Bachelard*. Paris: Vrin.

Lévy, P. (1992). *De la programmation comme un des beaux-arts*. Paris: Editions La DÉcouverte.

Lévy, P. (1995). *Qu'est-ce que le virtuel?* Paris: Editions La DÉcouverte.

Leyens, S. (2000). La conscience imaginée. Sur l'éliminativisme de Daniel Dennett. *Revue philosophique de Louvain, 98*(98–4), 761–782.

Manovich, L. (2013). *Software takes command*. New York: Bloomsbury Publishing.

Missonnier, S., and Lisandre, H. (Eds.). (2003). *Le Virtuel: la présence de l'absent*. Paris: EDK.

Moggridge, B. (2007). *Designing interactions*. Cambridge, MA: The MIT Press.

Mumford, L. (1934). *Technics and civilization* (2010th ed.). Chicago: The University of Chicago Press.

Nagel, T. (1974). What is it like to be a bat? *The Philosophical Review, LXXXIII*(4), 435–450.

Petit, V. (2013). Le concept de milieu en amont et en aval de Simondon. *Cahiers Simondon*, n°5, Paris: L'Harmattan.

Quéau, P. (1993). *Le Virtuel: vertus et vertiges*. Seyssel: Champ Vallon, coll. Milieux.

Simondon, G. (1958). *Du Mode d'existence des objets techniques*. Paris: Aubier. English edition: Simondon, G. (2016). *On the Mode of Existence of Technical Objects* (trans: Malaspina, C.). Minneapolis: Univocal Publishing.

Simondon, G. (1960–1961). Psycho-sociologie de la technicité. In Simondon, G. (2014). *Sur la technique*. Paris: Presses Universitaires de France.

Tisseron, S. (2008). *Virtuel, mon amour: penser, aimer, souffrir, à l'ère des nouvelles technologies*. Paris: Albin Michel.

Tisseron, S. (2012). *Rêver, fantasmer, virtualiser. Du virtuel psychique au virtuel numérique*. Paris: Dunod.

Turkle, S. (1995). *Life on the screen. Identity in the age of the internet*. New York: Simon & Schuster Paperbacks.

Uexküll Von, J. (2010 [1934]). *A foray into the world of animals and humans: With a theory of meaning*. Minneapolis: University of Minnesota Press.

Vial, S. (2013). *L'être et l'écran: comment le numérique change la perception*. Paris: Presses Universitaires de France.

Vial, S. (2014a). Pour introduire le "playsir": pourquoi les appareils numériques sont "ludogènes". *Interfaces numériques, 3*(1), 149–161.

Vial, S. (2014b). Ce que le numérique change à autrui: introduction à la fabrique phénoménotechnique de l'altérité. *Hermès, 1*(68), 151–157.

Made in the USA
Monee, IL
16 September 2026

63227619R00227